Factor Analytic Studies

1971 – 1975

Factor Analytic Studies

1971 – 1975

by

Suki Hinman

and

Brian Bolton

The Whitston Publishing Company
Troy, New York
1979

Library of Congress Catalog Card Number 78-069873

ISBN 0-87875-165-3

Printed in the United States of America

TABLE OF CONTENTS

Table of Contents

Table of Contents

FOREWORD

By a coincidence, this valuable volume and my own *Scientific Use of Factor Analysis* have appeared at virtually the same time, with the fortunate feasibility of mutual supplementation in instruction.

To develop the best use of factor analysis it needs to be viewed in the context of multivariate experimental research design, as in the *Handbook of Multivariate Experimental Psychology*. It has taken psychologists an entire generation to realize that the bivariate model, as it descended from Wundt in perception and Pavlov in reflexology, is far from being the whole of experimental psychology. Psychological causation is invariably multiple and the clinician, especially, has to deal with complex patterns rather than single symptoms and single causes.

The factor analyst's specification equation, expressing an act as the complex resultant of a multi-trait personality profile in a multi-dimensional situation, is the best model we have for both research and applied psychology. Unfortunately, many factor analyses have not been carried out with the refinements necessary for this model and have been a mere device for condensing a mass of data in an arbitrary way. The present writer has suggested six criteria of a factor analytic study essential for its contribution to the scientific model as commonly understood, and these criteria begin with a highly strategic choice of variables and samples. As of this decade one still meets all too often the situation where a student or researcher brings to a multivariate experimentalist a would-be factor analytic resolution for help at the last stage, the study being marred from the beginning by a wrong step at about half the choice points in planning and technique. This is perhaps the price we pay for the computer, which came at a most fortunate time for factor analysis, but which is now grossly misused simply because it is available, by researchers who have no training in multivariate design and want only to push a button on a big machine.

For some reason perhaps as many as one-third of factor analyses that are accepted for publication by editors are patently deficient in the half dozen main requirements of an adequate factor analysis. They start a chain of misleading ideas, being quoted by authors of surveys with no more discrimination than the editors, and contributing in the end only to the view of onlookers that factor analysts cannot get mutually consistent results. It is no remedy for this situation to have editors swing to the opposite extreme, as occurs in some journals, e.g., in perception, learning, and developmental psychology, by rejecting all factor analytic experiments on the grounds that they are too complex for their readers.

Basically we are facing the need for education, of editors and of journal readers, which will come about only when it is recognized that no psychologists are qualified in this complex behavioral science until they understand at least the logic of the factor analytic model, and hopefully some of the algebra of multivariate research. *Factor Analytic Studies: 1971-1975* should contribute substantially to this aim. Moreover, it promises to do so in an unusual and highly effective fashion by providing summaries of more than one thousand studies which exemplify alike good and bad technical procedures, the tremendous variety of applications to which factor analysis has been put, and the difficult problems that occur in attempting to integrate the results of disparate investigations. Used in conjunction with an up-to-date text of basic factor analytic principles it is likely to be a salient contributor to the scientific education of clinicians, as well as future experimental psychologists.

Raymond B. Cattell
Distinguished Research
Professor Emeritus, University
of Illinois
Visiting Professor, University
of Hawaii, Manoa
March 15, 1978

PREFACE

This volume provides a comprehensive survey of factor analytic studies reported in the psychological literature from 1971 to 1975. Each study is summarized and indexed using the *Psychological Abstracts* table of contents. The summaries follow a standard format: purpose, subjects, variables, method, results, and conclusions. Previous volumes, available from the publisher, have reviewed the literature from 1941 to 1970.

This volume contains summaries of the factor analytic studies reported between 1971 and 1975, inclusive, in the following 36 psychological journals: *Journal of Applied Psychology, Journal of Abnormal Psychology, Journal of Consulting and Clinical Psychology, Journal of Counseling Psychology, Journal of Educational Psychology, Journal of Experimental Psychology, Journal of Personality and Social Psychology, Psychometrika, Multivariate Behavioral Research, Educational and Psychological Measurement, Journal of Clinical Psychology, Journal of Personality, Archives of General Psychiatry, Psychological Reports, Perceptual and Motor Skills, Journal of Social Psychology, Personnel Psychology, Child Development, American Journal of Mental Deficiency, American Journal of Psychology, Australian Journal of Psychology, Canadian Journal of Psychology, British Journal of Psychology, British Journal of Medical Psychology, British Journal of Social and Clinical Psychology, Genetic Psychology Monographs, Organizational Behavior and Human Performance, Journal of Nervous and Mental Disease, Journal of General Psychology, American Educational Research Journal, Journal of Experimental Education, Journal of Educational Research, Journal of Educational Measurement, Journal of Special Education,* and the *Journal of Learning Disabilities.*

Preparation of this publication was supported by Grant No. 16-P-56812/RT-13 from the Rehabilitation Service Administration, Office of Human Development, Department of HEW

Factor Analytic Studies 1971-1975

to the Arkansas Rehabilitation Research and Training Center. Appreciation is expressed to Diane Graham for typing the manuscript, to James Duncan for his preliminary literature review, and to Patty George and Nell Spencer for preparing the indexes.

FACTOR ANALYSIS: AN OVERVIEW

Brief History

Factor analysis was invented by a psychologist, Charles Spearman, almost 70 years ago. He advocated a particular approach that produced a large general factor, the most notable example being general intelligence or G. Factor analysis, which achieved prominence as a psychometric technique in the late 1920s and in the 1930s in the fields of differential and educational psychology, primarily through the efforts of Truman Kelly, Godfrey Thompson, and L. L. Thurstone. In the 1940s and 1950s factorial methods were extended to the realm of personality, chiefly by R. B. Cattell and H. J. Eysenck. During the 1950s and 1960s factorial methods were increasingly employed in fields other than psychology—anthropology, economics, physiology, political science, etc.

Maximum research benefit probably has not been realized from this flexible statistical technique. The primary reason for the less than optimal utilization of factorial methods is that factor analysis has never been entirely accepted by the mainstream community of behavioral researchers. The development of factorial methods has been laced with pseudo-controversies and exaggerated claims. Extreme proponents of the technique have offered it as a universal panacea, while no lesser psychologists than Gordon Allport and Ann Anastasia have denied factors any psychological reality, referring to them as artifactors. (See Cureton [1939] for an amusing and insightful portrayal of the early community of factor analysts.)

Role of Factor Analysis

Factor analysis is, in simplest terms, a data reduction procedure. The primary objectives of factor analysis are the achievement of descriptive parsimony by the elimination of redundancy and the enhancement of reliability and generalizability of psychometrically defined constructs by the isolation and definition of major dimensions of variation.

An initial step in the sytematization of knowledge in any discipline or area of investigation is the discovery and operational definition of basic theoretical constructs. Skinner (1961) has made this point in no uncertain terms: "First it must be made clear that the formal properties of a system of variables can be profitably treated only after the dimensional problems have been solved" (p. 252). Any doubts that factor analysis should play a major role in the solution of dimensional problems have been dispelled by Underwood (1957): "But it is only recently that systematic attempts have been made to break down complex psychological dimensions into their component dimensions. In general, some form of factor analysis or derivative therefrom is being used most successfully in this very important work" (p. 44).

Factor Analytic Procedures

There are two basic factor analytic designs: the clustering of persons (Q) and the reduction of items or variables (R). The total variance of a variable or person is partitioned into *common* variance (that which is shared with other variables) and *unique* variance (which includes unreliable or error variance). The common variance is the basis for separating variables or persons into homogeneous clusters which are assumed to reflect underlying patterns or constructs. Factor analytic procedure follows a two stage sequence: the variable or person set is *condensed* into a minimum number of mathematical composite dimensions and then the composite dimensions are *rotated* to achieve a psychologically interpretable final solution.

The process of designing and conducting a factorial investigation can be separated into four major phases:

(1) Definition of the behavioral domain. The behavioral domain should meet three criteria: (1) the behaviors comprising the domain should be precisely specifiable; (b) a process for sampling and measuring the relevant behaviors should be available and feasible; and (c) the results should have reasonable probability of generalization beyond the behavioral sample used.

(2) Measurement of the behavioral domain. The optimal approach to measurement of the behavioral domain begins with a systematic sampling plan and uses multiple measurement methods where necessary. If a Q analysis is planned, an ipsative (forced-choice) measurement procedure is recommended. However, in R designs experimentally dependent variables should not be used.

(3) Reduction of the correlation matrix. After the domain has been measured, the variables (or persons) are intercorrelated. The square matrix which summarizes the relationships among the variables is reduced to a minimum number of independent composite variables, or underlying dimensions of variation. These new composite dimensions are then rotated to a meaningful final position, according to the "simple structure" criterion.

(4) Validation and interpretation. There are three general approaches to the validation and interpretation of the results of factor analytic investigations: (a) interpretation of each factor of cluster by rational analysis of the salient items or variables, (b) correlation of the factors with (or comparison of the person-clusters on) variables external to the original factor analysis, and (c) evaluation of the factorial structure in terms of hypotheses, theory, or previous research. Most factor analytic studies conclude with an interpretation by rational analysis. However, investigations which do not use external validation or evaluation of hypotheses cannot provide confirming evidence of the structure of behavioral domains because the final solution often depends on the judgment of the investigator.

The Meaning of Factors

An issue of long-standing debate in factor analysis concerns the psychological status of factors. One extreme position maintains that factor analysis provides a parsimonious description of a set of variables and argues that no psychological meaning can be assigned to the factors. This position is supported by studies such as that of Armstrong and Soelberg (1968), who factor analyzed random data, and still were able to "interpret" the results in a meaningful fashion.

At the other pole are those who regard factor analysis as a discovery technique that unearths basic psychological traits. Research on Osgood's semantic differential technique, which has spanned a period of 25 years and resulted in more than 1,000 publications, supports this position. The reality of the Evaluation-Potency-Activity structure of human semantic systems can hardly be doubted. Osgood (1969) has recently argued that the E-P-A factor structure reflects an underlying affective meaning system which is innately determined.

As with so many issues, there is no final answer to the question of the proper psychological interpretation of factors. For any particular investigation, several points must be considered: (a) the extent of previous theorizing, (b) the consistency of previous research results, (c) the nature of the variables measured, and (d) the adequacy of variable sampling procedures. (The interested reader is referred to articles by Coan [1964] and Royce [1963] that discuss the issue in depth.)

Additional Reading

Three recent books are recommended for the non-mathematically oriented reader: Comrey (1973), Gorsuch (1974), and Nunnally (1978). The mathematically sophisticated reader is referred to texts by Harman (1976) and Cattell (1978) and to Cattell's (1966) comprehensive review of multivariate methodology. The following articles contain briefer introductions to factor analysis: Cattell (1965a; 1965b), Eysenck (1952; 1953), Guilford (1952), McNemar (1951), and Weiss (1970; 1971).

REFERENCES

Armstrong, J. A., & Soelberg, P. On the interpretation of factor analysis. *Psychological Bulletin,* 1968, *70,* 361-364.

Cattell, R. B. Factor analysis: An introduction to essentials. *Biometrics,* 1965, *21,* 190-210 (a).

—. Factor analysis: An introduction to essentials. *Biometrics,* 1965, *21,* 405-435 (b).

—. (Ed.). *Handbook of multivariate experimental psychology.* Chicago: Rand-McNally, 1966.

—. *The scientific use of factor analysis.* New York: Plenum, 1978.

Coan, R. W. Facts, factors, and artifacts: The quest for psychological meaning. *Psychological Review,* 1964, *71,* 123-140.

Comrey, A. L. *A first course in factor analysis.* New York: Academic Press, 1973.

Cureton, E. E. The principal compulsions of factor-analysts. *Harvard Educational Review,* 1939, *9,* 287-295.

Eysenck, H. J. Uses and abuses of factor analysis. *Applied Statistics,* 1952, *1,* 45-49.

—. The logical basis of factor analysis. *American Psychologist,* 1953, *8,* 105-114.

Gorsuch, R. L. *Factor analysis.* Philadelphia: Saunders, 1974.

Guilford, J. P. When not to factor analyze. *Psychological Bulletin,* 1952, *54,* 26-37.

Harman, H. *Modern factor analysis* (Third edition). Chicago: University of Chicago Press, 1976.

McNemar, Q. The factors in factoring behavior. *Psychometrika,* 1951, *16,* 353-359.

Nunnally, J. C. *Psychometric theory* (Second edition). New York: Mc-Graw-Hill, 1978.

Osgood, C. E. On the whys and wherefores of E.P.A. *Journal of Personality and Social Psychology,* 1969, *12,* 194-199.

Royce, J. R. Factors as theoretical constructs. *American Psychologist,* 1963, *18,* 522-528.

Skinner, B. F. The flight from the laboratory. In *Cumulative Record* (Second edition). New York: Appleton-Century-Crofts, 1961.

Underwood, B. J. *Psychological research.* New York: Appleton-Century-Crofts, 1957.

Weiss, D. J. Factor analysis and counseling research. *Journal of Counseling Psychology,* 1970, *17*(5), 477-485.

—. Further considerations in applications of factor analysis. *Journal of Counseling Psychology,* 1971, *18*(1), 85-92.

III.   EXPERIMENTAL PSYCHOLOGY

Shay, C. B., Zimmerman, W. S., & Michael, W. B. The factorial validity of a rating scale for the evaluation of research articles. Educational and Psychological Measurement, 1972, 32, 453-457.
Purpose: To identify dimensions underlying a 25-characteristic rating scale for the evaluation of research articles
Subjects: 125 educational research articles from 39 journals, rated by a nominated expert
Variables: 25 item scores
Method: Principal components condensation with normal varimax rotation
Results: Eight factors emerged: I Method of analysis; II Design; III Sampling; IV Rigor; V Significance; VI Hypothesis; VII Exposition; VIII Objectivity.

A. Perception

Lehman, H. S. A multivariate model of synesthesia. Multivariate Behavioral Research, 1972, 7, 403-439.
Purpose: To identify clusters of color and adjective space descriptors
Subjects: 1 & 2) 90 college students, 3) subgroups of 50 who "saw" and 50 who did not "see" colors
Variables: 1) 30 adjective and 30 color ratings for (18) pieces of music, 2) adjective and color descriptions separately, 3) color and adjective descriptions for seers separately from nonseers
Method: Cluster analysis, independently for each subject-variable set
Results: Three color clusters (I Yellows vs. nonspectral grays and light olive; II Blue and purple; III Black and dark gray vs. white) and three adjective clusters (IV Happy, cheerful, merry, non-ponderous; V Yielding, delicate, non-emphatic, tranquil, tender; VI Lofty, majestic) emerged in the first analysis. In the second analyses, four composition clusters were defined by colors (I Light and fast; II Strong, heavy with brass, slight religious overtones; III Calm, quiet, soothing; IV Religious) and four highly similar ones were defined by adjectives. For seers, four color clusters (I Yellow green; II Purple, blue; III Nonspectral; IV Orange, orange-red, dark orange) and six adjective clusters (I Happy, cheerful, humorous, non-solemn; II Yielding, soothing, tranquil, tender; III Lofty, martial, majestic, triumphant; IV Exhilarated; V Good; VI Non-emphatic, delicate) emerged. For nonseers, three color clusters (I Yellow, gray; II Blue-purple; III Black-white) and three adjective clusters (I Happy; II Tender; III Lofty, majestic) emerged.

1. Illusion
2. Time

B. Vision
  Gronwall, D. M. A., & Sampson, H. Ocular dominance: A test
    of two hypotheses. British Journal of Psychology, 1971,
    62, 175-185.
    Purpose: To identify dimensions of ocular dominance
    Subjects: 50 college students
    Variables: Scores on 12 rapid tests of eye dominance, 3
      stereoscopic rivalry tests, phi movement, and 2 psy-
      chophysical methods scores
    Method: Cluster analysis
    Results: Six clusters emerged: I Nuclear tests; II Eye-
      closing; III Hold card; IV Two eye-muscle; V Sensory;
      VI Alignment.
    Conclusions: These data do not provide unequivocal support
      to the notion of one kind of eye dominance, although 14
      of the 18 tests appeared to be measuring some common
      property.
  1. Perception
  Coles, G. J., & Stone, L. A. A new methodological revision
    of Ekman's "content" model of multidimensional simi-
    larity analysis. Multivariate Behavioral Research,
    1972, 7, 85-107.
    Purpose: To identify dimensions of similarity judgments
      (by reanalyzing 14 data sets for various types of
      stimulus objects)
    Study One (Ekman, 1954)
    Subjects: Unspecified
    Variables: (Correlational) similarities of 14 colored
      stimuli
    Method: Principal components condensation with varimax
      and oblique rotation
    Results: Three factors emerged: I Blue-yellow; II Green-
      red; III Luminosity.
    Study Two (Kunnapas, Malhammar, & Svenson, 1964)
    Subjects: Unspecified
    Variables: (Correlational) similarities of 1) 7 parallelo-
      grams, 2) 7 heterogeneous geometrical figures
    Method: Principal components condensation with varimax
      rotation, by variable set
    Results: Three factors emerged for parallelograms: I
      Concentrated vs. extended area; II Horizontal extension;
      III Vertical extension. Three factors emerged in the
      second analysis: I Triangularity and roundness; II
      Square form; III Triangularity vs. cross tendency.
    Study Three (Ekman, 1965)
    Subjects: Unspecified
    Variables: (Correlational) similarities for 1) 9 Swedish
      words denoting emotional states, 2) 6 olfactory stimuli,
      3) 7 constant-brightness color stimuli, 4) 10 Swedish
      words denoting personality traits
    Method: Same as above

2

Results: Two emotional states factors emerged: I Happiness vs. sadness; II Unlabeled. Two olfactory factors emerged: I-II Unlabeled. Two color factors emerged: I Green-yellow; II Intensity or luminosity. Two personality factors emerged: I Shyness vs. self-centration; II Unsociable vs. independent.

Study Four (Kunnapas, 1966)
Subjects: Unspecified
Variables: (Correlational) similarities for 9 capital letters
Method: Same as above
Results: Three factors emerged: I Rectangularity and roundness; II Vertical linearity; III A.

Study Five (Kunnapas, 1967)
Subjects: Unspecified
Variables: (Correlational) similarities (visual memory) for 9 capital letters
Method: Same as above
Results: Three factors emerged: I Rectangularity and roundness; II Vertical linearity; III A.

Study Six (Kunnapas, 1968)
Subjects: Unspecified
Variables: (Correlational) similarities of 9 letters in terms of 1) acoustic perception, and 2) acoustic memory
Method: Same as above
Results: Three perception and three memory factors emerged: I O and F; II J; III E (singlet).

Study Seven (Kunnapas & Janson, 1969)
Subjects: Unspecified
Variables: (Correlational) similarities for 28 lower-case (Swedish) letters in terms of visual perception
Method: Same as above
Results: Seven factors emerged: I Vertical linearity and vertical linearity with dot vs. angularity open upward; II Roundness attached to vertical linearity; III Roundness; IV Parallel vertical linearity; V Roundness attached to a hook; VI Vertical linearity with crossness; VII Zigzaggedness.

Study Eight (Magnusson & Ekman, 1968)
Subjects: Unspecified
Variables: (Correlational) similarities for 11 personality traits
Method: Same as above
Results: Four factors emerged: I Not labeled; II Gay, candid vs. dependable, careful; III Helpful, kindhearted vs. ambitious, eager to learn; IV Ambitious, eager to learn vs. harmonious, gray.

Study Nine (Bratfisch & Ekman, 1969)
Subjects: Unspecified
Variables: (Correlational) similarities for 10 factor tests from a standardized intelligence test battery (Delta Battery)

3

Method: Same as above
Results: Two factors emerged: I Verbal comprehension,
    perceptual speed vs. reasoning ability, spatial ability;
    II Number computations.
Conclusions: Each of these solutions is contrasted with
    that originally reported.
Levin, J. A comparison of multidimensional scaling and fac-
    tor analysis of Ekman's color similarity matrix. Journal
    of General Psychology, 1972, 87, 111-115.
Purpose: To identify dimension of a rescaled matrix repre-
    senting Ekman's color similarity matrix
Subjects: Unspecified subjects used by Ekman (1954)
Variables: Rescaled matrix representing similarity ratings
    for 14 colors
Method: Principal components condensation with varimax
    rotation
Results: Three unlabeled (but similar to Stone & Cole's
    1971) factors emerged.
Sailor, P. J. Perception of line in clothing. Perceptual
    and Motor Skills, 1971, 33, 987-990.
Purpose: To identify dimensions underlying perception of
    line in clothing
Subjects: 29 undergraduate and 8 graduate students, plus
    9 faculty from a Textiles, Clothing, and Design Depart-
    ment
Variables: Q-sort of 46 line drawings of street dress for
    women
Method: Principal components condensation with orthogonal
    rotation (inverse)
Results: Four factors emerged: I Lines; II Fashion; III
    Silhouette; IV Detailing.
Smith, O. W., Smith, P. C., & Koutstaal, C. W. Apparent
    parallelism: Uni- or multidimensional? Perceptual
    and Motor Skills, 1972, 34, 834.
Purpose: To identify dimensions of apparent parallelism
Subjects: 52 individuals (unspecified here); raw data from
    Smith & Smith (1963)
Variables: 48 measures of apparent parallelism and 2 paper-
    and-pencil spatial visualization test scores
Method: Principal components condensation with varimax
    rotation
Results: An unspecified number of factors were extracted;
    however, it was concluded that apparent parallelism is
    not unitary.
Spring, C. Perceptual speed in poor readers. Journal of
    Educational Psychology, 1971, 62, 492-500.
Purpose: To identify dimensions of letter confusion
Subjects: Unspecified number (here) of preschoolers used
    by Gibson, Osser, Schiff, & Smith (1963)
Variables: Confusion-frequency matrix for 26 upper-case
    letters (Gibson et al.'s two matrices added together)
Method: Principal components condensation with varimax
    rotation

Results: Eight factors emerged: I Concave up and down;
        II Oblique lines; III Maximum number of points of inter-
        section with any horizontal line is three or more; IV
        Curved lines; V Vertical lines; VI Enclosed areas; VII
        Maximum number of points of intersection with any hori-
        zontal and any vertical line is one, disregarding serifs;
        VIII Concave up or down, but not both.
2. Size and Distance and Depth Perception
3. Color Vision
4. Form and Pattern Discrimination
5. Eye Movement
6. Brightness and Contrast Discrimination
C. Audition
  1. Perception
  2. Speech Discrimination
  3. Audiometry
D. Chemical Senses
    Gregson, R. A. M. Odour similarities and their multidimen-
        sional metric representation. Multivariate Behavioral
        Research, 1972, 7, 165-174.
    Purpose: To identify dimensions of odor similarities
    Subjects: 18 college students
    Variables: Similarity judgments (45) for 10 diverse odors
    Method: Inverse principal components condensation without
        rotation
    Results: The extracted factors were not labeled.
E. Somesthesia
F. Environmental Effects
    Holzman, P. S., & Rousey, C. Disinhibition of communicated
        thought: Generality and role of cognitive style. Journal
        of Abnormal Psychology, 1971, 77, 263-274.
    Purpose: To identify dimensions of cognitive control under
        white noise masking (WNM) conditions
    Subjects: 80 males (average age=34.8 years) and 80 females
        (38.6), all middle-class whites
    Variables: 9 cognitive control scores, age, 4 Davis Read-
        ing Test scores, 2 Digit substitution scores, 4 TAT
        scores
    Method: Principal axes condensation with normal varimax
        rotation, by sex
    Results: Seven factors emerged: I Field articulation;
        II Uninterpreted; III Scanning; IV-V Uninterpreted; VI
        Disinhibitory tendency under WNM; VII Leveling-sharpen-
        ing.
G. Sleep and Fatigue and Dreams
H. Hypnosis and Suggestibility
    Coe, W. C., & Sarbin, T. R. An alternative interpretation to
        the multiple composition of hypnotic scales: A single
        role-relevant skill. Journal of Personality and Social
        Psychology, 1971, 18, 1-8.
    Purpose: To identify dimensions underlying the Harvard
        Group Scale of Hypnotic Susceptibility (HGSHS)
    Subjects: 168 Ss, apparently college students (data from
        Coe & Sarbin, 1966)

Variables: 19 HGSHS scores
Method: Unspecified condensation
Results: Three factors emerged: I General hypnotic fac-
tor (high-scorer); II Moderate difficulty; III Easiest
items.
Conclusions: Results from several other studies are re-
viewed for comparison.
Curran, J. D., & Gibson, H. B. Critique of the Stanford
Hypnotic Susceptibility Scale: British usage and
factorial structure. Perceptual and Motor Skills, 1974,
39, 695-704.
Purpose: To identify dimensions underlying a modified
Stanford Hypnotic Susceptibility Scale (SHSS-A)
Subjects: 43 college students
Variables: 12 SHSS-A scores
Method: Principal components condensation with unspeci-
fied rotation
Results: Four factors emerged: I Loss of voluntary con-
trol over the musculature; II Hallucination; III Active
motor compliance within hypnosis; IV Active motor com-
pliance in waking state.
Gheorghiu, V. A., Hodapp, V., & Ludwig, C. M. Attempt to
construct a scale for the measurement of the effect of
suggestion on perception. Educational and Psychological
Measurement, 1975, 35, 341-352.
Purpose: To identify dimensions underlying a scale designed
to measure the effect of suggestion on perception.
Subjects: 112 eleventh- and twelfth-graders
Variables: Scores on 12 tests of the effect of suggestion
on perception (tactual, auditory, visual)
Method: Principal components condensation with varimax
rotation
Results: One unlabeled factor emerged; other factors
unique to sensory modality or type of suggested were
not identified.
Perry, C. Imagery, fantasy, and hypnotic susceptibility:
A multidimensional approach. Journal of Personality
and Social Psychology, 1973, 26, 217-221.
Purpose: To identify the relationship between imagery,
fantasy, and hypnotic suggestibility
Subjects: 62 college students
Variables: 23 scores on imagery, fantasy, and hypnotic
susceptibility measures
Method: Unspecified condensation with varimax rotation
Results: Five factors emerged: I Dream recall; II Bett's
QMI; III Distortion indexes; IV Hypnotic susceptibility;
V Not interpreted.
Conclusions: These results suggest that imagery, fantasy,
and hypnotic susceptibility are independent dimensions.
Tellegen, A., & Atkinson, G. Openness to absorbing and
self-altering experiences ("absorption"), a trait re-
lated to hypnotic suggestibility. Journal of Abnormal
Psychology, 1974, 83, 268-277.

6

Purpose: To identify dimensions underlying hypnotic sug-
    gestibility as measured by the Research Questionnaire
    (Q3)
Subjects: 1) 481 female college students, 2) subsample of
    142, 3) subsample of 171
Variables: 1) 71 Q3 item scores, 2 & 3) 11 Q3 factors
    from 1) plus Ego Resiliency and Ego Control scale scores
Method: Principal axes condensation with normal varimax
    rotation, independently for the three subject-variable
    sets
Results: Eleven Q3 factors emerged in the first analysis:
    I Reality absorption; II Fantasy absorption; III Dis-
    sociation; IV Sleep automatism; V Openness to experience;
    VI Devotion and trust; VII Autonomy-skepticism; VIII
    Optimism-placidity; IX Aloofness-reserve; X Caution
    vs. impulsiveness; XI Relaxation. Three factors
    similar across analyses, emerged for the second and
    third analyses: I Openness to absorbing and self-
    altering experiences, or absorption; II Stability;
    III Introversion.

I. Motivation and Emotion

Frazier, J. R. An exploratory attempt to define the compon-
    ents of activation. Psychological Reports, 1974, 34,
    1137-1138.
Purpose: To identify dimensions underlying a battery of
    physiological and behavioral measures of activation
Subjects: 40 male college students with satisfactory
    vision and generally good health
Variables: Scores from a 16-test battery of physiological
    and behavioral measures: physiological (4), standard
    anxiety questionnaires (3), WAIS subtests (3), Flicker
    Fusion and Tachistoscopic Threshold measures, PA task,
    RT, verbal fluency, concept-formation task
Method: Principal factors condensation with varimax
    rotation
Results: Three of seven factors extracted, were interpre-
    table: I Concentration-visualization; II Intensity A;
    III Intensity B.

Wiederanders, M. R. Effects of failure experiences on con-
    figural properties of the aspiration level concept.
    Psychological Reports, 1975, 37, 371-377.
Purpose: To identify dimensions of aspiration under pre-
    test and post-failure conditions
Subjects: 113 college students
Variables: Scores on 9 aspiration items, administered at
    pre-task and post-failure
Method: Principal factors condensation with varimax rota-
    tion; oblique cluster analysis (each by administration)
Results: Two pre-task factors (I Realism/fears; II Future
    pretentions) and three post-failure factors (I Realism/
    fears; II Intermediate hopes; III Future pretentions)
    emerged in both types of analyses.

7

J. Attention and Expectancy and Set
  Sack, S. A., & Rice, C. E. Selectivity, resistance to dis-
    traction and shifting as three attentional factors.
    Psychological Reports, 1974, 34, 1003-1012.
  Purpose: To identify dimensions of attention processes
  Subjects: 164 eighth-graders
  Variables: Scores on 3 measures of selectivity of atten-
    tion, 3 measures of distraction, 2 measures of shifting,
    and Speed of Color Discrimination Test
  Method: Principal components condensation with orthogonal
    and oblique transformations to approximate the hypothesized
    factor structure
  Results: The orthogonal solution was not considered a good
    fit for the data. Three oblique factors emerged: I
    Selectivity; II Resistence to distraction; III Shifting.
K. Motor Performance
  Briggs, P. F., & Tellegen, A. Development of the Manual
    Accuracy and Speed Test (MAST). Perceptual and Motor
    Skills, 1971, 32, 923-943.
  Purpose: To identify dimensions underlying a battery of
    manual accuracy and speed tests
  Subjects: 60 males, and 60 females, aged 16-39 years
  Variables: Scores on Tapping, Large Peg Placement, Small
    Peg Placement, Nails Pickup, Holes Total, True Travel,
    plus age
  Method: Principal axes condensation with varimax rotation,
    by sex
  Results: Seven specific factors (corresponding to the
    seven tests), emerged.
  Firth, C. D. Strategies in rotary pursuit tracking. British
    Journal of Psychology, 1971, 62, 187-197.
  Purpose: To identify dimensions of rotary pursuit tracking
  Subjects: 30 male volunteers, aged 20-35 years
  Variables: Scores derived from a 5 min. session of rotary
    pursuit tracking: 5 measures of hit-and-miss distri-
    bution, frequency of rest pauses, 2 measures of occur-
    rence of rhythmicities, total time on target, number of
    hits at maximum of curve; plus E and N scores from EPI
  Method: Principal components condensation with unspecified
    rotation
  Results: Two factors emerged: I Level of attainment; II
    Strategy.
  Geddes, D. Factor analytic study of perceptual-motor attri-
    butes as measured by two test batteries. Perceptual
    and Motor Skills, 1972, 34, 227-230.
  Purpose: To identify relationships between Perceptual-
    motor Attributes of Mentally Retarded Children and
    Youth battery (PAMRCY) and the Purdue Perceptual-motor
    Survey (PPS) with normal children
  Subjects: 80 public school first- and second-graders
  Variables: 6 PAMRCY and 22 PPS scores
  Method: Principal components condensation with varimax
    rotation

8

Results: Ten factors emerged: I Visual tracking; II Visual discrimination and copying of forms; III Visual discrimination and copying of rhythmic patterns; IV Verbal body image; V Dynamic balance; VI Spatial body perception; VII Postural maintenance; VIII Visual discrimination and copying of motor patterns; IX Gross agility; X Uninterpretable.

Meikle, S., Lonsbury, B., & Gerritse, R. Factor analysis of the Motor Steadiness Battery. Perceptual and Motor Skills, 1973, 36, 779-783.

Purpose: To identify dimensions underlying the Motor Steadiness Battery (MSB)

Subjects: 50 8- to 12-year-olds

Variables: 20 MSB-derived scores

Method: Principal components condensation with varimax rotation

Results: Ten factors emerged: I Simple kinetic movement; II Rate with which Ss can oscillate their feet; III Static steadiness; IV Vertical finger oscillation speed; V Complex kinetic steadiness; VI Errors and contact time from the Horizontal Groove Steadiness Test; VII Time for Foot Tapping Test; VIII-X Not interpreted.

Palmer, R. D. Dimensions of differentiation in handedness. Journal of Clinical Psychology, 1974, 30, 545-552.

Purpose: To identify dimensions of intraindividual differentiation in the development of handedness

Subjects: 60 male college students

Variables: Scores on the Embedded Figures Test, plus a handedness questionnaire and 19 manual skills tasks

Method: Principal components condensation with normal orthogonal varimax rotation

Results: Four factors emerged: I Focused application of power or strength; II Precision-control in fine hand movements; III-IV Uninterpretable.

Poole, C., & Stanley, G. A factorial and predictive study of spatial abilities. Australian Journal of Psychology, 1972, 24, 317-320.

Purpose: To identify dimensions of spatial abilities

Subjects: 163 first-year engineering students

Variables: Scores on 7 spatial abilities tests, plus an academic selection quota score and 7 criterion measures (final exam marks, etc.)

Method: Principal axes condensation with varimax rotation

Results: Three factors emerged: I Academic performance; II Visualization and manipulation of images; III Scanning and path finding.

L. Reaction Time

Stone, G. C. Individual differences in information processing: Comparison of simple visual stimuli. Perceptual and Motor Skills, 1971, 33, 395-414.

Purpose: To identify dimensions of RT to simple visual stimuli

Subjects: 40 young adults
Variables: RT scores to simple visual stimuli: 1) 48
odd and 48 even trials, 2) 10th and 50th percentile
scores for 2 sessions
Method: Principal components condensation with varimax
rotation, independently by odd and even trials, by
percentile scores
Results: The trials (tests) analyses produced five factors: I General; II Fast on first and relatively slow
on second and third tests; III Relative speed in
oddity vs. slowness in matching tasks; IV Changes due
to practice, matching-oddity difference, form-color
difference; V Not labeled. Three factors emerged for
percentile scores: I General; II Form-color difference;
III Matching-oddity differentiation.
M. Learning
   Harris, E. L., Lemke, E. A., & Rumery, R. E. Generalized
   learning curves and their ability and personality
   correlates. Multivariate Behavioral Research, 1974,
   9, 21-26.
   Purpose: To identify dimensions underlying learning curves
   for a 4-choice probability learning task
   Subjects: 118 college students
   Variables: 18 trial scores
   Method: Principal components condensation with graphic
   orthogonal rotation
   Results: Three factors emerged: I Consistent with Estes'
   stimulus sampling theory; II Pure strategy of Weir and
   Kendlers; III Consistent with Weir's hypothesis testing
   strategy.
   1. Conditioning
   2. Verbal Learning
   Berry, F. M., Detterman, D. K., & Mulhern, T. Stimulus en-
   coding as a function of modality: Aural versus visual
   paired-associate learning. Journal of Experimental
   Psychology, 1973, 99, 140-142.
   Purpose: To identify dimensions of trial-by-trial error
   for subjects presented CCCs aurally or visually
   Subjects: 1) 72 Ss using visual modality, 2) 72 Ss auditory
   modality, and 3) 144 Ss from both conditions, all college
   students
   Variables: Number of PA learning errors on Trials 1-10
   Method: Principal components condensation with varimax
   rotation, independently for each S group
   Results: Two similar factors emerged in all analyses:
   I Early-trial error; II Late-trial error.
   Conclusions: These results support a two-stage conceptualization of PA learning.
   Koh, S. D., Kayton, L., & Schwarz, C. The structure of word
   storage in the permanent memory of nonpsychotic schizophrenics. Journal of Consulting and Clinical Psychology,
   1974, 42, 879-887.

Purpose: To identify the storage structure of common
    English words in the memory of young nonpsychotic
    schizophrenics
Subjects: 1) 17 schizophrenic and 12 nonschizophrenic
    psychiatric patients, 16 normals, aged 17-29 years;
    2) 20 schizophrenics and 20 normals, aged 18-28
Variables: Q-sort of 40 common English words 1) without
    and 2) with time pressure
Method: Hierarchical cluster analysis, independently for
    each subgroup within each experiment
Results: In Experiment 1, eight clusters representing man-
    made objects (I Clothing; IV Furniture), natural re-
    sources (II Fuels; III Metals), lifelike creatures (V
    Shost-shadow; VI Humans), and nature--its associations
    (VII Earth formations; VIII Wheat-butter-market) emerged.
    In Experiment 2, clusters were interpreted with the
    aid of post-sort pleasantness ratings: I Pleasantness;
    II Unpleasantness; III Affectively neutral.
Koutstaal, C. W., Smith, O. W., & Knops, L. Correlates of
    Greenberg and Jenkins' S-scale across languages.
    Perceptual and Motor Skills, 1971, 32, 683-688.
Purpose: To identify dimensions underlying indices of
    CCVCs' distance from English, effort (subvocalization)
    ratings, strangeness ratings, and phonetic contrast
    (PCS) ratings
Subjects: 25 native speakers of English (USA college stu-
    dents), 17 Swedish high school students, 21 and 23
    Dutch high school students, 18 Dutch college students,
    22 Flemish college students
Variables: Scores on effort (subvocalization) ratings (6),
    S scale, strangeness values (4), PCS ratings (4) for
    CCVCs
Method: Principal components condensation with varimax
    rotation
Results: Two factors emerged: I S-scale; II Dutch asso-
    ciation.
Lippman, L. G. Item effects as indicators of subjective
    organization in serial learning. American Journal of
    Psychology, 1974, 87, 699-705.
Purpose: To examine the constancy of item effects across
    groups of subjects learning a short list of (moderately
    difficult) CVCs
Subjects: Serial learning performance data: at least 180
    unspecified (presumably college student) subjects;
    data reported by Lippman, 1971, 1974; and Lippman and
    Denny, 1964
    Ease of learning ratings: 53 introductory experimental
    psychology students
Variables: Nine sets of serial learning performance mea-
    sures, rating of ease of learning (the same CVCs)
Method: Unspecified condensation with varimax rotation
Results: Two factors emerged: I Item effects (uniform
    despite task differences); and II Rated ease of learning.

Conclusion: Rated ease of learning was only weakly re-
lated to item effects.
Meyers, L. S., & Boldrick, D. Memory for meaningful con-
nected discourse. Journal of Experimental Psychology:
Human Learning and Memory, 1975, 1, 584-591.
Purpose: To identify dimensions of recall data.
Subjects: 168 college students
Variables: 6 recall (of prose passage) scores
Method: Principal components condensation with varimax
rotation
Results: Three factors emerged: I General coherency of
output; II Sheer bulk of output; III Associative.
3. Verbal Paired Associate Learning
Butollo, W. H. A logistic test model approach to changes
of the factorial structure during learning. Journal
of General Psychology, 1972, 86, 189-200.
Purpose: To identify performance dimensions of learning
trials
Subjects: 80 college students
Variables: Scores on 27 5-item paired associate acquisition trials
Method: Matrix reduction condensation, apparently without rotation
Results: Four factors were reported: I Increase according
to average learning performance; II One clear change of
direction; III-IV Increasing numbers of direction
changes.
Conclusions: Analysis with Rasch's Logistic Test Model
did not yield changes in factorial structure of learning
performance, as the factor analytic results did.
4. Reinforcement
N. Memory
Aliotti, N. C. Note on validity and reliability of the
Bannatyne Visuo-spatial Memory Test. Perceptual and
Motor Skills, 1974, 38, 963-966.
Purpose: To identify relationships among a visual-motor
and motor abilities test battery and the Bannatyne
Visuo-spatial Memory Test (BVMT), a non-motoric visual
memory test
Subjects: 95 first- and second-graders
Variables: 27 scores from the Bender-Gestalt Visual-motor
Test, Rutgers Drawing Test, Memory-for-Designs,
Bannatyne Test, and BVMT
Method: Principal components condensation with varimax
rotation
Results: Three factors emerged: I Visual-motor copy
performance; II Visual memory; III Finger tapping
activity.
Blaine, D. D., & Dunham, J. L. Effect of availability on
the relationship of memory abilities to performance
in multiple-category concept tasks. Journal of
Educational Psychology, 1971, 62, 333-338.
Purpose: To identify dimensions underlying six memory
tests

Subjects: 60 college students
Variables: Scores on 6 memory tests (Guilford type)
Method: Principal axes condensation with varimax rotation
Results: Two factors emerged: I Rote memory; II Classes
memory.
   1. Short Term and Immediate Memory
O. Thinking
Adcock, C. J., & Webberley, M. Primary mental abilities.
Journal of General Psychology, 1971, 84, 229-243.
Purpose: To identify dimensions of primary abilities
using Guilford's content categories (except unexplored
psychological category) for practically significant
cognitive factors
Study One
Subjects: 89 college students
Variables: Scores on 12 Kit of Reference Tests (KIT)
tests, 9 other standard measures (e.g., selected 16PF
factors), 16 new measures, all theoretically measuring
8 operational and 3 content areas
Method: Principal-factor condensation with varimax and
promax rotation
Results: Twelve factors emerged: I Insight; II Fluency;
III Spatial ability; IV Perceptual speed; V Reasoning;
VI Word Gestalt completion; VII Sophistication; VIII
Alert impulsiveness; IX Perceptual closure; X Associa-
tive memory; XI Specific to 16PF; XII Impulsiveness?
Study Two
Subjects: 176 college students
Variables: Scores on 10 new tests, plus 3 DAT scores
Method: Principal factor condensation with varimax and
promax rotation
Results: Eleven factors emerged: I General intelligence;
II Symbol span; III Figural span; IV Figural fluency;
V Rote learning; VI Associative memory; VII English
language; VIII Perceptual Closure; IX Verbal fluency;
X Perceptual speed; XI 16PF intelligence.
Study Three
Subjects: 196 college students
Variables: Scores on 25 tsets for figural and semantic
abilities
Method: Principal-factor condensation with varimax and
promax rotation
Results: Ten factors emerged: I Flexibility of semantic
closure; II Examination success; III Perceptual span;
IV Perceptual closure; V Insight or general intelli-
gence; VI Figural memory scanning; VII Memory span;
VIII Study incentive; IX Fluency; X Reasoning.
Ashton, R., & White, K. Factor analysis of the Gordon Test
of Visual Imagery Control. Perceptual and Motor Skills,
1974, 38, 945-946.
Purpose: To identify dimensions underlying the Gordon Test
of Visual Imagery Control (GRVIC)

13

Subjects: 1562 first-year psychology students (Australian)
Variables: 12 GTVIC scores
Method: Principal components condensation with varimax rotation
Results: Three unlabeled factors emerged.
Conclusions: The GTVIC does not have one factor called Image controllability.

Das, J. P. Structure of cognitive abilities: Evidence for simultaneous and successive processing. Journal of Educational Psychology, 1973, 65, 103-108.
Purpose: To identify dimensions of information integration
Subjects: 1) 60 middle- and low-SES Canadian, and 2) 90 middle- and low-income Indian fourth-grade males
Variables: Scores on 1) six cognitive and five achievement tests, 2) six cognitive tests
Method: Principal components condensation with varimax rotation, independently for the two subject-variable sets
Results: Four factors emerged for Canadian boys: I Successive mode; II School achievement; III Simultaneous mode; IV Speed. Three factors emerged for Indian boys: I Simultaneous processing; II Speed; III Successive mode (tentative).

Das, J. P., & Molloy, G. N. Varieties of simultaneous and successive processing in children. Journal of Educational Psychology, 1975, 67, 213-220.
Purpose: To identify dimensions of information integration in children
Subjects: 60 high- and low-SES boys each in first and fourth grade (IQs in dull-normal range)
Variables: Scores on 1) 8 cognitive tests, 2) plus L-T Digit Span and IQ
Method: Principal components condensation with varimax rotation, independently by grade for the original battery and for Grade 4 for the expanded battery
Results: Three highly similar factors emerged, across grades for the 8-test battery and for Grade 4's expanded battery: I Successive mode; II Simultaneous mode; III Speed.

Leibovitz, M. P., London, P., Cooper, L. M., & Hart, J. T. Dominance in mental imagery. Educational and Psychological Measurement, 1972, 32, 679-703.
Purpose: To identify dimensions underlying measures of mental imagery
Subjects: 126 females, aged 16-61 years
Variables: Scores on 46 variables derived from individual and group imagery tests
Method: Principal components condensation with varimax rotation
Results: Thirteen factors emerged: I Secondary imagery; II Primary imagery; III Ss' pre-test reports; IV Kinesthetic imagery; V Tactile; VI Susceptibility test specific; VII Group test specific; VIII Nonspecific

imagery; IX Hearing imagery; X EEG alpha specific; XI
Second choice touch; XII Secondary taste-smell; XIII
Age.
White, K., Ashton, R., & Lan, H. Factor analyses of the
shortened form of Betts' Questionnaire Upon Mental
Imagery. Australian Journal of Psychology, 1974, 26,
183-190.
Purpose: To identify dimensions underlying a shortened
form of Betts' Questionnaire Upon Mental Imagery (QMI)
Subjects: 962 female and 600 male college students
Variables: 35 QMI item scores
Method: 1) Principal components condensation with varimax
rotation; 2) image analysis (principal components con-
densation with varimax rotation); and 3) principal com-
ponents condensation with oblimin rotation, all indepen-
dently for males, females, and total group
Results: Two factors emerged for all analyses: I General
imagery ability; II Separation of sensory modalities.
1. Problem Solving
Aaron, P. G. Epigenetic factors in information processing.
Psychological Reports, 1972, 31, 407-412.
Purpose: To identify performance dimensions underlying
a perceptual coding task
Subjects: 106 first-, eighth-, and ninth-graders
Variables: 10 perceptual coding task performance scores
Method: Principal axes condensation, apparently without
rotation
Results: Four factors emerged: I Integration; II Direc-
tionality; III Redundancy; IV Inhibition.
Coates, G. D., & Alluisi, E. A. Reliability and correlates
of a three-phase code transformation task (3P-COTRAN).
Perceptual and Motor Skills, 1971, 32, 971-985.
Purpose: To identify dimensions underlying measures of
three-phase code transformation task (3P-COTRAN) per-
formance and measures of intellectual ability and
personality
Subjects: 84 college students
Variables: Scores on 43 measures of 3P-COTRAN performance,
and 32 intellectual abilities and personality measures
(paper-and-pencil)
Method: Principal axes condensation with varimax rotation
Results: Eight factors emerged: I General time per solution; II General error rate; III Time per Phase III
solutions, or relative time in problem solving; IV
Accuracy in Phase II; V Error rate in Phase III solutions,
or relative errors in problem solving; VI Verbal intelligence; VII Social-religious vs. political-economic
values; VIII Internal vs. external motivation.
Morgan, B. B. Jr., & Alluisi, E. A. Acquisition and performance of a problem-solving skill. Perceptual and Motor
Skills, 1971, 33, 515-523.
Purpose: To identify dimensions underlying three-phase
code transformation task (3P-COTRAN) performance

15

Subjects: 20 unspecified individuals
Variables: Scores on 9 3P-COTRAN performance measures
(averaged across problems) for each of 8 sessions
Method: Principal axes condensation with varimax rotation,
by session
Results: Five factors, stable across sessions, emerged:
I General time per solution; II General error rate;
III Time per Phase III solution, or relative time in
problem-solving; IV Error rate in Phase III, or rela-
tive errors in problem solving; V Accuracy in Phase II.
Platt, J. J., & Spivack, G. Unidimensionality of the Means-
Ends Problem-Solving (MEPS) Procedure. Journal of
Clinical Psychology, 1975, 31, 15-16.
Purpose: To identify dimensions of the MEPS
Subjects: 72 male and 66 female psychiatric patients;
476 male adolescent reformatory inmates
Variables: Unspecified number of scores derived from the
nine MEPS stories
Method: Unspecified condensation with orthogonal varimax
rotation, separately for male and female patients and
delinquents
Results: One factor was extracted in all three analyses:
I Means-ends cognition.
Conclusion: This study demonstrated the unidimensionality
of the MEPS.
Sciortino, R. Effects of a two-stage solutions test, production and revision stages administered under neutral
and criteria-cued instructions, on psychological
processes. Journal of General Psychology, 1973, 88,
191-203.
Purpose: To identify dimensions of Sciortino's Inventory
of Psychological Processes (IPP-3) when administered
under two conditions
Subjects: 225 college students
Variables: 50 item scores from the IPP-3 each for post-
production and post-revision Solutions Test stages
(i.e., 450x50 matrix)
Method: Principal components condensation with varimax
rotation
Results: A six-factor solution was considered most acceptable: I Divergent thinking; II Frustration; III
Modification; IV Excitement; V Evaluated synthesis;
VI Incubated thinking.
Sciortino, R. Psychological processes in a three-stage
solutions test--preparation, production, revision--
administered under warm-up, neutral, and criteria-
cued instructions, respectively. Journal of General
Psychology, 1974, 90, 3-15.
Purpose: To identify dimensions of Sciortino's Inventory
of Psychological Processes (IPP-3) when administered
under three conditions
Subjects: 217 college students

Variables: 50 item scores from the IPP-3 each for post-
preparation, post-production, and post-revision
Solutions Test stages (i.e., 651x50 matrix)
Method: Principal components condensation with varimax
rotation
Results: Six factors emerged: I Cross fertilization; II
Frustration; III Evaluated refinement; IV Excitement;
V Motivated thinking; VI Freewheeled thinking.
2. Concepts
P. Decision and Choice Behavior
(Abstract Omitted)
G. Sleep and Fatigue and Dreams
Gillin, J. C., Buchsbaum, M. S., Jacobs, L. S., Fram, D. H.,
Williams, R. B. Jr., Vaughan, T. B. Jr., Mellon, E.,
Snyder, F., & Wyatt, R. J. Partial REM sleep depriva-
tion, schizophrenia and field articulation. Archives
of General Psychiatry, 1974, 30, 653-662.
Purpose: To identify dimensions of sleep changes from
baseline to post REM deprivation for a mixed sample
Subjects: 16 psychiatric inpatients (participating in
longitudinal, all-night sleep studies in their rooms)
Variables: Change scores (from baseline mean to post-
REM-deprivation recovery mean) on 7 sleep variables
Method: Principal components condensation, apparently
without rotation
Results: Four factors emerged: I REM compensation; II
Change in minutes of total sleep; III Change in REM
latency; IV Change in Stages III and IV sleep.

17

IV. PHYSIOLOGICAL PSYCHOLOGY

Martin, W. S., Fruchter, B., & Mathis, W. J. An investigation of the effect of the number of scale intervals on
principal components factor analysis. Educational and
Psychological Measurement, 1974, 34, 537-545.
Purpose: To identify dimensions underlying several physiometric measures (when several different numbers of
scale intervals were used)
Subjects: 110 tenth-grade males
Variables: 11 physiometric variables, scored on 9-, 7-,
5-, 4-, 3-, and 2-interval systems
Method: Principal components condensation with varimax
rotation, independently for each interval scoring
system
Results: Three unlabeled factors emerged in each analysis;
variations in size of eigenvalues and loadings were
discussed.
A. Neurology
    1. Neuroanatomy
B. Lesions
    1. Brain Lesions
    2. Brain Hypothalamic and Hippocampal Lesions
C. Brain Stimulation
    1. Chemical Stimulation
    2. Electrical Stimulation
D. Electrical Activity
    1. Electroencephalography and Evoked Potentials
E. Sensory Physiology
    Wolff, B. B. Factor analysis of human pain responses: Pain
    endurance as a specific pain factor. Journal of Abnormal
    Psychology, 1971, 78, 292-298.
    Purpose: To identify dimensions of pain response for
    cutaneous and deep somatic stimuli
    Subjects: 60 chronic arthritis patients (pre-surgical)
    Variables: Scores on 66 measures of cutaneous pain
    response and 54 measures of deep somatic pain response
    Method: Centroid condensation with varimax rotation
    Results: Four factors emerged: I Cutaneous sensitivity;
    II Gluteal sensitivity A; III Pain endurance; IV Gluteal
    sensitivity B.
F. Biochemistry
    1. Hormones
    2. Drug Effects - Human
    Crane, G. E., & Naranjo, E. R. Motor disorders induced by
    neuroleptics: A proposed new classification. Archives
    of General Psychiatry, 1971, 24, 179-184.
    Purpose: To identify factors of frequently occurring motor
    disorders attributed to neuroleptic drugs
    Subjects: 97 hospitalized (minimum of 2 years) schizophrenics
    free from physical disorders (70 receiving neuroleptics,
    others no drug)
    Variables: Scores on 14 symptoms attributed to neuroleptics
    Method: Principal components condensation with varimax
    rotation   18

Results: Three factors emerged: I Diffused slowing of
    movements and tremor; II Hyperactive; III "Other"
    postural disorder.
Nash, H., & Stone, G. C. Psychological effects of drugs: A
    factor analytic approach. Journal of Nervous and Mental
    Disease, 1974, 159, 444-448.
Purpose: To identify dimensions of psychological drug ef-
    fects (change scores reported by Nash, 1962)
Subjects: 1 and 2) 161 male penitentiary inmates partici-
    pating in a six-condition double-blind drug study;
    3) 201 (same subjects)
Variables: Scores on 18 objective indices of intellectual
    capacity, Sentence Completion, Writing Expansiveness,
    and 2 self-report ratings, at 1) pretest, 2) posttest,
    3) 22 change scores
Method: Principal components condensation with varimax
    rotation
Results: Seven pre- and six posttest factors emerged but
    were not interpreted; the results were similar, and
    neither ANOVA nor Tucker coefficients indicated the
    presence of any drug effects. Ten difference factors
    emerged but were not interpreted; however, two were
    significantly related to drug treatment.
  3. Drug Effects - Animal
G. Cardiovascular Processes
H. Environment and Stress
I. Genetics
J. Personality Correlates
K. Nutrition and Gastrointestinal Processes

V. ANIMAL PSYCHOLOGY

Dudzinski, M. L., Norris, J. M., Chmura, J. T., & Edwards,
   C. B. H. Repeatability of principal components in
   samples: Normal and non-normal data sets compared.
   Multivariate Behavioral Research, 1975, 10, 109-117.
   Purpose: To identify dimensions underlying selected
      rabbit behaviors for four samples, to illustrate
      repeatability of principal components
   Subjects: 1) 224 "dominant" rabbits (non-normal); 2)
      146 rabbits (supplementary); 3) 166 (homogeneous);
      4) 205 (mixed)
   Variables: Log transformations of time in seconds spent
      performing chinning, sniffing, digging, and eating
      behaviors
   Method: Principal components condensation, independently
      by sample
   Results: Two factors emerged in each case but were un-
      labeled.
A. Comparative Psychology
B. Natural Observation
C. Early Experience
D. Instincts
E. Motivation and Emotion
   Royce, J. R., & Poley, W. Invariance of factors of mouse
      emotionality with changed experimental conditions.
      Multivariate Behavioral Research, 1975, 10, 479-487.
   Purpose: To identify dimensions of mouse emotionality
      across conditions
   Subjects: Unspecified numbers of mice: 1) six strains
      (Royce, Poley, & Yeudall, 1973), 2) two strains (Egan
      & Royce, 1973), 3) two strains
   Variables: Scores on 19 emotionality variables
   Method: Alpha condensation with promax rotation, indepen-
      dently for each subject group
   Results: Six factors emerged across studies: I Autonomic
      balance; II Motor discharge; III Acrophobia; IV Terri-
      toriality; V Tunneling 1; VI Tunneling 2.
F. Learning
   1. Conditioning
   2. Discrimination
   3. Avoidance and Escape
   4. Reinforcement
   5. Reinforcement Schedule
   6. Punishment and Extinction
G. Social and Sexual Behavior
H. Sensory Processes

A. Infancy
B. Childhood
    1. Learning
        Adkins, D. C., & Ballif, B. L. A new approach to response
            sets in analysis of a test of motivation to achieve.
            Educational and Psychological Measurement, 1972, 32,
            559-577.
        Purpose: To identify dimensions of elementary school
            children's motivation to achieve, as assessed by
            Gumpgookies (Gu)
        Subjects: 2,313 children (2,063 4-year-olds, 250 first-
            and second-graders) for interpreted results (several
            other samples analyzed)
        Variables: 75 Gu item scores
        Method: Unspecified factor analysis yielding orthogonal
            factors
        Results: Five factors emerged: I Instrumental activity;
            II School enjoyment; III Evaluative; IV Self-confidence;
            V Purposive.
        Berry, F. M., Baumeister, A. A., & Detterman, D. Free-
            learning among intellectually average children and
            mentally retarded individuals: A study of response
            integration. American Journal of Mental Deficiency,
            1971, 76, 118-124.
        Purpose: To identify errors-over-trials dimensions for
            a free-learning task
        Subjects: 96 intellectually average third- through sixth-
            graders, and 96 retarded ("learners") residents of a
            state institution
        Variables: Number of free-learning errors for each of 13
            trials
        Method: Principal components condensation with varimax
            rotation, independently for intellectually average and
            retarded subjects
        Results: Three factors emerged for both analyses, though
            they differed somewhat in definition: I Response
            learning; II Associative learning; III Unexpected,
            uninterpreted.
        Vernon, P. E., & Mitchell, M. C. Social-class differences
            in associative learning. Journal of Special Education,
            1974, 8, 297-311.
        Purpose: To identify dimensions underlying a battery of
            learning measures
        Subjects: 94 "white collar" and 94 "blue collar" children,
            range in age from 9-2 to 12-11 years
        Variables: Scores on 3 subject variables (SES, sex, "youth"),
            3 intelligence tests, 3 verbal and achievement tests, 5
            teacher assessments, 7 learning measures, 3 gain scores,
            and 5 retention scores
        Method: Principal components condensation with varimax
            rotation, independently for two SES groups and total

21

Results: Five factors emerged in each analysis: I g;
II Teacher halo; III Verbal and span memory; IV
Associative learning; V Retention. A sixth factor
was extracted for the total group: VI SES-general
information.
White, W. F., & McConnell, J. Affective responses and
school achievement among eighth grade boys and girls.
Perceptual and Motor Skills, 1974, 38, 1295-1301.
Purpose: To identify dimensions of motivation in eighth-
graders
Subjects: 267 eighth-graders
Variables: 50 Junior Index of Motivation item scores
Method: Principal components condensation with varimax
rotation
Results: Six factors were interpreted: I General adequacy;
II Pessimism; III Dependency; IV Mental rigor as opposed
to creativity; V School atmosphere; VI Teacher dominance.
2. Concepts and Language
Crockett, D. J. Component analysis of within correlations
of language-skill tests in normal children. Journal
of Special Education, 1974, 8, 361-375.
Purpose: To identify the dimensionality of language skills
present in childhood
Subjects: 353 children, ranging in age from 5-5 to 13-5
years
Variables: 38 scores from the Neurosensory Centre Comprehensive Examination for Aphasia
Method: Principal components condensation with varimax
rotation, independently for within-cell and unadjusted
correlations
Results: Seven within-cell factors emerged: I Ability to
associate encoded and decoded material; II Memory for
verbal material; III Name-finding ability; IV Auditory-
language comprehension; V Ability to verbally produce
grammatically correct and meaningful sentences, and
syntactic fluency; VI Ability to record and recall
digits and ability to perform simple cognitive operations; VII Ability to understand, retain, and recall
digits. Four factors emerged for the unadjusted-
correlation analysis: I Ability to associate encoded
and decoded material; II Memory for verbal material;
III Ability to perform simple cognitive operation and to
understand, retain, record, and recall digits; IV Name-
finding ability.
Lacoursiere-Paige, F. Development of right-left concept in
children. Perceptual and Motor Skills, 1974, 38, 111-117.
Purpose: To identify dimensions of right-left concept
development in children
Subjects: 80 normal children, aged 4-11 years
Variables: 26 scores derived from S-B Draw-A-Diamond,
Draw-A-Man, Benton's Right-Left Discrimination Battery,
Raven's Progressive Matrices, WISC Vocabulary, SRS
Space, Harris Test of Lateral Dominance.

Method: Principal components condensation with varimax
    rotation
Results: Eight factors emerged: I General maturation;
    II Directional orientation; III Lateralization; IV-
    VIII Not interpreted.
Naylor, F. D. The characteristics of semantic space in a
    sample of young children. Australian Journal of
    Psychology, 1971, 23, 161-165.
Purpose: To identify dimensions of semantic space in young
    children
Subjects: 120 6- to 8-year-olds
Variables: 20 bipolar semantic scale ratings summed across
    5 person concepts
Method: Principal components condensation with varimax
    rotation
Results: Eight factors emerged: I Evaluation; II Potency;
    III Activity; IV-VI Not interpreted; VII Denotative
    (poor-rich); VIII Denotative (wet-dry).
Saltz, E., Dunin-Markiewicz, A., & Rourke, D. The develop-
    ment of natural language concepts. II. Developmental
    changes in attribute structure. Child Development,
    1975, 46, 913-921.
Purpose: To identify semantic structure at four age levels
Subjects: 40 six-year-olds, 40 nine-year-olds, 40 12-year-
    olds, 80 college students
Variables: Scores on 16 semantic differential scales
    (ratings for 4 ficticious animals as stimuli)
Method: Multiway principal components condensation
    (CANDECOMP, Carroll & Chang, 1970) without rotation
Results: A progressive five-factor solution emerged: I
    General evaluative (all ages); II Potency (all ages);
    III Activity (absent for 6-year-olds); IV Form vs.
    substance evaluation (absent for 6- and 9-year-olds);
    V Social interactions - introversion-extraversion mea-
    sure (present for college students only).
3. Abilities
Ball, D. W., Payne, J. S., & Hallahan, D. P. Factorial com-
    position of the Peabody Picture Vocabulary Test with
    Head Start children. Psychological Reports, 1973, 32,
    12-14.
Purpose: To identify dimensions underlying the Peabody
    Picture Vocabulary Test (PPVT) for a sample of Head
    Start children
Subjects: 354 Head Start children (mean age=4.4 years,
    PPVT IQ=75.73)
Variables: 44 PPVT item scores
Method: Principal components condensation without rotation
    (varimax rotation yielded uninterpretable factors)
Results: Eight factors emerged: I General; II & IV Nouns
    & Progressive verbs, respectively (probably error and
    specific variance); III, V-VIII Uninterpreted.

23

Bergan, J. R., Zimmerman, B. J., & Ferg, M. Effects of variations in content and stimulus grouping on visual sequential memory. Journal of Educational Psychology, 1971, 62, 400-404.
Purpose: To identify relationships among sequential memory subtests, intelligence, and achievement
Subjects: 43 male and 60 female fifth-graders
Variables: Scores on 9 sequential memory subtests, plus IQ and achievement scores
Method: Principal axes condensation with varimax rotation
Results: Three factors emerged: I Memory for multiple units of information; II Memory for single units of information; III Intelligence and achievement.
Blaha, J., Wallbrown, F. H., & Wherry, R. J. Hierarchical factor structure of the Wechsler Intelligence Scale for Children. Psychological Reports, 1974, 35, 771-778.
Purpose: To identify the hierarchical factor structure of the Wechsler Intelligence Scale for Children (WISC)
Subjects: 200 children each at ages 7½, 10½, and 13½ years (from standardization sample)
Variables: 12 WISC subtest scores
Method: Hierarchical factor condensation
Results: Three factors were obtained at each age level: I General (g); II Verbal-numerical-educational (v:ed); III Practical-mechanical-spatial-physical (k:m).
Chissom, B. S. A factor-analytic study of the relationship of motor factors to academic criteria for first- and third-grade boys. Child Development, 1971, 42, 1133-1143.
Purpose: To identify the factor structure of motor skills in young boys
Subjects: 79 first- and 90 third-grade boys
Variables: 9 motor skills test scores
Method: Principal components condensation with orthogonal varimax rotation, independently by grade level
Results: Three factors emerged in both analyses: I Balance; II Dynamic strength; III Gross motor coordination.
Coates, S. Field independence and intellectual functioning in preschool children. Perceptual and Motor Skills, 1975, 41, 251-254.
Purpose: To identify factors common to a measure of field independence and aspects of intellectual functioning in young children
Subjects: 88 girls and 81 boys (mean age=4.7 years)
Variables: Preschool Embedded-figures test score, and 11 WPPSI scores
Method: Principal components condensation with varimax rotation, independently for each sex
Results: Four factors emerged in each analysis: I Perceptual analytic; II Verbal comprehension; III-IV Uninterpreted (but not identical for the two groups).

Conclusion: The field independence measure loaded, for both sexes, on the first factor, Perceptual analytic.
Coates, S., & Bromberg, P. M. Factorial structure of the Wechsler Preschool and Primary Scale of Intelligence between the ages of 4 and 6½. Journal of Consulting and Clinical Psychology, 1973, 40, 365-370.
Purpose: To identify dimensions underlying the Wechsler Preschool and Primary Scale of Intelligence (WPPSI)
Subjects: 200 children each at age 4, 4½, 5, 5½, 6, and 6½ years (standardization sample)
Variables: WPPSI subtest scores (11)
Method: Principal components condensation with varimax rotation, independently by age group
Results: Four factors emerged: I Verbal comprehension (common across age); II Perceptual organization (common across age); III Quasi-specific (similar to Cohen); IV Not interpretable because not matchable across age.
Conger, A. J., & Conger, J. C. Reliable dimensions for WISC profiles. Educational and Psychological Measurement, 1975, 35, 847-863.
Purpose: To identify (reliable) dimensions of WISC profiles across age groups
Subjects: 200 children each at 7½, 10½, and 13½ years of age (WISC manual)
Variables: 11 WISC scores
Method: Canonical condensation with varimax rotation, by age group
Results: Two dimensions which were readily interpretable (I Total IQ; II Verbal performance differences), plus four others which were similar across age (III Verbal comprehension; IV Relevance; V Perceptual organization; VI Mazes-specific) emerged.
Corman, L., & Budoff, M. Factor structures of retarded and nonretarded children on Raven's Progressive Matrices. Educational and Psychological Measurement, 1974, 34, 407-412.
Purpose: To identify dimensions underlying Raven's Progressive Matrices (PM) for normal and EMR children
Subjects: 243 low-income 7- to 12-year-old normals, 379 6- to 11-year-old normals; 399 7- to 15-year-old and 174 5½ to 14-year-old EMRs
Variables: 36 PM item scores
Method: Principal components condensation with varimax rotation, independently for the 4 groups of Ss
Results: Four similar factors emerged across analyses: I Continuity and reconstruction of simple and complex structures; II Discrete pattern completion; III Reasoning by analogy; IV Simple continuous pattern completion (less consistent for EMR groups)
Corman, L., & Budoff, M. Factor structures of Spanish-speaking and non-Spanish-speaking children on Raven's Progressive Matrices. Educational and Psychological Measurement, 1974, 34, 977-981.

25

Purpose: To identify dimensions underlying Raven's Progressive Matrices (PM) for Spanish- and English-speaking children

Subjects: 228 Spanish-speaking 6- to 17-year-olds, 243 English-speaking 7- to 12-year-olds

Variables: 36 PM item scores

Method: Principal components condensation with varimax rotation, by group

Results: Four factors emerged for English-speaking children: I Continuity and reconstruction of simple and complex structures; II Discrete pattern completions; III Reasoning by analogy; IV Simple continuous pattern completion. The first three Spanish-speaking factors (I-III) were matched with the first analysis, and the fourth (IV) was not labeled.

Das, J. P. Patterns of cognitive ability in nonretarded and retarded children. American Journal of Mental Deficiency, 1972, 77, 6-12.

Purpose: To identify dimensions of cognitive (reasoning and memory) ability for EMR and non-retarded children

Subjects: 60 EMR and 60 nonretarded children, matched on MA

Variables: Scores (6) from Raven's Progressive Matrices, Memory for Designs (errors), Cross-modal Coding, a visual short-term memory task, and auditory free- and serial recall tasks, plus IQ

Method: Principal axes condensation with varimax rotation, independently by group

Results: Two essentially similar factors emerged for EMR and nonretarded children: I Successive information processing mode; II Simultaneous information processing mode.

DeVries, R. Relationships among Piagetian, IQ, and achievement assessments. Child Development, 1974, 45, 746-756.

Purpose: To identify the relationships of Piagetian and psychometric assessments of intelligence, and school achievement

Subjects: 1) 122 bright, average, and retarded public school children (bright and average CA=5-7 years; retarded CA=6-12 years, MA=5-7 years); 2) subgroup of 50 bright and average children

Variables: 1) Stanford-Binet MA plus scores on 15 Piagetian tasks; 2) Stanford-Binet MA, 15 Piagetian tasks, CTMM Language and Non Language IQ, 4 Metropolitan Achievement Test subscores

Method: Principal components condensation with orthogonal rotation, independently for the two variable sets

Results: Five factors emerged for the first analysis: I Conservation; II Piagetian variables plus MA; III Identity; IV-V Unlabeled, defined by Sibling egocentrism, Length transivity, and Object sorting. Six factors emerged for the second analysis: I Conservation; II Identity; III MAT subtest and Dream; IV Object sorting and identity; V Guessing game; VI Length transivity.

Conclusions: The results of a principal-factor condensation

with oblique rotation also supported these results
which indicate some overlap between Piagetian, IQ,
and achievement tests but also that they measure
different aspects of cognitive functioning.

Doughtie, E. B., Wakefield, J. A. Jr., Sampson, R. N., &
Alston, H. L. A statistical test of the theoretical
model for the representational level of the Illinois
Test of Psycholinguistic Ability. Journal of Educational Psychology, 1974, 66, 410-415.

Purpose: To identify dimensions underlying the Illinois
Test of Psycholinguistic Abilities (ITPA) for eight
age levels

Subjects: 107 children aged 2-7 to 3-1, 116 3-7 to 4-1,
115 4-7 to 5-1, 128 5-7 to 6-1, 124 6-7 to 7-1, 123
7-7 to 8-1, 127 8-7 to 9-1, 122 9-7 to 10-1 (from
Paraskevopoulos & Kirk, 1969)

Variables: 12 ITPA scores

Method: Principal components condensation with varimax
rotation, by age level

Results: Five factors were extracted at each age level;
placement of the subtests in factor space was un-
related to the model for 3- and 4-year-olds but showed
correspondences for all other ages.

Eckert, H. M., & Eichorn, D. H. Construct standards in
skilled action. Child Development, 1974, 45, 439-445.

Purpose: To identify dimensions of skilled action across
age

Subjects: 27 boys and 31 girls tested at ages 4.5, 5.5,
6.5, 7.5, and 8.5 years (Berkeley Growth Study)

Variables: Scores on 6 eye-hand coordination tasks for
all but second occasion, 7 tasks for 2nd occasion

Method: Unspecified condensation with normalized varimax
rotation, independently by sex and occasion (age)

Results: The number of factors extracted reduced from
three to two for girls between 4.5 and 5.5 years and
for boys between 5.5 and 6.5; these factors were
tentatively identified as I Patterning, and II Speed.

Feigenbaum, K. A study of the relationship between physical
transivity and social transivity. Genetic Psychology
Monographs, 1974, 90, 3-42.

Purpose: To identify dimensions of physical and social
transivity

Subjects: 39 boys and 33 girls from one kindergarten

Variables: Trial scores on 2 social discrimination
tasks (2, 8 trials), 2 social seriation, 2 social
transivity (2, 8 trials), 2 physical seriation, 2
physical transivity (3, 8 trials) tasks

Method: Unspecified condensation with varimax rotation,
independently for boys and girls

Results: Four factors, of 11 extracted, were interpreted
for boys: I Transivity of photographs; II Transivity
of colored sticks; III Length discrimination; IV Discrimination of photographs. Three factors, of 11 extracted, were interpreted for girls: I Transivity of
stick figures; II Length discrimination; III Transivity
of photographs.

27

Feldhussen, J. F., Houtz, J. C., & Ringenbach, S. The Perdue
Elementary Problem-Solving Inventory. Psychological
Reports, 1972, 31, 891-901.
Purpose: To identify dimensions underlying the Perdue
Elementary Problem-solving Inventory (PEPSI)
Subjects: 1073 second-, fourth-, and sixth-graders
Variables: 1) Total scores from PEPSI, Logical Thinking,
Concept Formation, Lorge-Thorndike Intelligence Test,
tests of reading comprehension, Perceptual Abilities;
2) 49 PEPSI problem scores
Method: 1) Principal factors condensation; 2) unspecified
condensation
Results: An unspecified number of factors were extracted
in the first analysis. Six factors emerged for PEPSI
alone: I Evaluative ability; II Ability to sense problems; III Ability to define problems; IV Ability to
analyze critical details; V Ability to see implications;
VI Ability to make unusual associations.
Fulgosi, A., & Guilford, J. P. Factor structures with
divergent- and convergent-production abilities in groups
of American and Yugoslavian adolescents. Journal of
General Psychology, 1972, 87, 169-180.
Purpose: To identify dimensions of divergent- and convergent-
production abilities with Yugoslavian adolescents
Subjects: 131 17- to 18-year-old Zagreb, Yugoslavia students
Variables: Scores on 8 controlled-association and 16 marker
tests, all of which had been translated into Croatian
Method: Principal axes condensation with varimax and
orthogonal graphical rotation
Results: Five factors emerged: I DMU; II NMU; III DMR;
IV NMR; V Uninterpreted.
Conclusions: The factor structure was compared with that
for an American adolescent sample.
Hallahan, D. P., Ball, D. W., & Payne, J. S. Factorial
composition of the short form of the Stanford-Binet
with culturally disadvantaged Head Start children.
Psychological Reports, 1973, 32, 1048-1050.
Purpose: To identify dimensions underlying the short form
of the Stanford-Binet with Head Start children
Subjects: 363 culturally disadvantaged Head Start children
(mean age=4.3 years, IQ=91.6)
Variables: 15 Stanford-Binet (1960 revision, short form)
subtest scores
Method: Unspecified condensation with varimax rotation
Results: Three factors emerged: I General knowledge; II
Visual and visual-motor; III Verbal.
Hardy, M., Smythe, P. C., Stennett, R. G., & Wilson, H. R.
Developmental patterns in elemental reading skills:
Phoneme-grapheme and grapheme-phoneme correspondences.
Journal of Educational Psychology, 1972, 63, 433-436.
Purpose: To identify dimensions of knowledge of sound-
symbol relationships in Grades 1-3

28

Subjects: 149 first- through third-graders
Variables: 33 phoneme-grapheme (P-G) and 33 grapheme-phoneme (G-P) scores from an association test
Method: Unspecified factor analysis, independently for all, P-G, and G-P scores
Results: The first analysis yielded 12 factors; when analyzed separately, P-G scores yielded 6 factors and G-P 5.
Conclusions: The factors were not labeled; however, the test is obviously not measuring a unitary ability.
Harris, C. W., & McArthur, D. L. Another view of the relation of environment to mental abilities. Journal of Educational Psychology, 1974, 66, 457-459.
Purpose: To identify the relationship between environmental measures and mental abilities
Subjects: Unspecified here (data reported by Marjoribanks, 1972)
Variables: Scores on 4 SRA subtests, 8 press measures, and 2 global status indicators
Method: Interbattery (Tucker, 1958) factor condensation
Results: One unlabeled factor emerged.
Hick, T. L., & Santman, M. C. Test of a strategy to increase the predictability of first grade reading skills from letter naming abilities in kindergarten. Journal of Educational Research, 1971, 65, 147-150.
Purpose: To identify dimensions of letter-naming development
Subjects: 41 kindergarten children at a campus school
Variables: Letter-naming scores for the lower-case Charles E. Merrill letters (number unspecified)
Method: Unspecified condensation with varimax rotation
Results: Five factors emerged: I e i k r s t x y z; II Letter shape (a b c f g h j k m n p q r u v w); III l; IV j o; V d p.
Conclusions: Results of retesting 35 children in the first grade with another instrument were also reported.
Hollenbeck, G. P. A comparison of analyses using the first and second generation Little Jiffy's. Educational and Psychological Measurement, 1972, 32, 45-51.
Purpose: To identify dimensions underlying the McCarthy Scale of Children's Abilities (MSCA)
Subjects: 142 5- and 5½-year-olds (standardization sample)
Variables: 24 MSCA scale scores
Method: 1) Little Jiffy-I (principal components condensation with varimax rotation), 2) Little Jiffy-II (orthoblique rotation)
Results: Seven and six factors were extracted by LJ-I and LJ-II, respectively; they were not labeled.
Hollenbeck, G. P., & Kaufman, A. S. Factor analysis of the Wechsler Preschool and Primary Scale of Intelligence (WPPSI). Journal of Clinical Psychology, 1973, 29, 41-45.

29

Purpose: To identify dimensions underlying the Wechsler
Preschool and Primary Scale of Intelligence (WPPSI)
Subjects: WPPSI standardization sample, grouped into 4-
and 4½-year-olds (N=400), 5- and 5½-year-olds (400),
and 6- and 6½-year-olds (400)
Variables: Scores on 11 WPPSI subtests
Method: 1) Principal components condensation with varimax
rotation; 2) second-generation Little Jiffy with orthoblique
rotation; 3) principal factors condensation with varimax
rotation, with oblimax rotation, and with oblimin
(quartimin, biquartimin, covarimin) rotations
Results: Two factors emerged at all three age levels in all
analyses: I Verbal; II Performance.
Hollos, M. Logical operations and role-taking abilities in
two cultures: Norway and Hungary. Child Development,
1975, 46, 638-649.
Purpose: To identify relationships between Piagetian
measures of conservation, classification, role-taking,
and verbal egocentrism for Hungarian children
Subjects: 45 first-, 45 second-, and 45 third-graders from
Hungarian farm settlements, village, or town
Variables: Scores from 9 Piagetian measures of conservation, classification, and role-taking, plus a measure
of verbal egocentrism
Method: Principal components condensation, apparently
without rotation
Results: Two interpretable factors emerged: I Logical
operations; II Role taking.
Hollos, M., & Cowan, P. A. Social isolation and cognitive
development: Logical operations and role-taking abilities
in three Norwegian social settings. Child Development,
1973, 44, 630-641.
Purpose: To identify dimensions of conservation, classification, and role-taking in children in three Norwegian
settings
Subjects: 16 seventh-, 16-eighth-, and 16 ninth-graders
each from a farm community, a village, and a town in
Norway
Variables: Scores on 3 tests each for conservation, clas-
sification, and role-taking
Method: Principal components condensation with orthogonal
rotation
Results: Three factors emerged: I Logical operations; II
Role taking; III Uninterpreted (perspective singlet).
Hopkins, K. D., & Bracht, G. H. Ten-year stability of verbal
and nonverbal IQ scores. American Educational Research
Journal, 1975, 12, 469-477.
Purpose: To obtain an understanding of the meaning of
verbal and non-verbal IQ scores across time
Subjects: Students in two high school graduating classes
tested at grades 1 (CTMM, N=446), 2 (CTMM, N=995), 4
(N=1629), 7 (LT, N=2284), 9 (LT, N=2166), and 11 (LT,
1859)

Variables: Verbal and non-verbal IQ scores at each grade
   level
Method: Principal axes condensation with oblique rotation
Results: Four factors emerged: I Verbal (evident at
   grade 4 on); II Non-verbal (grade 7 and after); III
   Transitory maturational or test-taking (primarily grade
   1, some grade 2); IV Uninterpreted transitory (grades
   2 and 4) factor.
Humphreys, L. G., & Taber, T. Ability factors as a function
   of advantaged and disadvantaged groups. Journal of
   Educational Measurement, 1973, 10, 107-115.
   Purpose: To identify dimensions underlying the Project
   Talent abilities measures for advantaged and disadvan-
   taged groups
   Subjects: Ninth-grade boys: 4977 high quartile intelli-
   gence and SES, 939 high quartile IQ and low quartile
   SES, 1336 low quartile IQ and high quartile SES, 4491
   low quartile IQ and SES
   Variables: 21 standard aptitude and achievement test
   scores
   Method: Unspecified condensation with varimax and ortho-
   gonal procrustes rotation, independently for the four
   groups
   Results: Six factors emerged in all four analyses: I
   Academic achievement; II Verbal comprehension; III
   Spatial visualization; IV Clerical speed; V Rote
   memorization; VI Fluency.
   Conclusions: Differences in factor loadings between groups
   were examined.
Jensen, A. R. Level I and Level II abilities in three ethnic
   groups. American Educational Research Journal, 1973,
   10, 263-276.
   Purpose: To identify dimensions of Level I and II abilities
   Subjects: 237 white fourth-, 242 fifth-, 219 sixth-graders;
   189 black fourth-, 198 fifth-, 169 sixth-graders; 239
   Mexican fourth-, 211 fifth-, 218 sixth-graders
   Variables: Scores on 2 Lorge-Thorndike subtests, Raven's
   Colored Progressive Matrices, Figure Copying Test, 7
   SAT subtests, 3 Memory for Numbers subtests, Listening-
   Attention Test, 2 speed and persistence tests
   Method: Principal components condensation with oblique
   promax rotation, independently for each grade level
   within each race
   Results: Three factors emerged in all nine analyses: I
   Crystalized intelligence; II Fluid intelligence; III
   Memory.
   Conclusions: Differences between mean factor scores for
   the groups were discussed.
Kaufman, A. S. Piaget and Gesell: A psychometric analysis
   of tests built from their tasks. Child Development,
   1971, 42, 1341-1360.
   Purpose: To identify dimensions of Piagetian and Gesellian
   tasks

Subjects: 103 kindergarteners
Variables: Scores on 1) 13 Piagetian tasks, 2) 11 Gesell
    tasks, and 3) 13 Piagetian and 11 Gesell tasks plus 3
    Thorndike-Lorge subtests
Method: Principal axes condensation with varimax rotation
    for each set of variables
Results: Three Piaget factors emerged: I Number; II Logic;
    III Cognition of size relationships.   Three Gesell fac-
    tors emerged: I Paper-and-pencil coordination; II
    Awareness of part-whole relationships; III Academic
    achievement.   Four factors emerged from the joint
    analysis: I Paper-and-pencil coordination; II Number;
    III Abstract thinking; IV Academic achievement.
Kaufman, A. S.   Factor analysis of the WISC-R at 11 age
    levels between 6½ and 16½ years.   Journal of Consulting
    and Clinical Psychology, 1975, 43, 135-147.
Purpose: To identify dimensions underlying the WISC-R
    across age groups
Subjects: 200 children each aged 6½, 7½, 8½, 9½, 10½,
    11½, 12½, 13½, 14½, 15½, and 16½ years (standardization
    sample)
Variables: 12 WISC-R subtest scores
Method: 1) Principal components condensation with varimax
    rotation; 2) principal factors condensation with varimax
    and 3) oblimax and 4) biquartimin rotation, all by age
    group
Results: The three-factor varimax (principal factors) solu-
    tion (I Verbal comprehension; II Perceptual organization;
    III Freedom from distractability) was considered most
    sensible for nine of the groups; a four-factor solution
    (IV Quasi-specific) was required for ages 6½ and 14½.
Kaufman, A. S.   Factor structure of the McCarthy Scales at
    five age levels between 2½ and 8½.   Educational and
    Psychological Measurement, 1975, 35, 641-656.
Purpose: To identify dimensions underlying the McCarthy
    Scales of Children's Abilities (MSCA) for five age
    levels
Subjects: 102 2½-, 204 3- to 3½-, 206 4- to 4½-, 206 5-
    to 5½-, and 314 6½-, 7½-, 8½-year-olds (from standardi-
    zation sample)
Variables: 16 MSCA scale scores
Method: Principal factors condensation with varimax rota-
    tion, at each age level
Results: Four factors emerged for 2½: I Verbal; II Motor;
    III General cognitive; IV Memory.   Six factors emerged
    for 3-3½: I General cognitive; II Motor; III Verbal;
    IV Memory; V Drawing; VI Perceptual-performance.   Six
    factors emerged for 4-4½: I Drawing; II Motor; III
    Memory; IV Verbal; V Perceptual-performance; VI Semantic
    memory.   Five factors emerged for 5-5½: I General cog-
    nitive/verbal; II Motor; III Perceptual-performance;
    IV Memory; V Quantitative.   Six factors emerged for 6½-
    8½: I General cognitive/verbal; II Motor; III Percep-
    tual performance; IV Memory; V Reasoning; VI Quantitative.

Kaufman, A. S., & Hollenbeck, G. P. Factor analysis of the
    standardization edition of the McCarthy Scales. Journal
    of Clinical Psychology, 1973, 29, 358-362.
Purpose: To identify dimensions underlying the McCarthy
    Scales of Children's Abilities (MSCA)
Subjects: 99 3- to 3½-year-olds, 132 5- to 5½-year-olds,
    and 142 7½- to 8½-year-olds (60% of standardization
    sample: half male, 11-13% nonwhite)
Variables: 24 scores from (18) MSCA tests (standardization
    edition)
Method: 1) Principal components condensation with varimax
    rotation, 2) principal-factors condensation with varimax
    rotation, 3) Little Jiffy II with orthoblique rotation,
    all independently by age group
Results: The results were similar across methods; the
    principal-factors solution is reported. Three factors
    emerged consistently across age levels (I General
    cognitive; II Memory; III Motor), and three additional
    factors which varied across age were extracted (IV
    Verbal; V Quantitative; VI Perceptual-performance).
Kaufman, A. S., & Hollenbeck, G. P. Comparative structure
    of the WPPSI for blacks and whites. Journal of Clinical
    Psychology, 1974, 30, 316-319.
Purpose: To identify WPPSI factor structure for blacks
    and whites
Subjects: 156 black and 1032 white 4- to 6½-year-olds
    from standardization sample
Variables: 11 WPPSI subtest scores
Method: 1) Principal components condensation with varimax
    rotation, 2) principal factors condensation with varimax
    rotation, 3) Little Jiffy IV with orthoblique rotation,
    all independently by race
Results: Two factors, like those for the 4-4½, 5-5½, and
    6-6½ age groups in the total standardization sample,
    emerged in all cases: I Verbal; II Performance.
Klonoff, H. Factor analysis of a neuropsychological battery
    for children aged 9 to 15. Perceptual and Motor Skills,
    1971, 32, 603-616.
Purpose: To identify dimensions underlying a neuropsycho-
    logical battery for children
Subjects: 200 normal children, aged 9-15 years
Variables: Scores on 29 Halstead-Reitan, 2 Benton, 1 Reitan-
    Klove, 16 Klove, and 12 WISC variables
Method: Principal components condensation with varimax
    rotation
Results: Nineteen factors emerged: I Directional sequencing
    of visual stimuli; II Verbal fluency; III Static motor
    steadiness; IV Directed motor steadiness; V Tactile dis-
    crimination speed; VI Coordinated motor speed; VII
    Undirected motor speed; VIII Manipulative dexterity;
    IX Patterned critical discrimination; X Form reproduction
    accuracy; XI Crossed modality consistency; XII Set for

provisional solutions; XIII Cue resultant shifting of
attention; XIV Tactile retention; XV Alternation between
conceptual realms; XVI Directed motor speed; XVII
Analytic-synthetic visual-motor ability; XVIII Form
reproduction speed; XIX Auditory recognition.
Lee, L. C. The concomitant development of cognitive and
moral modes of thought: A test of selected deductions
from Piaget's theory. Genetic Psychology Monographs,
1971, 83, 93-146.
  Purpose: To identify dimensions underlying six Piagetian
  cognitive tasks
  Subjects: 195 boys, equally represented at ages 5-17 years
  Variables: Scores on tests of mass and liquid conserva-
  tion, lateral discrimination, projected space, balance,
  and shadow projection
  Method: Hierarchical factor condensation, apparently with-
  out rotation
  Results: Three orthogonal factors emerged: I General
  cognitive; II Negation; III Formal operations.
  Conclusions: These results support Piaget's notion of two
  distinct cognitive operations.
Lyle, J. G., & Johnson, E. G. Analysis of WISC Coding: 5.
Prediction of coding performance. Perceptual and Motor
Skills, 1974, 39, 111-114.
  Purpose: To identify relationships between several pre-
  dictors and WISC Coding performance
  Subjects: 44 male and 41 female fourth-graders
  Variables: Scores on Coding, Writing Speed, Associate
  Learning, verbal IQ, non-verbal IQ, plus age and sex
  Method: Maximum likelihood condensation with varimax
  rotation
  Results: Three factors emerged: I General intelligence;
  II Slowness in writing; III Poor associate learning.
Madaus, G. F., Woods, E. M., & Nuttall, R. L. A causal
model analysis of Bloom's taxonomy. American Educational
Research Journal, 1973, 10, 253-262.
  Purpose: To identify the factorial structure of cognitive
  Kit of Reference Tests (KIT) (to determine the effect of
  "g" on taxonomic structure)
  Subjects: 1,128 total: 303 ninth-, 267-303 tenth-, 267-
  303 eleventh-, 267 twelfth-graders
  Variables: Scores on an unspecified number of KIT tests
  for cognitive factors
  Method: Principal components condensation without rotation, for each grade level and total sample
  Results: An unspecified number of factors emerged; the
  first factor (I. "g") was congruent across solutions
  and was used in subsequent examination of taxonomic-
  type tests.
McLaughlin, J. A., & Stephens, B. Interrelationships among
reasoning, moral judgment, and moral conduct. American
Journal of Mental Deficiency, 1974, 79, 156-161.

Purpose: To identify interrelationships between reasoning, moral judgment, and moral conduct
Subjects: 75 retarded and 75 nonretarded subjects, tested
during first and third years of 8-year longitudinal
study
Variables: 1) Unspecified number of scores from measures
of reasoning, WISC or WAIS subscales, and WRAT, from
Wave 1; 2) unspecified number of scores from measures
of reasoning, moral judgment, and moral conduct, from
Wave 1; 3) 53 scores from measures of reasoning, moral
judgment, moral conduct, WISC and WRAT subscales
Method: Unspecified condensation with oblique rotation,
independently for Wave 1 first set of variables (combined group), Wave 1 second set of variables by group,
and Wave 2 data by group
Results: Five interpretable factors emerged from the first
analysis: I Wechsler; II WRAT; III Conservation; IV
Classification; V Spatial imagery. Thirty factors
emerged for retardates in Waves 1 and 2; though unlabeled,
factor structures were compared across time. For non-
retardates, 26 factors emerged for Wave 1 and 30 for
Wave 2 data; they were likewise compared.
Conclusions: Intelligence and reasoning appear to remain
independent over time; however, reasoning, moral judgment,
and moral conduct tend to become more related over time
for both retarded and nonretarded individuals.
Mendels, G. E. The predictive validity of the Lorge-Thorndike
Intelligence Tests at the kindergarten level. _Journal_
of Educational Research, 1973, _66_, 320-322.
Purpose: To identify factors common to the Lorge-Thorndike
Intelligence Tests (LTIT), Metropolitan Achievement
Tests (MAT), and other selected variables
Subjects: 79 children administered the LTIT at the end of
kindergarten and the MAT mid-first grade
Variables: 1 LTIT, 2 MAT, 2 teacher's grade, 1 reading
level score, plus age, father's occupation, sex, and
ability level of school
Method: Unspecified condensation with varimax rotation
Results: Four factors, of the seven extracted, were
interpretable: I General ability; II Age; III School
achievement and/or socioeconomic status; IV Sex.
Neeman, R. L. Perceptual-motor attributes of normal school
children: A factor analytic study. _Perceptual and_
Motor Skills, 1972, _34_, 471-474.
Purpose: To identify dimensions underlying the Perdue
Perceptual-Motor Survey (PPMS)
Subjects: 200 school children of average intelligence
and "normal" perceptual-motor development (data col-
lected by Roach & Kephart, 1966)
Variables: 30 PPMS item scores
Method: Unspecified condensation with orthogonal rotation
Results: Nine factors emerged: I Walking board; II Chalk-
board; III Ocular pursuits; IV Intelligence; V Rhythmic

writing; VI Form Perception; VII Obstacle course; VIII
Kraus-Weber; IX Uninterpretable.

Newcomer, P., Hare, B., Hammill, D., & McGettigan, J. Construct validity of the Illinois Test of Psycholinguistic
Abilities. Journal of Learning Disabilities, 1975, 8,
220-231.
Purpose: To identify factors common to the ITPA and selected
criterion tests
Subjects: 167 children similar to the 1968 revision standardization sample
Variables: Scores from 8 short-form ITPA and 4 intact ITPA
subtests, plus a total of 20 external criterion tests
which varied either in content or channel
Method: Principal components condensation with orthogonal
rotation
Results: Ten factors emerged: I Reading and writing
ability; II Auditory association; III Meaningful receptive language; IV Auditory memory; V Oral language
usage; VI Expression of function; VII Visual closure;
VIII Auditory closure; IX Motor free visual closure;
X Meaningful association.
Conclusions: Nine subtests measured discrete abilities,
and the level and process dimensions of the underlying
model were substantiated.

O'Sullivan, M., & Guilford, J. P. Six factors of behavioral
cognition: Understanding other people. Journal of
Educational Research, 1975, 12, 255-271.
Purpose: To identify factors of behavioral cognition
Subjects: 240 eleventh-graders
Variables: Scores on 11 behavioral cognition tests, and
24 marker tests
Method: Principal factor condensation with orthogonal
rotation
Results: Six behavioral cognition factors (I Cognition
of behavioral units; II Cognition of behavioral classes;
III Cognition of behavioral relations; IV Cognition of
behavioral systems; V Cognition of behavioral transfor-
mations; VI Cognition of behavioral implications) as
well as the 12 reference factors (CMU, CMC, CMR, CMS,
CMI, CFU, CFR, NMU, NFT, DMU, DMT, NMS) emerged.

Proger, B. B., McGowan, J. R., Bayuk, R. J. Jr., Mann, L.,
Trevorrow, R. L., & Massa, E. The relative predictive
and construct validities of the Otis-Lennon Mental
Ability Test, the Lorge-Thorndike Intelligence Test,
and the Metropolitan Readiness Test in grades two and
four: A series of multivariate analyses. Educational
and Psychological Measurement, 1971, 31, 529-538.
Purpose: To identify relationships among the Otis-Lennon
Mental Ability Test (O-L), Lorge-Thorndike Intelligence
Test (L-T), Metropolitan Readiness Test (MRT), teacher
ratings (TR), and Stanford Achievement Test (SAT) for
Grades 2 and 4
Subjects: 322 second- and 316 fourth-graders

Variables: 1 O-L, 3 L-T, 1 MRT, 3 TR, and 9 SAT scores
Method: Principal components condensation, apparently
  without rotation, by grade
Results: Two second-grade (I Verbal; II Science and
  social studies) and two fourth-grade factors (I Verbal;
  II Numerical) emerged and were compared with canonical
  results.
Quereshi, M. Y. Patterns of intellectual development during
  childhood and adolescence. Genetic Psychology Monographs,
  1973, 87, 313-344.
Purpose: To identify patterns of intellectual development
  during childhood and adolescence
Subjects: 514 5- through 17-year-olds, divided into either
  4 or 7 age groups
Variables: Test and retest scores on 11 WISC (age 5-14)
  or WAIS (age 16-17) subtests (plus 4 pivot "sum score"
  variables for square-root analyses); correlations
  either uncorrected or corrected (for attenuation,
  differential variability)
Method: 1) Square-root condensation, independently for
  testing by age grouping(s) by correlation type; 2) principal axes condensation with varimax rotation, independently for testing by age grouping(s) by correlation
Results: Data based on the Wechsler scales do not corroborate
  the differentiation hypothesis, whether or not the correlations are corrected, at the various age levels. The
  factorial structure of WISC and WAIS are similar for the
  two testings: I Verbal; II Perceptual organization; III
  Freedom from distractability; IV Perceptual speed and
  accuracy.
Schiff, W., & Dytell, R. S. Deaf and hearing children's
  performance on a tactual perception battery. Perceptual
  and Motor Skills, 1972, 35, 683-706.
Purpose: To identify dimensions underlying a tactile battery performance for deaf and hearing children
Subjects: 179 deaf and 121 hearing children and adolescents
Variables: Scores on a 15-test tactile battery
Method: Principal components condensation with varimax
  rotation, independently for deaf and hearing Ss
Results: Seven factors emerged in each analysis, and three
  were found to be common: I Perceptual speed; II Pattern
  perception; III Vibratory sensitivity.
Silverstein, A. B. Factor structure of the Wechsler Intelli-
  gence Scale for Children for three ethnic groups.
  Journal of Educational Psychology, 1973, 65, 408-410.
Purpose: To identify dimensions underlying the Wechsler
  Intelligence Scale for Children (WISC) for three ethnic
  groups
Subjects: 505 English-speaking caucasians, 318 blacks,
  487 Mexican-Americans, all aged 6-11 years
Variables: 11 WISC subtest scores
Method: Principal factors condensation with maxplane con-
  densation, by group

Results: Two factors emerged in each group: I Verbal comprehension; II Perceptual organization.
Conclusions: These data suggest factorial invariance across groups.
Speedie, S. M., Houtz, J. C., Ringenbach, S., & Feldhussen, J. F. Abilities measured by the Perdue Elementary Problem-solving Inventory. Psychological Reports, 1973, 33, 959-963.
Purpose: To identify dimensions underlying the Perdue Elementary Problem-solving Inventory (PEPSI) for a second-grade sample
Subjects: 364 second-graders
Variables: 46 PEPSI problem scores
Method: Principal axes condensation with varimax rotation
Results: Seven factors emerged: I Ability to select the best solution to a problem; II Ability to sense that a problem exists; III Ability to define the problem specifically; IV Ability to notice details; V Ability to see implications; VI Ability to make remote associations; VII Uninterpreted.
Stennett, R. G., Smythe, P. C., Hardy, M., & Wilson, H. R. Developmental trends in letter-printing skills. Perceptual and Motor Skills, 1972, 34, 183-186.
Purpose: To identify dimensions of letter-printing skill
Subjects: Approximately 100 kindergarteners and first-graders
Variables: Binary scores (acceptable copy or not) for 26 upper- and lower-case letters
Method: Principal factors condensation with varimax rotation, independently for upper- and lower-case letters
Results: Seven unlabeled factors emerged in each analysis; several possibilities were suggested for rationales to use in interpretation.
Taylor, L. J. The Peabody Picture Vocabulary Test: What does it measure? Perceptual and Motor Skills, 1975, 41, 777-778.
Purpose: To identify dimensions common to the Peabody Picture Vocabulary Test (PPVT), WPPSI, and Illinois Test of Psycholinguistic Abilities (ITPA).
Subjects: 65 boys and 68 girls in kindergarten or first grade
Variables: PPVT score, plus 1) 13 WPPSI and 2) 13 ITPA scores
Method: Unspecified condensation with oblique rotation
Results: The total number of factors extracted was not reported; however, the PPVT loaded on the same factor as the verbal tests in the WPPSI analysis. In the ITPA analysis, the factor on which the PPVT loaded was not interpreted since it was composed of tests measuring many different aspects of linguistic ability.

Toussaint, N. A. An analysis of synchrony between concrete-operational tasks in terms of structural and performance demands. Child Development, 1974, 45, 992-1001.
   Purpose: To identify dimensions of performance on tasks presumed to require equivalent logical competence when performance demands are equated
   Subjects: 32 first- and 32 second-graders
   Variables: 9 scores from tasks tapping multiplication of classes, multiplication of relations, seriation, and transivity
   Method: Unspecified ("classical") condensation with oblique rotation, independently by grade and combined groups
   Results: Two factors (I Operative; II Figurative) emerged for all analyses, although Grade 2 produced a third uninterpreted factor.
Wallbrown, F. H., Blaha, J., & Wherry, R. J. The hierarchical factor structure of the Wechsler Preschool and Primary Scale of Intelligence. Journal of Consulting and Clinical Psychology, 1973, 41, 356-362.
   Purpose: To identify dimensions underlying the Wechsler Preschool and Primary Scale of Intelligence (WPPSI)
   Subjects: 200 children each at age 4, 4½, 5, 5½, 6, and 6½ years (standardization sample)
   Variables: 11 WPPSI subtest scores
   Method: Hierarchical factor condensation, by age group
   Results: One general factor (I General intelligence) and two subgeneral factors (II Verbal-educational; III Spatial-perceptual) emerged across age.
Wallbrown, F. H., Wallbrown, J. D., & Wherry, R. J. Sr. The construct validity of the Wallach-Kogan creativity test for inner-city children. Journal of General Psychology, 1975, 92, 83-96.
   Purpose: To identify dimensions common to intelligence, creativity, control, and criterion variables for inner-city children
   Subjects: 73 third- and fourth-graders from an inner-city parochial school
   Variables: Scores on 11 WISC subtests, internality (Bialer), 5 Wallach-Kogan tests, 8 judges' product ratings, and 8 demographic indices
   Method: Hierarchical factor condensation
   Results: Ten first-order factors emerged: I Creativity-visual; II Creativity-verbal; III Method variance, or judges' bias, related to crayon drawings; IV Freedom from distractibility; V Precision in verbal expression; VI Perceptual organization; VII Quasi-specific (Coding); VIII-X Not interpreted. Two higher-order factors were extracted: I General intelligence; II General creativity.
Wexley, K., Guidubaldi, J., & Kehle, T. An evaluation of Montessori and day care programs for disadvantaged children. Journal of Educational Research, 1974, 68, 95-99.

Purpose: To identify dimensions common to eight cognitive tests
Subjects: 25 Montessori program, 25 day care center, 19 disadvantaged model cities neighborhood, and 19 disadvantaged non-model cities neighborhood children
Variables: 3 WPPSI, 1 WRAT, 2 CATB, 1 Color Recognition Test, and 1 Weight Gradation Test scores
Method: Principal components condensation without rotation
Results: A single factor (I Composite) accounted for 48% of the common variance
Conclusion: Factor scores were computed and used in further investigation.

Williams, T. H. The Wechsler scales: Parents and (male) children. Journal of Educational Measurement, 1975, 12, 119-128.
Purpose: To identify factor structures of Wechsler tests for parents and (male) children
Subjects: 100 children (males) about age 10, 124 parents (67 mothers, 57 fathers)
Variables: 11 subtest scores from the WISC or WAIS as appropriate (Mazes excluded)
Method: Principal factor condensation, independently for children, parents, mothers, and fathers, with varimax rotation
Results: Two factors emerged in all four analyses: I Verbal; II Performance.
Conclusions: Factor structures were compared across generations and sex.

Winkelmann, W. Factorial analysis of children's conservation task performance. Child Development, 1974, 45, 843-848.
Purpose: To identify dimensions of children's conservation task performance
Subjects: 281 5- to 8-year-olds
Variables: Scores on 31 (12 concrete and 19 paper) conservation items
Method: Unspecified condensation with varimax rotation, independently for all, concrete only, and paper only items
Results: Four factors emerged in all analyses: I Conservation of substance equality or identity; II Conservation of substance inequality; III Conservation of number equality; IV Conservation of number inequality.
Conclusions: These results were replicated with a second sample.

4. Perception
5. Personality
Ali, F. Dimensions of competence in young children. Genetic Psychology Monographs, 1973, 88, 305-328.
Purpose: To identify dimensions of competence in young children

Subjects: 50 disadvantaged black four- and five-year-olds
Variables: Scores on 15 exploration, 10 manipulation, and
    12 individual-dispositional variables (observed-scored)
Method: Principal axes condensation with oblique binormamin
    rotation
Results: Three factors emerged: I General competence; II
    Emotional freedom; III Intellectual competence.
Bates, J. E., Bentler, P. M., & Thompson, S. K. Measurement
    of deviant gender development. Child Development, 1973,
    44, 591-598.
Purpose: To identify dimensions of the parent-report
    Gender Behavior Inventory for Boys (GBIB)
Subjects: (mothers of) 175 normal and 15 gender-referral
    (clinic) boys, aged 5-12 years
Variables: 173 GBIB item scores
Method: Principal axes condensation with clustran (ortho-
    gonal?) rotation
Results: Four factors emerged: I Feminine behavior; II
    Extraversion; III Behavior disturbance; IV Mother's boy.
Conclusions: Results from a validation and from a develop-
    mental trends study are also reported.
Bledsoe, J. C. Factor invariance in the measurement of
    children's manifest anxiety. Psychological Reports,
    1975, 36, 12.
Purpose: To identify dimensions underlying the Children's
    Manifest Anxiety Scale (CMAS)
Subjects: 611 fourth- through seventh-graders
Variables: Scores on 53 CMAS items
Method: Principal components condensation with orthogonal
    rotation
Results: Two factors emerged: I Anxiety; II Lie scale.
Busby, W. A., Fillmer, H. T., & Smittle, P. Interrelation-
    ships between self-concept, visual perception, and
    reading disabilities. Journal of Experimental Education,
    1974, 42, 1-6.
Purpose: To identify the relationship between self-concept,
    visual perception, and reading disabilities
Subjects: 50 seventh- and 50 ninth-graders
Variables: 15 Tennessee Self Concept Scale, 3 language
    achievement test, 2 Spatial Visualization Section
    (Multiple Aptitude Test) test, and 2 intelligence
    test scores.
Method: Principal axes condensation with varimax rotation
Results: Three of the six factors extracted were inter-
    preted: I Intelligence and language achievement; II
    Self-concept; III Visual perception.
Buss, A. H., Plomin, R., & Willerman, L. The inheritance of
    temperaments. Journal of Personality, 1973, 41, 513-524.
Purpose: To identify dimensions of mother-rated personality
    for twins
Subjects: 127 pairs of white, same-sexed twins, rated by
    their mothers (70 male; 58 female)

Variables: Scores on 20 temperament questionnaire items
Method: Unspecified condensation with varimax rotation,
   by sex
Results: Four factors emerged in each case: I Emotionality;
   II Activity; Sociability; III Impulsivity.
Conclusions: Factor scores were examined for differences
   between monozygotic and dizygotic pairs.
Carlier, M.   Flexibility, a dimensional analysis of a modality
   of divergent thinking.   Perceptual and Motor Skills,
   1971, 32, 447-450.
Purpose: To identify dimensions underlying a 26-test
   flexibility battery
Subjects: 78 boys and 107 girls, average age 18 years
Variables: Scores on 26 verbal and nonverbal flexibility
   tests
Method: Unspecified condensation with unspecified rotation, by sex
Results: Six factors emerged for both analyses: I General
   flexibility; II Associative flexibility; III Ideational
   flexibility; IV Graphic flexibility; V-VI Not interpretable.
Case, R., & Globerson, T.   Field independence and central
   computing space.   Child Development, 1974, 45, 772-778.
Purpose: To identify the relationship between field
   independence and central computing space in young
   children
Subjects: 43 8-year-olds
Variables: Scores on 4 analytic field measures, 2 measures
   of central computing space, 1 attention-concentration
   item demanding central computing space
Method: Common-factor (and principal components and image
   analytic) condensation, all with orthoblique rotation
Results: Two factors emerged in all analyses: I Field
   independence; II Central computing space.
Cattell, R. B., & Dreger, R. M.   Personality structure as
   revealed in questionnaire responses at the preschool
   level.   Child Development, 1974, 45, 49-54.
Purpose: To identify the factorial structure of the
   Preschool Personality Questionnaire (PSPQ)
Study One (Cattell & Peterson, "advance report," 1958)
Subjects: 32 male and 48 female 4-5-year-olds
Variables: 104 preliminary - PSPQ item scores
Method: Centroid condensation with oblique rotation
Results: Thirteen (unlabeled here) factors emerged; the
   number of items was reduced to 90.
Study Two
Subjects: 115 6-year-olds
Variables: 26 PSPQ parcel scores (2 parcels for each
   factor), plus scores on 20 ability (fluid and crystallized)
   measures
Method: Principal axes condensation with oblique maxplane
   and blind rotoplot rotation

Results: Four ability (I-IV) and 14 personality ("unlabeled"
until future confirmation possible) factors emerged:
V Desurgency; VI-VII Unknown; VIII Excitability; IX
Premsia; X Dominance; XI Guilt proneness; XII Affecto-
thymia; XIII Ego strength; XIV Parmia; XV Autia; XVI
Self-sufficiency; XVII Ergic tension; XVIII Self-
sentiment (negative).
Study Three
Subjects: 180 4-, 5-, and 6-year-olds
Variables: Scores on 13 PSPQ parcels (from (1)), 12 PSPQ
parcels (from (2)), plus 95 PSPQ single items
Method: Same as Study Two
Results: Twenty-two factors emerged, of which 9 are "new,"
7 are satisfactory matches for the original factors,
and 6 are less satisfactory matches.
Cattell, R. B., & Klein, T. W. A check on hypothetical per-
sonality structures and their theoretical interpretation at 14-16 years in T-data. British Journal of
Psychology, 1975, 66, 131-151.
Purpose: To identify dimensions underlying a battery of
95 tests
Subjects: 394 14- to 16-year-olds
Variables: 95 scores from a battery of tests selected to
measure 12 personality factors defined in previous
researches
Method: Principal axes condensation with 1) promax analytic,
hand, and maxplane, 2) oblique procrustes, blind rotoplot, and 3) hand, procrustes, blind rotoplot, and
maxplane rotations
Results: Seventeen factors emerged and were generally
consistent across the three rotation strategies: I
Assertive ego; II Independence; III Comention; IV
Exuberance; V Regression; VI Anxiety; VII Realism vs.
tensidia; VIII Premature and rigid superego development; IX Exvia-invia; X Dismay; XI-XVII Not interpreted.
DeVito, P. J. An analysis of selected behavioral characteristics of disadvantaged students. Journal of Educational Research, 1975, 68, 178-181.
Purpose: To identify dimensions underlying a behavior
characteristics inventory
Subjects: 503 educationally disadvantaged children
(kindergarten-third grade)
Variables: 32 item scores from a behavior characteristics
inventory
Method: Common factor condensation with varimax rotation
Results: Six factors emerged: I Creativity; II Aggression-
trouble-maker; III Dependence on teacher; IV Achievement
motivation; V Comfort in school; VI Friendliness.
Conclusions: Factor scores were used as dependent variables
in further analyses.
Eaves, L. J. The structure of genotypic and environmental
covariation for personality measurements: An analysis
of the PEN. British Journal of Social and Clinical
Psychology, 1973, 12, 275-282.

43

Purpose: To identify genotypic and environmental dimen-
sions of variation on personality measures
Subjects: 101 pairs of monozygotic twins
Variables: Within-pairs covariance matrix for PEN scores
Method: Principal components condensation, with normalized
rotation
Results: Three factors of environmental variation emerged
but were not labeled.
Conclusions: Twenty-one canonical roots, or genotypic
factors were extracted, and promax rotation rendered
three interpretable: I Sex difference on psychoticism;
II Component of E; III Neuroticism (also related to P).
Egeland, B., & Halperin, S. A factor analysis of the ele-
mentary level of the California Test of Personality.
Journal of Clinical Psychology, 1971, 27, 105-108.
Purpose: To identify dimensions underlying California
Test of Personality (CTP) scales
Subjects: 60 fifth-grade boys selected by their teachers
as difficult to manage in the classroom
Variables: 12 CTP scale scores
Method: 1) Principal components condensation with varimax
rotation; 2) alpha analysis; 3) image analysis
Results: Principal components and alpha analyses both
yielded two factors, whereas image analysis produced
seven. Two factors were interpretable and common to
all three solutions: I Child's self-esteem and perception of ability to adjust to his environment; II
Social conformity, and child's ability to comprehend
rules/standards of certain social behaviors.
Conclusions: While these factors could be labeled as
suggested in the CTP manual, their composition does
not agree with the manual.
Gaudry, E., & Poole, C. A further validation of the state-
trait distinction in anxiety research. Australian
Journal of Psychology, 1975, 27, 119-125.
Purpose: To identify dimensions underlying anxiety items
from the State-Trait Anxiety Inventory for Children
(STAI-C) under various conditions
Subjects: 472 eighth-graders
Variables: 80 STAI-C item scores from test and (1 day
later) retest (done by scale, following completion
of hard or easy task)
Method: Principal axes condensation with varimax and
oblique promax rotation, varimax rotation of second-
order solution
Results: Eleven first-order factors emerged: I-II State
scale on two different administrations; III Extreme
statements of anxiety; IV-IX, XI Linkage of subsets
of trait scale items; X Test situation. Two second-
order factors emerged: I Trait; II State.
Hargreaves, D. J., & Bolton, N. Selecting creativity tests
for use in research. British Journal of Psychology,
1972, 63, 451-462.

Purpose: To identify dimensions underlying divergent and
non-divergent tests
Subjects: 117 10- to 11-year-olds
Variables: 44 scores from a battery of 11 divergent and
4 non-divergent creativity tests plus 2 IQ measures
as markers
Method: Principal components condensation with varimax
rotation
Results: Nine factors emerged: I Creativity; II IQ; III
Divergent equivalent of a (narrow) bipolar group factor of general intelligence; IV Verbal/pictorial content
differences between tests; V Nonsense words (task-
specific); VI-IX Not interpreted (also task-specific).
Katzenmeyer, W. G., & Stenner, A. J. Strategic use of random
subsample replication and a coefficient of factor repli-
cability. Educational and Psychological Measurement,
1975, 35, 19-29.
Purpose: To identify dimensions underlying children's
affective behavior, as assessed by the (primary level
of the) Self Observation Scales (SOS)
Subjects: 4 random subsamples of 1,575 first-, second-,
and third-graders each
Variables: 45 SOS item scores
Method: Unspecified condensation with varimax and maxplane rotation, by subsample
Results: Four matched factors emerged across samples:
I Self acceptance; II Social maturity; III Self security;
IV School affiliation.
Kazelskis, R., Jenkins, J. D., & Lingle, R. K. Two alternative definitions of creativity and their relationships
with intelligence. Journal of Experimental Education,
1972, 41, 58-62.
Purpose: To identify the factor structure of traditional
and Wallach-Kogan creativity tests and measures of
verbal and non-verbal intelligence
Subjects: 111 tenth- and eleventh-graders from a rural,
predominantly black area
Variables: 5 verbal and 3 non-verbal subtest scores from
the Lorge-Thorndike Intelligence Test, 5 traditional
creativity measures (2 Torrance, 3 Getzels-Jackson
scales), 5 Wallach-Kogan creativity battery subtest
scores
Method: Principal axes condensation with oblique maxplane
rotation
Results: Three factors emerged: I Intelligence; II Traditional creativity; III Wallach-Kogan creativity.
Kogan, N. A clarification of Cropley and Maslany's analysis
of the Wallach-Kogan Creativity Tests. British Journal
of Psychology, 1971, 62, 113-117.
Purpose: To identify a rotated solution for Cropley &
Maslany's (1969) data
Subjects: Unspecified (here) number of fifth-graders

Variables: Scores on 5 Wallach-Kogan creativity tests
and 6 Thurstone PMA tests
Method: (Principal axes condensation, done by Cropley &
Maslany) with promax rotation
Results: Three factors emerged: I Creativity; II Intel-
ligence (numerical-spatial, plus letter series); III
Verbal-semantic intelligence.
Conclusions: Creativity emerged as independent of the
two (related) intelligence factors when rotation was
employed.

Kohn, M., & Rosman, B. L. Cross-situational and longitudinal
stability of social-emotional functioning in young
children. Child Development, 1973, 44, 721-727.
Purpose: To identify dimensions of the Test Behavior
Inventory (TBI), a measure of preschool test-taking
behavior
Subjects: 287 five-year-old boys attending public kinder-
gartens or day care centers
Variables: Ratings on 38 TBI items, summed over five
test-taking sessions
Method: Unspecified condensation with varimax rotation
Results: Two factors were considered interpretable: I
Confident-friendly vs. anxious-withdrawn; II Attentive-
cooperative vs. distractible-disruptive.
Conclusions: Factor scores were used in the larger study
reported in this paper.

Kreitler, S., Kreitler, H., & Zigler, E. Cognitive orienta-
tion and curiosity. British Journal of Psychology, 1974,
65, 43-52.
Purpose: To identify dimensions of curiosity in children
Subjects: 84 lower- and middle-class children
Variables: Scores on 19 measures of curiosity
Method: Unspecified condensation with normalized varimax
rotation
Results: Five factors emerged: I Manipulatory curiosity;
II Perceptual curiosity; III Conceptual curiosity;
IV Curiosity about the complex; V Adjustive-reactive
curiosity.
Conclusions: These results provided the basis for the
major study reported.

Labouvie, E. W., & Schaie, K. W. Personality structure as
a function of behavioral stability in children. Child
Development, 1974, 45, 252-255.
Purpose: To identify dimensions underlying children's
personality, as assessed by the Children's Personality
Questionnaire (CPQ) for "stable" and "unstable" children
Subjects: 275 high-stable (on basis of 1-year CPQ sta-
bilities) and 284 low-stable children in grades 3, 4, and
5 at first time of measurement
Variables: 90 CPQ item scores at test and 1-year retest
Method: Principal axes condensation with unspecified
rotation, independently for each stability level and
each occasion

Results: Two factors were found to be relatively invariant
across all analyses: I Unsocialized aggression versus
friendly cooperation; II Outgoing assertiveness versus
social withdrawal.
Conclusions: Analyses of variance were reported for esti-
mated factor scores and general intelligence scores.

Landis, D., Hayman, J. L. Jr., & Hall, W. S. Multidimensional
analysis procedures for measuring self-concept in poverty
area classrooms. Journal of Educational Psychology, 1971,
62, 95-103.
Purpose: To identify dimensions of fourth-graders' self-
image
Subjects: 487 poverty area fourth-graders from 16 classrooms
Variables: 12 similarity-in-concept-structure scores from
the Most-Like Questionnaire
Method: Unspecified condensation with equamax rotation,
independently for each classroom
Results: Three factors emerged across classrooms: I
Racial; II Educated vs. noneducated, personalized vs.
institutional self, activity; III Upper-track group
from white segment of community.

Lerner, R. M., & Korn, S. J. The development of body-build
stereotypes in males. Child Development, 1972, 43,
908-920.
Purpose: To identify dimensions underlying body-build
stereotypes in males
Subjects: 30 chubby and 30 average males each aged 5.2-
6.9 years, 14.0-15.9 years, and 19.1-20.9 years
Variables: Percentage agreement (group) scores for 56
verbal checklist (body-build stimulus attributes) items
Method: Principal components condensation with varimax
rotation, independently by age group
Results: Three factors, highly stable across age, emerged:
I Mesomorph; II Endomorph; III Ectomorph.

Michael, W. B., Smith, R. A., & Michael, J. J. The factorial
validity of the Piers-Harris Children's Self-Concept
Scale for each of three samples of elementary, junior
high, and senior high school students in a large metro-
politan school district. Educational and Psychological
Measurement, 1975, 35, 405-414.
Purpose: To identify dimensions underlying the Piers-
Harris Children's Self-Concept Scale (PHCSCS) for three
samples
Subjects: 299 elementary, 302 junior high, 300 senior
high school pupils
Variables: 80 PHCSCS item scores
Method: Principal components condensation with varimax
rotation, by group
Results: Three essentially invariant factors emerged across
samples: I Physical appearance; II Socially unacceptable
(bad) behavior; III Academic or school status. Two
other elementary school factors emerged: IV Self-
depreciation (abasement); V Anxiety involving withdrawing

47

behavior. Six other factors emerged for junior high:
IV Abasement or self-depreciation; V Self-depreciation
involving alienation and self-pity; VI Self-contentment
(happiness); VII Self-dissatisfaction (unhappiness)
involving anxiety and instability; VIII Popularity;
IX (Perceived) psychomotor coordination. Three other
factors appeared for senior high: IV Self-dissatisfaction reflecting alienation and anxiety; V Popularity;
VI (Perceived) psychomotor coordination.

Miller, L. C., Barrett, C. L., Hampe, E., & Noble, H. Factor
structure of childhood fears. Journal of Consulting and
Clinical Psychology, 1972, 39, 264-268.
Purpose: To identify dimensions underlying a fear inven-
tory for children
Subjects: 179 6- to 16-year-olds
Variables: Scores on 60 fear items, rated by parents
Method: Principal components condensation with varimax
rotation
Results: Three factors emerged: I Fear of physical injury;
II Fear of natural events; III Fear of psychic stress.

Orloff, H. Thematic Apperception Test card rejection in a
large sample of normal children. Multivariate Behavioral
Research, 1973, 8, 63-70.
Purpose: To identify relationships between TAT card rejec-
tion and other selected variables with normal children
Subjects: 505 male and 496 female 6- to 11-year-old normals
Variables: Scores on 2 school and 2 parent ratings; WISC
BD and V, DAM, WRAT RS and AS, age, sex, and 8 TAT
variables
Method: Criterion factor condensation with varimax rota-
tion, for total and by sex
Results: Five similar factors emerged across analyses:
I Academic ability; II Linguistic deficit; III Creativity;
IV (Conceptual Maturity and Card Rejection); V "Malad-
justment."

Plass, H., Michael, J. J., & Michael, W. B. The factorial
validity of the Torrance Tests of Creative Thinking for
a sample of 111 sixth-grade children. Educational and
Psychological Measurement, 1974, 34, 413-414.
Purpose: To identify dimensions underlying the Torrence
Tests of Creative Thinking (TTCT)
Subjects: 111 sixth-graders
Variables: 30 TTCT scores
Method: Unspecified condensation with varimax rotation
Results: Seven factors emerged and were interpreted as
descriptions of each task (measure) rather than the
hypothesized process for which it was scored.

Richmond, B. O., & White, W. F. Sociometric predictors of
the self concept among fifth and sixth grade children.
Journal of Educational Research, 1971, 64, 425-429.
Purpose: To identify dimensions of 1) self-esteem and
2) peer concepts in the classroom

48

Subjects: 204 fifth- and sixth-graders who were either
advantaged white children or economically deprived
black children
Variables: 1) 58 item scores from the Coopersmith Self-
esteem Inventory (SEI); 2) 12 semantic-differential
scores from each of approximately 8 peers
Method: 1) Principal axes condensation with varimax rotation; 2) principal components condensation with varimax
rotation
Results: Five SEI factors emerged: I Self rejection; II
Parental approval; III Rejection by authority; IV Lie
scale; V Social and self-acceptance. Three peer-concepts
factors emerged: I Evaluation; II Potency; III Activity.
Rubin, K. H. Egocentrism in childhood: A unitary construct?
Child Development, 1973, 44, 102-110.
Purpose: To identify the factorial structure of egocentrism
in childhood
Subjects: 10 boys and 10 girls each in kindergarten and
Grades 2, 4, 6
Variables: CA, MA, and scores on measures of cognitive,
spatial, role-taking, and communicative-recommunicative
(2) egocentrism, conservation, and popularity
Method: Unspecified condensation with varimax rotation
Results: Two factors emerged: I Decentration; II Popu-
larity (singlet).
Seitz, V. R. Multidimensional scaling of dimensional pre-
ferences: A methodological study. Child Development,
1971, 42, 1701-1720.
Purpose: To identify clusters of children with similar
preferences
Subjects: 144 Head Start children (average age 72.4
months)
Variables: Comparative distances of preference judgments
for 12 triadic stimulus sets
Method: Cluster analysis
Results: Two clusters emerged: I Form; II Color
Conclusions: Multidimensional scaling produced Logicality
and Color-form preference axes.
Thompson, S. K. Gender labels and early sex role develop-
ment. Child Development, 1975, 46, 339-347.
Purpose: To identify relationships between demographics
and children's performance on tests of gender labeling
and sex-role stereotypy
Subjects: 11 females and 11 males each aged 24-25, 30-31,
and 36-37 months
Variables: Scores on 19 demographic variables and 10 tests
of gender labeling/sex-role stereotyping
Method: Principal axes condensation with varimax rotation
Results: Three factors emerged: I Cognitive-developmental;
II Parental status; III Playmate.
White, W. F., & Bashaw, W. L. High self-esteem and identification with adult models among economically deprived
children. Perceptual and Motor Skills, 1971, 33, 1127-
1130.

Purpose: To identify dimensions underlying the Self
   Social-constructs Scale (SSCS) for economically de-
   prived children
Subjects: 120 economically deprived kindergarteners and
   first-graders
Variables: 8 SSCS scale scores
Method: Principal components condensation with varimax
   rotation
Results: Three factors emerged: I Self-esteem; II Self-
   realism; III Dependency.

6. Social Behavior

Nias, D. K. B. The structuring of social attitudes in children. Child Development, 1972, 43, 211-219.
Purpose: To identify dimensions underlying English children's
   social attitudes
Subjects: 217 male and 224 female English 11- and 12-year-
   olds
Variables: Scores on 50 3-point social attitude items
Method: Principal components condensation with varimax
   and promax rotation, independently by sex
Results: Four similarly labeled factors emerged for boys
   and girls: I Religion; II Ethnocentrism; III Punitive-
   ness; IV Sex.

Piche, G. L., Michlin, M. L., Rubin, D. L., & Johnson, F. L.
   Relationships between fourth graders' performances on
   selected role-taking tasks and referential communication accuracy tasks. Child Development, 1975, 46, 965-
   969.
Purpose: To identify dimensions underlying fourth-graders'
   role-taking and referential communication accuracy
   task performance
Subjects: 20 fourth-graders
Variables: 3 role-taking and 2 referential communication
   accuracy task scores
Method: Principal components condensation with varimax
   rotation
Results: Two factors emerged: I Visual recoding; II
   Social role-taking.

Richards, H. C., & McCandless, B. R. Socialization dimensions among five-year-old slum children. Journal of
   Educational Psychology, 1972, 63, 44-55.
Purpose: To identify dimensions of socialization among
   two samples of slum children
Subjects: 1) 86 male and 95 female, and 2) 74 male and
   female 4- and 5-year-old slum children
Variables: Scores on 1) 43 demographic and parent-child,
   peer popularity, sex role reference, self-social construct, intensity of task involvement, PPVT, and teacher
   rating measures; 2) 38 of the above measures
Method: Principal components condensation with varimax
   rotation, independently by sample and by sex within
   the first sample

Results: For the first sample, seven factors emerged for
    total sample and females while six emerged for males;
    for the second sample, both seven- and eight-factor
    solutions were acceptable. Overall, six factors were
    considered meaningful: I Verbal facility; II Coping
    by withdrawal; III Coping by aggression; IV Alienation;
    V Sex; VI Remaining environmental handicaps.
Severy, L. J., & Davis, K. E. Helping behavior among normal
    and retarded children. Child Development, 1971, 42,
    1017-1031.
  Purpose: To identify 1) dimensions of helping behaviors
    among normal and retarded children, and 2) clusters of
    similar children
  Subjects: 16 normal and 13 retarded 3- to 5-year-olds,
    and 14 normal and 12 retarded 8- to 10-year-olds
  Variables: Scores on 55 helping behaviors (or helping-
    behavior-derived indices)
  Method: 1) Minimum residual condensation with varimax
    rotation; 2) inverse minimum residual condensation
    with varimax rotation
  Results: Four dimensions of helping behavior emerged:
    I Task helping (opportunity controlled); II Showing
    concern; III Helping activity; IV Psychological helping
    (opportunity controlled). Five factors representing
    groups of subjects emerged: I Engaged in very little
    helping behavior and not warm; II Immature pattern
    (concern but little skill to act); III Task (not psy-
    chological) helpers; IV Sophisticated helpers.
Waldrop, M. F., & Halverson, C. F. Jr. Intensive and exten-
    sive peer behavior: Longitudinal and cross-sectional
    analyses. Child Development, 1975, 46, 19-26.
  Purpose: To identify dimensions of peer behaviors in
    7½-year-olds
  Subjects: 35 male and 27 female 7½-year-olds
  Variables: Scores on 12 measures of peer behavior
  Method: Principal components condensation, apparently
    without rotation, independently by sex
  Results: An unspecified number of factors were extracted;
    Factor I for boys was labeled Extensiveness, and
    Factor I for girls was labeled Intensiveness.
7. Parent-Child and Family Relations
Anderson, J. G., & Johnson, W. H. Stability and change
    among three generations of Mexican-Americans: Factors
    affecting achievement. American Educational Research
    Journal, 1971, 8, 285-308.
  Purpose: To identify dimensions underlying a subset of
    items from a questionnaire assessing family and child
    characteristics
  Subjects: 163 junior and senior high school students from
    Mexican-American families
  Variables: 30 item scores from a 140-item questionnaire
  Method: Unspecified condensation with orthogonal varimax
    rotation

Results: Nine factors emerged: I Language usage in the
family; II Student's desire to achieve in school; III
Participation in extracurricular activities; IV Parental stress on academic achievement; V Parental stress
on completing high school; VI Parental stress on
attending college; VII Parental assistance with school
work; VIII Self-concept of ability; IX Student's educational aspirations.

Arnold, L. E., & Smeltzer, D. J. Behavior checklist factor
analysis for children and adolescents. Archives of
General Psychiatry, 1974, 30, 799-804.
Purpose: To identify dimensions underlying a behavior
checklist for children and adolescents
Subjects: 185 young children (age 12 years or less) and
166 teenagers (13-18), rated by parents
Variables: Scores on 79 behavior checklist items
Method: Principal axes condensation with varimax rotation,
by age group
Results: Six factors emerged for the younger group: I
unsocialized aggression; II Inattentive unproductiveness; III Sociopathy; IV Hyperactivity; V Withdrawal-
depression; VI Somatic complaint. Seven teenage
factors emerged: I Unsocialized aggression; II Inattentive unproductiveness; III Sociopathy; IV Hyperactivity; V Withdrawal-depression; VI Somatic neuroticism;
VII Sleep disturbance.

Barton, K., Dielman, T. E., & Cattell, R. B. An item factor
analysis of intrafamilial attitudes of parents. Journal
of Social Psychology, 1973, 90, 67-72.
Purpose: To identify dimensions underlying the Family
Attitude Measure (FAM)
Subjects: Parents of 250 junior high school children
Variables: 108 FAM item scores (Estimates subtest);
unspecified number of item scores for Paired Words subtest
Method: Principal axes condensation with oblimax and
rotoplot-assisted graphic rotation, independently for
Estimates and Paired Words
Results: Twelve factors emerged for Estimates: I Family
cohesiveness (high gregariousness-protectiveness); II
Low hostility-pugnacity; III Low pugnacity-assertiveness;
IV High assertiveness-protectiveness (primarily interspouse); V High interspouse hostility; VI High pugnacity-
hostility; VII Low interspouse hostility; VIII Low
parent-child pugnacity; IX Low interspouse protectiveness-
gregariousness; X Low parent-child pugnacity-assertiveness;
XI Low parent-child gregariousness; XII Interspouse high
hostility-low gregariousness.
Twelve factors emerged for Paired Words: I Aggressiveness or hostility towards mate; II Protectiveness toward
mate; III High protectiveness toward child; IV High
protectiveness-low fear (parent-parent); V Assertion;
VI Uninterpretable; VII Gregarious; VIII High hostility-
high fear; IX Preference for mate vs child; X-XI High sex,

52

high assertion, low protection A & B; XII High protection-low hostility.

Beckwith, L. Relationships between attributes of mothers and their infants' IQ scores. Child Development, 1971, 42, 1083-1097.

Purpose: To identify dimensions underlying the Parental Attitude Response Inventory (PARI)

Subjects: 24 adoptive mothers

Variables: 1) 19 PARI scale scores; 2) scores on factors extracted in (1) and 13 observation measures

Method: 1) Centroid condensation with oblique rotation; 2) cluster analysis

Results: Three PARI factors emerged: I Assertive maternal control; II Impulse suppression; III Overpossessiveness. Two clusters of behavior emerged: I Stimulation; II Generalized restrictiveness.

Berzonsky, M. D. Some relationships between children's conceptions of psychological and physical causality. Journal of Social Psychology, 1973, 90, 299-309.

Purpose: To identify dimensions underlying children's conceptions of psychological and physical causality

Subjects: 22 kindergarteners and 19 first-graders

Variables: 11 scores from measures of psychological and physical causality, morality, justice, skepticism, animism, and ability to take other perspectives

Method: Principal components condensation with orthogonal rotation

Results: Four factors emerged: I Social judgment; II Psychological causality; III Physical causality; IV Conceptions of life.

Dielman, T. E., Barton, K., & Cattell, R. B. Cross-validational evidence on the structure of parental reports of child rearing practices. Journal of Social Psychology, 1973, 90, 243-250.

Purpose: To identify dimensions of parental reports of child-rearing practices

Subjects: 331 mothers and 307 fathers of junior high school children

Variables: Scores from a child-rearing practices questionnaire: 96 for mothers, 65 for fathers

Method: Principal axes condensation with oblimax and rotoplot-assisted graphic rotations, independently for mothers and fathers

Results: Seven factors of fifteen extracted for mothers, were considered good matches with those from a previous study: I Patriarchial family structure; II High use of physical punishment; III Mother's lack of self-confidence; IV Low involvement; V High use of reward in child rearing; Vi Preference for older children; VII Low use of discipline. Eight factors of the eleven extracted for fathers were considered good matches with those of Dielman, et al. (1971): I High use of reward in child rearing; II High use of physical punishment; III

53

Promotion of independence; IV Preference for younger
children; V Strict discipline; VI Low use of reasoning;
VII Wife responsible for child rearing; VIII Dissatis-
faction with home life.
Dielman, T. E., Cattell, R. B., & Lepper, C. Dimensions of
problem behavior in the early grades. Journal of
Consulting and Clinical Psychology, 1971, 37, 243-249.
Purpose: To identify dimensions of problem behavior in
the early grades
Subjects: 362 first-, second-, and third-graders, rated
by their teachers
Variables: 3-point frequency ratings of 62 behaviors
Method: Principal axes condensation with promax rotation
Results: Eight first-order factors emerged: I Hyper-
activity; II Disciplinary problems; III Sluggishness;
IV Paranoic tendencies; V Social withdrawal; VI Acting
out; VII Speech problem; VIII Antisocial tendencies.
Three second-order factors emerged: I Neuroticism;
II Sociopathic behavior; III Autism.
Dielman, T. E., Cattell, R. B., Lepper, C., & Rhoades, P.
A check on the structure of parental reports of child-
rearing practices. Child Development, 1971, 42, 893-903.
Purpose: To identify dimensions of parental reports of
child-rearing practices
Subjects: 156 mothers, and 133 fathers of 6- to 8-year-old
children
Variables: 101 item scores for mothers and 68 item scores
for fathers, from a questionnaire adapted from Sears,
et al. (1957, 1965) and Milton (1958)
Method: Principal axes condensation with promax condensation, independently for mothers and fathers
Results: Sixteen first-order factors emerged for mothers:
I Patriarchal family structure; II High physical punishment; III Mother's lack of self-confidence; IV Low
involvement; V High use of rewards in child rearing;
VI Strict discipline; VII Lack of affection toward child;
VIII Low child orientation; IX Low use of praise; X Dislike of intrusive child behavior; XI Promotion of independence; XII Low use of discipline; XIII Low goals -
feeling of inadequacy in mother; XIV Easily annoyed by
child; XV Early socialization; XVI Permissiveness.
Eight second-order factors emerged for mothers: I
Permissiveness associated with distaste for interacting
with child; II Lack of affection for child; III Inactive
father role; IV Mother's self-adequacy and higher goals/
early socialization of child; V Annoyability; VI Promotion of independence, perhaps from desire to be left
alone; VII Responsible child-rearing orientation; VIII
Low use of praise and low discipline.
Eighteen first-order factors emerged for fathers: I
High use of sanctions in directing child's behavior; II
High physical punishment; III Promotion of independence;
IV Not Labeled; V Strict discipline; VI Not labeled;
VII Permissive, low expectations; VIII High child

orientation; IX Low praise-high punishment for parental
aggression; X-XIV, XVI Not labeled; XV Use of threat
rather than physical punishment; XVII-XVIII Uninterpretable. Seven second-order factors emerged for fathers;
they were discussed in terms of definers but not labeled.
Epstein, A. S., & Radin, N. Motivational components related
to father behavior and cognitive functioning in preschoolers. Child Development, 1975, 46, 831-839.
  Study One (Radin & Epstein, 1974 unpublished manuscript)
  Purpose: To identify dimensions of father-child interaction
  Subjects: Unspecified fathers and sons, fathers and
  daughters
  Variables: Scores on 25 interaction categories
  Method: Unspecified factor analysis, independently for
  boys and girls
  Results: Four factors emerged for boys: I Positive
  response to child and cognitive stimulation; II Empathy
  and psychological manipulation; III Preventive and
  physical control; IV Verbal restrictiveness. Six factors
  emerged for girls: I Meeting and ignoring explicit needs;
  II Aversive and non-aversive control; III Verbal restric-
  tiveness and requesting; IV Empathy, psychological mani-
  pulation, and cognitive stimulation; V Attention to child's
  verbalizations; VI Physical restrictiveness.
  Study Two (Epstein, 1973 unpublished dissertation)
  Purpose: To identify dimensions underlying Piagetian tasks
  using a modified scoring system
  Subjects: Unspecified children
  Variables: Unspecified number (apparently 16) modified
  DeVries scores on 8 Piagetian tasks
  Method: Unspecified factor analysis
  Results: Five factors emerged: I Sex role constancy; II
  Conservation; III Interpersonal inferring; IV Class
  inclusion; V Uninterpreted.
  Study Three
  Purpose: To identify dimensions underlying the Stanford-
  Binet face-sheet items
  Subjects: 180 lower-, working-, or middle-class four-
  year-olds
  Variables: Scores on 13 Stanford-Binet face-sheet items
  Method: Principal components condensation with varimax
  rotation
  Results: Two interpretable factors emerged: I Person-
  oriented motivation (behavior); II Task-oriented
  motivation (behavior).
Gecas, V., Calonico, J. M., & Thomas, D. L. The development
  of self-concept in the child: Mirror theory versus
  model theory. Journal of Social Psychology, 1974, 92,
  67-76.
  Purpose: To identify dimensions of self-evaluation and
  evaluation of other (intact) nuclear family members

55

Subjects: Mother, father, college student offspring, and
high school student offspring in each of 219 intact
nuclear families
Variables: Scores on 10 5-point semantic differential
scales, applied by each respondent to himself and each
other family member
Method: Principal axes condensation with varimax rotation
Results: Three factors emerged: I Worth; II Power; III
Activity.
Hanson, R. A. Consistency and stability of home environmental
measures related to IQ. Child Development, 1975, 46,
470-480.
Purpose: To identify dimensions underlying home environmental measures
Subjects: 60 male and 50 female children
Variables: Ratings on 7 home environmental measures, for
periods when children were aged 0-3, 4-6, and 7-10
years
Method: Common factor condensation with varimax rotation,
independently for each time period
Results: Two unlabeled factors emerged for each time
period.
Harrison, P. R. A technique for analyzing the distance
between organisms in observational studies. Journal
of General Psychology, 1974, 91, 269-271.
Purpose: To identify dimensions of inter-organism distance
Subjects: 4 preschool children
Variables: Undirected distances between all possible pairs
of children, averaged over unspecified number of time
samples
Method: Centroid condensation without rotation
Results: Two factors emerged: I Three children maintain-
ing consistent distance from fourth but not each other
(or, one child maintaining consistent distance from
other three); II Uninterpreted.
Conclusions: The time sampling-grid technique proposed
may be useful in observational studies.
Huberty, C. J., & Swan, W. W. Preschool classroom experience
and first-grade achievement. Journal of Educational
Research, 1974, 67, 311-316.
Purpose: To identify dimensions underlying observable
behavioral symptoms measured by the Wray Behavior
Scale (WBS)
Subjects: 49 children who entered an experimental program
at age 3, and 58 who entered at age 6
Variables: 15 WBS item scores
Method: Principal components condensation with varimax
rotation
Results: Three factors emerged but were unlabeled.
Conclusions: The factors were interpreted as dimensions
of behavior common to the two groups, and factor scores
were used as input variables in further study of the
independent groups.

Marjoribanks, K. Environmental correlates of diverse mental abilities. Journal of Experimental Education, 1971, 39, 64-68.
Purpose: To identify dimensions underlying eight environmental forces sections of a questionnaire
Subjects: 90 middle-class and 95 lower-class 11-year-olds and their parents
Variables: Scores on 8 environmental forces sections of a home interview schedule
Method: Principal components condensation without rotation
Results: Two factors emerged: I Learning environment of the home; II Parental dominance.
Conclusions: Relationships between various indices of the environment and mental abilities were also examined.
Marjoribanks, K. Environment as a threshold variable: An examination. Journal of Educational Research, 1974, 67, 210-212.
Purpose: To identify dimensions of environment forces questionnaire items
Subjects: Mothers and fathers of 90 11-year-olds classified as middle class, and parents of 95 lower-class 11-year-olds in Canada
Variables: 8 scale scores from a questionnaire pertaining to the learning environment of the home
Method: Principal components condensation without rotation
Results: Two factors emerged: I Learning environment of the home; II Parental dominance.
Conclusions: Relationships between various performance variables on the children were also examined.
Miller, L. C. School Behavior Check List: An inventory of deviant behavior for elementary school children. Journal of Consulting and Clinical Psychology, 1972, 38, 134-144.
Purpose: To identify dimensions underlying a modified Pittsburg Survey (the School Behavior Checklist, SBC)
Subjects: 5,373 kindergarteners through sixth-graders (unspecified numbers of males and females for factor analyses)
Variables: Scores on 1) 96 SBC items, 2) 6 first-order SBC factors, plus 8 demographic variables and 6 teacher ratings
Method: Principal components condensation with varimax rotation, independently for total sample and sex subsamples for first analysis
Results: Six first-order factors emerged in all three analyses: I Low need achievement; II Aggression; III Anxiety; IV Academic disability; V Hostile isolation; VI Total disability. Three factors emerged in the "second-order" analysis: I High-achievement (bipolar) encompassing both cognitive and social skills; II Cognitive-race; III Age-grade.
Paulson, M. J., Lin, T., & Hanssen, C. Family harmony: An etiologic factor in alienation. Child Development, 1972, 43, 591-603.

Purpose: To identify dimensions of the Parental Attitude
  Research Instrument (PARI) when filled out by young
  adults in retrospect
Subjects: 34 male and 82 female anti-Establishment individuals and 40 male and 54 female (Establishment)
  college students, all mean age 20 years
Variables: Scores on 23 maternal-form PARI subscales;
  items responded to in terms of present (retrospective)
  recall of parents' earlier attitudes
Method: Principal factors condensation with varimax rotation
Results: Three factors emerged: I Parental authority; II
  Family harmony; III Involvement.
Conclusions: Factor scores were used as dependent variables
  in a Life style x Sex x Dominant parent ANOVA.
Peisach, E., Whiteman, M., Brook, J. S., & Deutsch, M. Interrelationships among children's environmental variables
  as related to age, sex, race, and socioeconomic status.
  Genetic Psychology Monographs, 1975, 92, 3-17.
Purpose: To identify dimensions underlying children's
  environmental and demographic variables as assessed
  via questionnaire and interview
Subjects: 127 first- and 165 fifth-graders
Variables: Scores on 8 social class, 3 living conditions,
  3 family composition, 4 aspiration, 8 parental interaction with child, 3 child's exposure to mass media,
  3 child's educational background, and 1 nutrition item
Method: Principal components condensation with unspecified
  rotation
Results: Six factors emerged: I Social class; II Interactive opportunity and experience for the child; III
  Maturity; IV Aspiration; V Family interaction; VI Amount
  of income.
Peters, E. N., Pumphrey, M. W., & Flax, N. Comparison of
  retarded and nonretarded children on the dimensions
  of behavior in recreation groups. American Journal
  of Mental Deficiency, 1974, 79, 87-94.
Purpose: To identify dimensions underlying children's
  behavior in recreation groups as measured by the Group
  Participation Report Form (GPRF)
Subjects: 1) 223 normal children (85 rated twice), and
  2) 70 EMR children (rated up to 5 times), all parti-
  cipants in adult-led recreation groups
Variables: 30 GPRF item scores
Method: Principal components condensation with varimax
  rotation, independently for normals and EMRs
Results: Three factors emerged for both analyses: I
  Aggressive acting-out; II Belongingness and comfort;
  III Leader-orientedness.
Prichard, A., Bashaw, W. L., & Anderson, H. E. Jr. A com-
  parison of the structure of behavioral maturity between
  Japanese and American primary-grade children. Journal
  of Social Psychology, 1972, 86, 167-173.

Purpose: To identify dimensions of behavioral maturity
Subjects: 100 second-graders in Takaoka, Japan
Variables: 18 item scores from a Japanese-adapted Child Behavior Scale
Method: Principal components condensation with varimax rotation
Results: Three factors emerged: I Academic maturity; II Interpersonal maturity; III Emotional maturity.
Conclusions: These results were compared with those for an American sample.

Shweder, R. A. How relevant is an individual difference theory of personality? Journal of Personality, 1975, 43, 455-484.

Purpose: To identify dimensions underlying the 44 theoretically significant scales proposed by Sears, Maccoby, and Levin (1957) to influence a mother's child-rearing practices
Subjects: 5 graduate students
Variables: Pooled similarity judgments for 44 declarative assertions representing the Sears et al. scales
Method: Principal axes condensation with varimax rotation
Results: Seven factors emerged, of which two accounted for 70% of the variance: I Permissiveness; II Self-confidence.
Conclusions: These results are discussed in conjunction with those of three others reported here.

Siegelman, M. Parent behavior correlates of personality traits related to creativity in sons and daughters. Journal of Consulting and Clinical Psychology, 1973, 40, 43-47.

Purpose: To identify dimensions of adults' retrospective reports of early parental behavior, as assessed by selected scales of the Parent-Child Relations Questionnaire, Short Form 1 (PCR-SF1)
Subjects: 144 male and 274 female college students
Variables: Scores on 6 PCR-SF1 scales, for each parent
Method: Principal components condensation with varimax rotation, independently by parent within each sex
Results: Three factors emerged consistently: I Love-reject; II Casual-demand; III Protect.

Solomon, D., Houlihan, K. A., Busse, T. V., & Parelius, R. J. Parent behavior and child academic achievement, achievement striving, and related personality characteristics. Genetic Psychology Monographs, 1971, 83, 173-273.

Purpose: To identify dimensions of 1) parental behavior during parent-child task-oriented interactions, and 2) indices of their children's general achievement, classroom behavior and task behavior
Subjects: 72 black fifth-graders, and their parents, rated by an observer

59

Variables: Scores on 1) 50 parental behavior measures
covering specific task ratings, global and "teacher
behavior" ratings, verbal and nonverbal interaction
scores; 2) 35 child measures of general achievement,
classroom and task behavior
Method: Principal axes condensation with oblique rotation,
independently for mothers, fathers, and children
Results: Four factors emerged for mothers: I Direct,
simple participation; II Encouragement of independent
achievement efforts; III Warmth; IV General interest.
Five paternal behavior factors emerged: I Encouragement of independent achievement efforts; II General
verbal participation; III Geniality; IV Hostility;
V Interest in situation and tasks. Six child achievement factors emerged: I General academic achievement;
II Achievement behavior in classroom - individual work
situations; III Achievement behavior in classroom -
recitation situations; IV Perserverance; V Divergent
task achievement behavior; VI Convergent task achievement behavior.
Stayton, D. J., Hogan, R., & Ainsworth, M. D. S. Infant
obedience and maternal behavior: The origins of
socialization reconsidered. Child Development, 1971,
42, 1057-1069.
Purpose: To identify dimensions underlying selected
maternal behavior and infant behavior variables
Subjects: 25 infant-mother pairs from white middle-class
families
Variables: Scores on 6 maternal behavior measures, plus
infant's age, sex, internalized controls, compliance
to commands
Method: Principal components condensation, apparently
without rotation
Results: Three factors emerged: I Maternal behaviors
promoting mother-infant harmony (quality of interaction);
II Floor freedom, physical intervention, IQ; III Verbal
commands, physical intervention, sex.
Thomas, D. R. Conservatism, authoritarianism and child-
rearing practices. British Journal of Social and
Clinical Psychology, 1975, 14, 97-98.
Purpose: To identify dimensions of child-rearing practices
Subjects: 56 Australian mothers (of children 4 to 6 years
old)
Variables: Unspecified number of variables from a modified
Sears et al. child-rearing practices interview
Method: Unspecified factor analysis
Results: Three factors emerged: I Authoritarianism; II
Orderliness-training; III Family adjustment.
Conclusions: Correlations between factors scores and C-
Scale data were reported.
C. Adolescence
Adcock, C. J., & Martin, W. A. Flexibility and creativity.
Journal of General Psychology, 1971, 85, 71-76.

60

Purpose: To identify the relationship of closure factors
   to creativity
Subjects: 188 tenth-graders
Variables: Scores on 16 (closure, creativity) tests plus
   age, sex, IQ, English and math exam scores
Method: Principal axes condensation with varimax and
   promax rotation
Results: Eight factors emerged: I Divergent speed of
   closure with ambiguous figural material; II Sex differences; III Convergent closure; IV Flexibility of
   semantic closure; V Academic achievement; VI Convergent
   figural closure; VII Age; VIII Figural flexibility.
Cattell, R. B., Shrader, R. R., & Barton, K. The definition
   and measurement of anxiety as a trait and a state in
   the 12- to 17-year range. British Journal of Social
   and Clinical Psychology, 1974, 13, 173-182.
   Study One
   Purpose: To identify dimensions of 162 anxiety items
   Subjects: 300 junior high and high school students, aged
      12-15 years
   Variables: 162 item scores (anxiety pool), marked with
      13 HSPQ primaries (B excluded)
   Method: Unspecified factor analysis
   Results: A general anxiety factor emerged; 60 items were
      selected from the 162-item pool.
   Study Two
   Purpose: To identify dimensions of anxiety items
   Subjects: 190 boys and girls aged 14-15 years
   Variables: 60 items selected in Study One, 100 extension
      items, 13 HSPQ factors
   Method: Unspecified factor analysis
   Results: A general anxiety factor emerged; 60 items were
      again selected for further item analysis.
   Study Three
   Purpose: To identify an anxiety state factor
   Subjects: 250 children, aged 12-17 years
   Variables: Test and retest (1 month apart) for 18 HSPQ
      scores, plus 80 anxiety items finally selected in
      prior steps
   Method: dR technique condensation with unspecified rota-
      tion
   Results: Nine second-order (and unspecified number of
      first-order) factors emerged: I Anxiety; II Exvia v.
      invia; III-IX Not interpreted.
   Conclusions: 40 items were selected as the High School
      Anxiety Scale, which scores separately on state and
      trait.
Chabassol, D. J. A scale for the evaluation of structure
   needs and perceptions in adolescence. Journal of
   Experimental Education, 1971, 40, 12-16.
   Purpose: To identify dimensions underlying the Wants
      Structure (WS) and Has Structure (HS) subscales of
      the adolescent structure inventory (CASI)

Subjects: 101 junior-senior secondary students in British
   Columbia
Variables: 20 item scores each from CASI subscales 1) WS,
   and 2) HS
Method: Principal axes condensation without rotation
Results: Five WS factors emerged: I General WS factor;
   II Supportive; III Clarity; IV Concern; V Uninterpreted.
   Five HS factors emerged: I General HS factor; II
   Restraint; III Depreciation; IV-V Uninterpreted.
Cooper, M. Factor analysis of measures of aptitude, intelli-
   gence, personality, and performance in high school
   subjects. Journal of Experimental Education, 1974,
   42, 7-10.
Purpose: To identify relationships between aptitude, in-
   telligence, personality, and performance in high school
   subjects
Subjects: 527 tenth-graders
Variables: Scores on 14 HSPQ subtests, 7 Differential
   Aptitude Tests, 2 Henmon Nelson Tests, and 7 school
   marks
Method: Unspecified condensation with oblimin rotation
Results: Five factors emerged: I Invia; II Anxiety; III
   Cortertia; IV Verbal ability; V Academic performance.
Conclusions: The extracted factors did not represent
   composites of variables from the four areas measured.
Delaney, J. O., & Maguire, T. O. A comparison between the
   structures of divergent production abilities of chil-
   dren at two levels of intellectual functioning.
   Multivariate Behavioral Research, 1974, 9, 37-45.
Purpose: To identify dimensions underlying divergent pro-
   duction tests for children at two levels of intellectual
   functioning
Subjects: 46 adolescents with average WISC IQ (full scale)
   =69.5, and 48 adolescents with average WISC IQ=104.5
Variables: Scores on 12 divergent production tests
Method: Image analytic condensation with orthogonal
   procrustes rotation, by group
Results: Six factors were extracted for both groups,
   though the subnormals' provided a slightly better fit:
   I Divergent production of figural units; II Divergent
   production of symbolic units; III Divergent production
   of symbolic relations; IV Divergent production of
   semantic units; V Divergent production of semantic
   classes; VI Divergent production of figural systems.
Ivinskis, A., Allen, S., & Shaw, E. An extension of Wechsler
   Memory Scale norms to lower age groups. Journal of
   Clinical Psychology, 1971, 27, 354-357.
Purpose: To identify relationships among Weschsler Memory
   Scale (WMS) and WISC/WAIS subtests
Subjects: 30 persons in the 10-14 year age range, and
   44 aged 16-18 years
Variables: Scores on WMS (11?) and WISC/WAIS (11) sub-
   tests plus sex

Method: Principal components condensation with quartimax
rotation
Results: Six factors emerged: I Verbal comprehension;
II Memory; III Nonverbal organization; IV Not labeled
(negative loadings for Orientation, sex, and Digit
Symbol); V Logical verbal memory (?); VI Not labeled
(Logical Memory and Associative Learning).
Jones, F. H. A 4-year follow-up of vulnerable adolescents:
The prediction of outcomes in early adulthood from
measures of social competence, coping style, and over-
all level of psychopathology. Journal of Nervous and
Mental Disease, 1974, 159, 20-39.
Purpose: To identify dimensions of social competence for
vulnerable adolescents
Subjects: 24 male-adolescents seen at the UCLA Family
Project clinic
Variables: Scores on 109 items from a standardized initial
interview, averaged across parents if both were present
to respond
Method: Cluster analysis
Results: Eleven clusters emerged: I School achievement;
II Career ambition; III Peer friendship; IV Dating;
V Family openness about sex; VI Assertiveness; VII
Separation from home; VIII Constructive relationships
within the family; IX Responsibility/autonomy struggles
and family distance; X Control and countercontrol in
communication; XI Frustration hostility.
Khan, S. B. Learning and the development of verbal ability.
American Educational Research Journal, 1972, 9, 607-613.
Purpose: To identify the factorial structure of a test
battery for two criterion, a control, and an experimental group
Subjects: 75 eleventh- and 95 ninth-graders (criterion);
48 seventh-graders (control); 63 seventh-graders
(experimental group)
Variables: Scores on Kit of Reference Tests sets (3 each)
for verbal comprehension, number facility, perceptual
speed, and spatial scanning
Method: Principal components condensation with normal
orthogonal varimax rotation, independently for each
grade level at pretest (all groups) and posttest
(control and experimental groups only)
Results: Four eleventh-grade factors emerged: I Verbal;
II Numerical; III Perceptual speed; IV Spatial. Four
ninth-grade factors emerged: I Verbal; II Numerical;
III Spatial ability; IV Uninterpreted. Four factors
each emerged for pre- and posttest for control and experimental groups; they were compared with criterion
group factors.
Conclusions: These comparative results indicated a tendency
toward specificity and differentiation as a result of
relevant learning.

Kourtrelakos, J. Perceived parental attitudes and demographic
variables as related to maladjustment. Perceptual and
Motor Skills, 1971, 32, 371-382.
Purpose: To identify dimensions of perceived parental
attitudes
Subjects: 80 students at a college for the performing
arts
Variables: 20 perceived parental attitude item scores
Method: Principal axes condensation with varimax rotation
Results: Four factors emerged: I Mother's trust in stu-
dent's judgment; II Father's trust in student's judgment;
III Parental encouragement of independent living; IV
Parental encouragement of independent effort.
Labouvie-Vief, G., Levin, J. R., & Urberg, K. A. The rela-
tionship between selected cognitive abilities and
learning: A second look. Journal of Educational
Psychology, 1975, 67, 558-569.
Purpose: To identify dimensions underlying selected mea-
sures of intelligence and memory
Subjects: 75 females and 75 males each at Grade 7 and 12
Variables: Scores on 8 memory and intelligence variables
(4 PMA, 4 KIT)
Method: Principal components condensation with varimax
rotation, for total group and by grade level
Results: The factor results were not specified, except
that independent factors of intelligence and memory
did not emerge in any of the analyses.
Lovegrove, S. A., & Hammond, S. B. A procedure for factoring an unlimited pool of items with results for the
C.P.I. using young Australian males. Australian Journal
of Psychology, 1973, 25, 29-43.
Purpose: To identify dimensions underlying an Australian
version of the California Psychological Inventory (CPI)
Subjects: 116 male 14-year-olds
Variables: 1) 450 item scores from an Australian version
of the CPI, divided into 3 groups of 150 each; 2) scores
on each factor extracted in the first stage (90 scores)
Method: Principal components condensation with varimax
rotation, independently for each 150-item set in the
first stage, and for the second stage.
Results: Thirty first-order factors were extracted from
each 150-item matrix; items not loading at least ≥ .30
on one factor were dropped in computing "factor scores"
for the second stage. Six "second-order" factors
emerged in the second stage: I Personal adequacy and
well-being; II Serious/flippant life attitude; III
Sociability and interpersonal competence; IV Community
alienation; V Rigidity of thought; VI Authoritarianism.
MacKenzie, A. J. A factor analysis of modified span memory
tests. Australian Journal of Psychology, 1972, 24,
19-30.
Purpose: To identify dimensions underlying span memory
tests

64

Subjects: 100 Junior Naval Recruits, aged 15.5-16.5 years
Variables: Scores on 33 modified span memory and standard
  reference marker tests, plus 2 random numbers
Method: Principal components condensation with varimax
  rotation
Results: Nine factors emerged: I Serial span memory; II
  Number facility; III Verbal comprehension; IV Simultan-
  eous span memory; V Random number specific; VI Associa-
  tive memory; VII Two-channel span memory; VIII General
  reasoning; IX Random number specific.
Meeker, M., & Meyers, C. E. Memory factors and school success
  of average and special groups of ninth-grade boys.
  Genetic Psychology Monographs, 1971, 83, 275-308.
Purpose: To identify dimensions of memory
Subjects: 90 ninth-grade boys (IQ 90-110)
Variables: Scores on 4 auditory and 4 visual tests of
  immediate memory, CTMM, 2 CAT tests, 8 DAT tests, grade
  and evaluation in math and English, best subject, and
  a reading test
Method: Principal components condensation with orthogonal
  varimax rotation
Results: Eight factors emerged: I General educational
  (possibly paper-pencil); II Auditory backward; III Nonverbal DAT plus math; IV Teacher grades and evaluations;
  V Visual backward; VI Auditory forward; VII-VIII Uninterpreted.
Conclusions: Three (II, V, VI) of the four posited memory
  factors emerged.
Nihira, K., Yusin, A., & Sinay, R. Perception of parental
  behavior by adolescents in crisis. Psychological
  Reports, 1975, 37, 787-793.
Purpose: To identify dimensions underlying adolescents-
  in-crisis' perceptions of parental behavior, as assessed
  by the Bronfenbrenner Parent Behavior Questionnaire
  (BPBQ)
Subjects: 86 12- to 18-year-olds in a crisis ward
Variables: 15 BPBQ subscale scores, plus sex for 1) mater-
  nal and 2) paternal behavior
Method: Principal components condensation with varimax
  rotation, independently for maternal and paternal ratings
Results: Four factors emerged in both analyses: I Nur-
  turing; II Punishing; III Firm control vs. lax control;
  IV Rejection (paternal) or Sex differences (maternal).
D. Adulthood and Old Age

VII.   SOCIAL PSYCHOLOGY

Beswick, D. G., & Tallmadge, G. K.   Reexamination of two
    learning style studies in the light of the cognitive
    process theory of curiosity.   Journal of Educational
    Psychology, 1971, 62, 456-462.
    Purpose: To identify dimensions underlying experimental
    treatments conditions
    Subjects: 11 experimental treatments, rated by 2 expert
    judges
    Variables: Scores on 8 arousal-of-curiosity scales
    Method: Principal components condensation with varimax
    rotation
    Results: Two factors emerged: I Ideational vs. substan-
    tive; II Implicitness vs. explicitness of subject
    matter.
Collins, B. E., & Hoyt, M. F.   Choice, aversive consequences,
    and the "truthtelling" potential of the situation as
    integrating concepts in forced compliance.   Psychologi-
    cal Reports, 1972, 30, 875-885.
    Purpose: To identify dimensions underlying S-judged
    ratings of experimental conditions
    Subjects: Unspecified number of college students
    Variables: Scores on 18 judgment scales for 1 of 6 ex-
    perimental conditions
    Method: Unspecified factor analysis
    Results: Four factors emerged: I (Collins') aversive
    consequences; II Choice, III Truth-evoking potential;
    IV Probability of compliance in a particular experi-
    mental situation.
French, J. W.   Application of T-technique factor analysis
    to the stock market.   Multivariate Behavioral Research,
    1972, 7, 279-286.
    Purpose: To identify dimensions of common stocks price
    change
    Subjects: 425 individual common stocks
    Variables: Normalized proportional changes in prices
    for 44 quarters
    Method: Principal components condensation with maxplane
    rotation
    Results: Six of the seven factors extracted appeared to
    have been generated by the nature of the CP matrix,
    while the seventh was interpreted as representing
    general stock market strengths/weaknesses.
Guertin, W. H.   Typing ships with transpose factor analysis.
    Educational and Psychological Measurement, 1971, 31,
    397-405.
    Purpose: To identify types of ships using transpose factor
    analysis
    Subjects: 29 ships (data provided by Cattell & Coulter)
    Variables: 1) 12 ship measures, 2) dissimilarities of 12
    measures

Method: Principal axes condensation with varimax rotation
    for 1) and 2)
Results: Four factors emerged in both analyses: I Carrier;
    II Destroyer; III Frigate 1; IV Frigate 2.
Lundberg, U., & Ekman, G. Subjective geographic distance: A
    multidimensional comparison. Psychometrika, 1973, 38,
    113-122.
Purpose: To identify factors of subjective judgments of
    geographic inter-distances
Subjects: 60 college students
Variables: (Cosine-transformed) interdistance judgments
    for 13 places
Method: Principal components condensation with unspecified
    rotation
Results: A three-factor solution that was similar to that
    for a Kruskal analysis emerged; however, a two-dimen-
    sional configuration could also describe the data.
Silverman, I., Shulman, A. D., & Wiesenthal, D. L. The
    experimenter as a source of variance in psychological
    research: Modeling and sex effects. Journal of
    Personality and Social Psychology, 1972, 21, 219-227.
Purpose: To identify dimensions of observer-rated experi-
    menter personality traits
Subjects: 224 college students (film observers)
Variables: Scores assigned to 22 personality trait scales
    for 6 experimenters
Method: Principal components condensation with varimax
    rotation
Results: Four factors emerged: I Cold-warm; II Competence-
    incompetence; III Extraversion-introversion; IV Vigorous-
    apathetic.
A. Culture and Social Processes
  1. Ethnology
  2. Social Structure and Social Role
  Gough, H. G. A cluster analysis of Home Index status items.
      Psychological Reports, 1971, 28, 923-929.
  Purpose: To identify dimensions of socioeconomic status,
      as assessed by the Home Index (HI)
  Subjects: 1,379 high school students
  Variables: 22 HI item scores
  Method: Cluster analysis with oblique rotation
  Results: Four factors emerged: I Social status; II
      Ownership; III Socio-civic involvement; IV Aesthetic
      involvement.
  3. Religion
  Coursey, R. D. Consulting and the Catholic crisis. Journal
      of Consulting and Clinical Psychology, 1974, 42, 519-528.
  Purpose: To identify dimensions of religious controversy
      in the Catholic church
  Subjects: 1) 275 parishioners from a moderately conserva-
      tive parish, 2) 403 from a moderately liberal parish
  Variables: Scores on 55 religious issues items

Method: Principal components condensation with varimax
   rotation, by parish and combined
Results: Six factors were reported for the combined sample:
   I Pious submissiveness; II The Catholic ghetto; III
   Marriage issues; IV Church rules; V Styles of worship;
   VI Social rights.
Wearing, A. J., & Brown, L. B. The dimensionality of religion.
   British Journal of Social and Clinical Psychology, 1972,
   11, 143-148.
   Purpose: To identify dimensions of religious belief and
      behavior
   Subjects: 420 college students, stratified across sex,
      course, and year in school
   Variables: 32 questionnaire-derived scores: strength of
      belief, theological conservatism vs. liberalism, Allport-
      Vernon-Lindzey religious score, dogmatism, authoritarian-
      ism, neuroticism, 9 areas of religious behavior, 8
      religious belief items, 8 determinants of career success
      ratings, sex
   Method: Principal components condensation with varimax
      rotation
   Results: Ten factors emerged: I Religious belief; II Career
      determinants; III Denomination - religious origins; IV
      Parental church attendance; V Anxious dogmatism; VI
      Ruthlessness opposed to hard work and ability in career
      success; VII Favorable stereotyped conventionalism to
      religious institutions; VIII Uninterpreted; IX Sex
      differences; X Christian defined in terms of faith in
      and salvation through Christ.
   Conclusions: Religious belief and practice seem to fall
      on one primary dimension in this study.
4. Cross Cultural Comparison
   Bagley, C., Boshier, R., & Nias, D. K. B. The orthogonality
      of religious and racialist/punitive attitudes in three
      societies. Journal of Social Psychology, 1974, 92,
      173-179.
   Purpose: To identify attitude dimensions as measured by
      the Conservatism (C) Scale for three societies
   Subjects: Unspecified number of male education students
      in England; unspecified number of individuals aged
      21-65 years, plus 79 "young professionals" in Holland
      (50% male); 350 continuing education students, mostly
      white-collar workers or their wives, in New Zealand
      (34% male)
   Variables: 50 C Scale item scores and total score, plus
      age, sex, class, education and 4-9 other markers
      related to religious behavior and affiliation
   Method: Unspecified condensation with promax rotation,
      independently for each nationality
   Results: Although 20 factors emerged for English students,
      12 for the Dutch, and 19 for the New Zealanders, only
      the first two (naturally orthogonal) factors were
      interpreted: I Religious values; II Racial and puni-
      tive values.

Bahr, H. M., & Chadwick, B. A. Conservatism, racial intolerance, and attitudes toward racial assimilation among
whites and American Indians. Journal of Social Psychology, 1974, 94, 45-56.
Purpose: To identify dimensions of conservatism
Subjects: 122 American Indians and 304 whites, all adults
from 1 urban area
Variables: Unspecified numbers of items from the Wilson-
Patterson Conservatism Scale and measures of racial
intolerance and support for assimilation
Method: Principal axes condensation with varimax rotation,
independently by ethnic group
Results: Sixteen factors emerged for whites and 18 for
Indians; the first unrotated factor appeared similar
across analyses: I General. Four rotated factors for
whites were discussed: I Family and motherhood; II
Religious; III Racial; IV Radical counterculture. Four
rotated factors for Indians were discussed: I Religious;
II Radical counterculture; III-IV Not interpreted.
Boshier, R. To rotate or not to rotate: The question of
the Conservatism Scale. British Journal of Social and
Clinical Psychology, 1972, 11, 313-323.
Purpose: To provide a rotated solution for Conservatism
(C) Scale data
Subjects: 122 male and 238 female New Zealanders in adult
education courses
Variables: 50 C Scale item scores
Method: Principal components condensation (presumably)
with varimax and oblique rotation
Results: Fifteen first-order factors emerged: I Religiosity;
II Anti-innovation; III Feminine repression; IV Hostility
to youth; V National and ethnic dominance; VI Hard
moralism; VII Retributive conservatism; VIII Age artifact;
IX Political (foreign affairs) conservatism; X Chaperones;
XI Moral fundamentalism; XII Horoscopes; XIII White lies;
XIV Social repression; XV Inborn conscience. Eight
second-order factors emerged: I Retributive - racial
toughness; II Superstition; III Retributive - political
toughness; IV Intolerance of youth; V Feminine repression;
VI Moral fundamentalism; VII Anti-innovation; VIII
Repressive religiosity. Four third-order factors emerged:
I Superstitious retribution; II Intolerance of youth;
III Socio-sexual fundamentalism; IV Socio-religious
rigidity.
Eysenck, H. J., & Coulter, T. T. The personality and attitudes
of working-class British Communists and Fascists.
Journal of Social Psychology, 1972, 87, 59-73.
Purpose: To identify dimensions of personality and attitudes
in working-class British Communists and Fascists (data
collected by the late Coulter, 1951-53)
Subjects: 43 Communists, 43 Fascists, and 86 soldiers
(controls), all working-class British non-Jewish males

69

Variables: Scores on Melvin's adaptation of R and T Scale,
  California Ethnocentrism Scale, California Fascism
  Scale, Dog-Cat Test of Intolerance of Ambiguity, Luchins
  water jar test of Einstelling rigidity, Rokeach social
  map test, California rigidity test, Intolerance of
  ambiguity questionnaire, TAT, Emphasis score (from
  R & T)
Method: Unspecified condensation with unspecified conden-
  sation, independently by group
Results: Four factors, fairly congruent across groups,
  emerged: I Tough-mindedness; II Rigidity; III Intolerance
  of ambiguity; IV Indirect aggression.
Feather, N. T. Explanations of poverty in Australian and
  American samples: The person, society, or fate?
  Australian Journal of Psychology, 1974, 26, 199-216.
  Purpose: To identify dimensions underlying Australians'
  attributions of poverty
  Subjects: 667 Adelaide respondents (heads, spouses, chil-
  dren, others dwelling in 328 randomly-selected households)
  Variables: 3-point ratings on 11 reasons for poverty
  Method: Principal-factor condensation with varimax rota-
  tion
  Results: Three factors emerged: I Structural, or socio-
  economic; II Individualistic, or personal responsibility;
  III Fatalistic, or personal misfortune.
Gardner, R. C., Kirby, D. M., & Arboleda, A. Ethnic stereotypes: A cross-cultural replication of their unitary
  dimensionality. Journal of Social Psychology, 1973,
  91, 189-195.
  Purpose: To identify dimensions of stereotypes cross-
  culturally
  Subjects: 250 Filipino students
  Variables: Scores on 40 stereotype differential scales
  for the concept Chinese in the Philippines, plus degree
  of contact and 3 social distance ratings
  Method: Unspecified condensation with normalized varimax
  rotation
  Results: Three factors emerged: I Evaluative; II Stereotype; III Social distance.
Gault, U., & Wang, A. M. Cultural variations in British and
  Australian personality differentials. British Journal
  of Social and Clinical Psychology, 1974, 13, 37-40.
  Purpose: To identify dimensions of a personality differ-
  ential
  Subjects: 80 Australians, equally distributed by sex, 4
  age groups, and 2 occupational groups
  Variables: Scores on 48 bipolar rating scales (for 25
  person-concepts)
  Method: Principal components condensation with varimax
  rotation
  Results: Six factors emerged: I, VI Evaluation; II, III
  Potency; IV, V Activity.
  Conclusions: These results are compared with those of Warr
  & Haycock (1970) for a British sample.

70

Gorsuch, R. L., & Smith, R. A. Changes in college students'
evaluations of moral behavior: 1969 versus 1939, 1949,
and 1958. Journal of Personality and Social Psychology,
1972, 24, 381-391.
Purpose: To identify dimensions underlying Crissman's 50-
item moral behavior scale for college samples at 5 times
Subjects: Unspecified (here) numbers of individuals
assessed by Rettig & Pasamanick: 1) college students
in 1958 (reported 1959); 2) students and alumni (1960);
3) unspecified (1961); 4) American and Korean college
students (1962); 5) settlers in Israel (1963)
Variables: Scores on 50 moral behavior items
Method: Unspecified new rotation for each group
Results: Apparently eight factors emerged consistently:
I Misrepresentation; II Irreligious hedonism; III Sexual
misbehavior; IV Non-philanthropic behavior; V Nonconser-
vative marriage pattern; VI Refusing to bear arms in a
war one believes to be unjust (singlet); VII Nations at
war using poison gas on the homes and cities of its
enemy behind the lines (singlet); VIII Total severity.
Conclusions: Eight scales were used in the present study
with 1969 data.
Hasleton, S. Permissiveness in Australian society. Australian
Journal of Psychology, 1975, 27, 257-267.
Purpose: To identify dimensions of permissive attitudes in
Australian society
Subjects: 1881 registered Australian voters
Variables: Scores on 10 5-point items related to permis-
siveness
Method : Principal components condensation with varimax
rotation, independently for total group and subgroups
selected on basis of age (3) and education (2)
Results: Two factors emerged in all six analyses: I
General permissiveness; II Abortion, prostitution,
divorce.
Henderson, R. W., Bergan, J. R., & Hurt, M. Jr. Development
and validation of the Henderson Environmental Learning
Process Scale. Journal of Social Psychology, 1972, 88,
185-196.
Purpose: To identify dimensions of the Henderson Environmental Learning Process Scale (HELPS)
Subjects: 60 lower-SES Mexican-American and 66 middle-SES
Anglo-American first-graders
Variables: 1) 25 HELPS item scores, 2) 25 HELPS item
scores, plus SES, ethnic group membership
Method: Principal components condensation with varimax
rotation, independently for the two variable sets
Results: Five factors emerged in both analyses: I Extended interests and community involvement; II Valuing
language and school related behavior (SES, ethnicity
loaded highly on this factor); III Intellectual guidance;
IV Providing a supportive environment for school learning; V Attention.

Kohn, P. M. Authoritarianism, rebelliousness, and their
    correlates among British undergraduates. British
    Journal of Social and Clinical Psychology, 1974, 13,
    245-255.
    Purpose: To identify dimensions underlying an anglicized
    version of Kohn's Authoritarianism-Rebellion Scale (ARS)
    Subjects: 234 British undergraduates
    Variables: 30 item scores from the British ARS
    Method: Principal components condensation with direct
    oblimin rotation
    Results: Five factors emerged: I Sociopolitical rebellious-
    ness; II General right-wing authoritarianism; III
    Authoritarian submission; IV Prepolitical authoritarian-
    rebellious; V Ethnocentrism.
Laosa, L. M., Swartz, J. D., & Moran, L. J. Word association
    structures among Mexican and American children. Journal
    of Social Psychology, 1971, 85, 7-15.
    Purpose: To identify dimensions of word associations for
    Mexican and American children
    Subjects: 75 Mexico City (M) and 75 Austin, Texas (A)
    children with mean age 9.7 years; 79 M and 77 A with
    mean age 12.7 years; 51 M and 51 A with mean age 15.7
    years; all matched within age group on age, sex, and
    SES
    Variables: Ratings, from an 80-word association list, on
    6 semantic and 7 grammatical variables
    Method: Principal components condensation with normalized
    varimax rotation, independently for each age x culture
    group within semantic and grammatical sets
    Results: Two semantic factors, similar across age and cul-
    ture, emerged: I Dimension vs. predication reference
    set; II Concept referent dimensions. Two grammatical
    factors, again similar across age and culture, emerged:
    I Noun; II Verb, adjective.
Larsen, K. S., Arosalo, U., Lineback, S., & Ommundssen, R.
    New Left ideology - a cross-national study. Journal of
    Social Psychology, 1973, 90, 321-322.
    Study One
    Purpose: To identify dimensions of New Left ideology with
    a cross-cultural sample
    Subjects: 117 Japanese, 152 Norwegian, and 136 American
    college students
    Variables: 78 New Left Scale item scores
    Method: Unspecified condensation with varimax rotation,
    independently by nation
    Results: Seven factors emerged for Japan: I Antiestablish-
    ment-action; II Bourgeoisie; III Law and order; IV
    Traditional-reactionary; V Cynical-machiavellianism;
    VI Cynicism antistructure; VII Revolutionary tactics.
    Two factors were interpretable for Norway: I Bourgoisie;
    II Peaceful reconstruction. Seven factors emerged for

the United States: I Free from structure; II Trust-
change; III "Let it be"; IV Bourgoisie; V Machiavellian-
ism; VI Concerned liberal; VII Suspicion.
Conclusions: These results were compared with those of
Christie, et al. (1970, see below).
Study Two Christie, Friedman, & Ross (mineographed paper,
1970)
Purpose: To identify dimensions of New Left ideology
Subjects: American university students (number unspecified
here)
Variables: (presumably) 78 New Left Scale item scores
Method: (Unspecified here) factor analysis
Results: Five factors emerged: I New Left philosophy; II
Machiavellian tactics; III Trust in others; IV Revolu-
tionary tactics; V Traditional moralism.
Lundberg, U., & Devine, B. Negative similarities. Educa-
tional and Psychological Measurement, 1975, 35, 797-807.
Purpose: To identify dimensions underlying emotional terms
Subjects: 1) 150 college students, 2) 150 college students
all Swedish
Variables: Similarity ratings of 15 emotional terms (Ekman,
1955) on 1) 5-point scale (no similarity to identity),
and 2) 9-point scale (complete opposites to identity)
Method: Principal components condensation with varimax
rotation, independently for the two subject-variable
sets
Results: Seven factors emerged in the first analysis: I
Fear; II General agitation; III-VII Not labeled. Six
factors emerged in the second analysis: I Fear; II
General agitation; III Discontentedness and contentedness; IV Passive repulsion; V-VI Not labeled.
Marjoribanks, K., & Josefowitz, N. Kerlinger's theory of
social attitudes: An analysis. Psychological Reports,
1975, 37, 819-823.
Purpose: To identify dimensions of social attitudes
Subjects: 370 male and 90 female British and Welsh secon-
dary students (17-year-olds)
Variables: 1) 32 Worldmindedness Scale (W-Scale) item
scores, 2) 12 Authoritarianism-Rebellion Scale (AR
Scale) item scores, 3) 50 Conservatism Scale (C-Scale)
item scores, 4) 41 items identified in 1-3
Method: Principal components condensation with varimax
rotation for each variable set
Results: Three W-Scale factors emerged: I Racial prejudice;
II Nationalism; III Patriotism. One AR Scale emerged:
I Social conservatism. Four C-Scale items emerged: I
Disrespect for authority; II Political activism; III
Support for modern art; IV Sexual freedom. Two second-
order factors emerged: I Conservative; II Liberal.
Mezei, L. Factorial validity of the Kluckhohn and Strodtbeck
Value Orientation Scale. Journal of Social Psychology,
1974, 92, 145-146.
Purpose: To identify dimensions underlying the Kluckhohn
and Strodtbeck Value Orientation Scale (KSVOS)

Subjects: 106 persons from Zuni, Navaho, Mormon, Spanish
American and Texas cultures in New Mexico (original data
collected by Kluckhohn & Strodtbeck, 1961)
Variables: Rank-ordered solutions for 40 KSVOS variables
Method: Principal axes condensation with varimax and
quartimax rotation
Results: The quartimax solution yielded two major bipolar
factors: I Activity-passivity; II Individualistic-
traditional. The varimax solution yielded seven factors:
I Economic relations; II Community relations; III Rela-
tional collaterality and harmony with (man-) nature; IV
Individualistic-traditional; V Activity-passivity; VI
Activity; VII Not reported.
Morse, S. J., Reis, H. T., Gruzen, J., & Wolff, E. The "eye
of the beholder": Determinants of physical attractiveness
judgments in the U. S. and South Africa. Journal of
Personality, 1974, 42, 528-542.
Purpose: To identify dimensions of other- and self-evaluated
physical attractiveness for an American sample (present)
and a South African sample (data from Morse, Gruzen, &
Reis, 1974 unpublished manuscript)
Subjects: 58 male and 74 female American college students,
59 male and 102 female South Africans
Variables: 15 10-point semantic differential ratings of
self-evaluated heterosexual attractiveness, 2) 5 (11 for
Africans) ratings for other-evaluation
Method: Principal components condensation with varimax
rotation, independently for self- and other-evaluations
by sex within nationality
Results: For Americans, five unlabeled male and four female
self-evaluation factors emerged; two unlabeled factors
emerged for both sexes for other-evaluations. The numbers of factors extracted for the South African analyses
were not reported.
Schuh, A. J. An alternative questionnaire strategy for con-
ducting cross-cultural research on managerial attitudes.
Personnel Psychology, 1974, 27, 95-102.
Purpose: To identify dimensions underlying a basic cultural
differences questionnaire
Subjects: 63 male and 28 female American, 101 male and 112
female Filipino college students
Variables: 30 binary item scores from a basic cultural
differences questionnaire
Method: Hierarchical condensation without rotation
Results: One general factor emerged.
Conclusions: Ten items defining the general factor were
selected for further use.
Scott, W. A., & Peterson, C. Adjustment, Pollyannaism, and
attraction to close relationships. Journal of Consulting
and Clinical Psychology, 1975, 43, 872-880.
Purpose: To identify dimensions of liking for objects in
three cultures
Subjects: 88 U. S., 80 Japanese, and 40 New Zealand college
students

Variables: Liking scores for an unspecified number of objects
Method: Cluster analysis, by culture
Results: Four self-role (I Sociable; II Domestic; III
    Responsible; IV Distasteful), four family relationship
    (V Intimate relations; VI Recreation; VII Skill display;
    VIII Hostile relations), three (of seven to nine) ac-
    quanitance (IX Close relatives; X Valued persons; XI
    Avoided persons), and nine nations clusters (XII-XXI
    Unlabeled, combining geographical and ideological con-
    siderations) which were common to the three cultures
    emerged.
Stacey, B. G., & Green, R. T. Working-class conservatism:
    A review and an empirical study. British Journal of
    Social and Clinical Psychology, 1971, 10, 10-26.
  Purpose: To identify dimensions of attitudes in the working
    class
  Subjects: 121 Conservative and 181 Labour supporters, all
    males in manual occupations
  Variables: Scores on 58 attitude items and 11 descriptive
    characteristics (markers)
  Method: Unspecified tri-level (markers correlated with
    extracted factors) condensation without rotation
  Results: Fourteen first-level factors, of 23 extracted,
    were retained: I Royalist; II Sympathy and sensitivity
    toward others; III Not labeled; IV Reform rather than
    punishment of criminals; V-VII Not labeled; VIII Firm-
    ness in discipline; IX-XV Not labeled; XVI Paternal
    support for party (Labour); XVII Not labeled; XVIII
    Egalitarian; XIX-XXI Not labeled; XXII Critical, ques-
    tioning; XXIII Unswerving loyalty to authority and
    leadership. Seven second-level factors emerged: I
    Patriotic; II Not labeled; III Egalitarianism, helpful
    sympathy, reform of criminals; IV-V Not labeled; VI
    Elitism and 'natural' inegalitarianism; VII Self-
    interest combined with firm convictions, toughness.
    Three third-level factors emerged: I Self-interest,
    inegalitarianism, lack of sympathy for underdog, per-
    sonal initiative emphasis, elite leadership orienta-
    tion, nationalistic outlook; II Anti-royalty, anti-
    aristocracy; III Discipline, toughness, submission to
    authority.
Stanley, G., & Vagg, P. Attitude and personality character-
    istics of Australian protestant fundamentalists. Journal
    of Social Psychology, 1975, 96, 291-292.
  Purpose: To identify dimensions of a set of attitude and
    personality items
  Subjects: 219 Australian college students
  Variables: Scores on 40 Eysenck attitude scale items and
    57 Eysenck Personality Inventory (Form B) items
  Method: Unspecified condensation with promax rotation
  Results: Three attitude (I Moral radicalism; II Political
    radicalism; III Socioethnic radicalism) and four per-
    sonality (IV Neuroticism; V-VII Not labeled) factors
    emerged.
75

Conclusions: Factor scores were compared between this
sample and one of bible college students.
Thomas, D. R. The relationship between ethnocentrism and
conservatism in an "authoritarian" culture. Journal
of Social Psychology, 1974, 94, 179-186.
Purpose: To identify dimensions underlying 12 social
distance items
Subjects: 77 mothers (Australian) participating in a
child-rearing study
Variables: 7-point ratings on 12 behavioral intentions
items (3 ethnic groups x 4 degrees of intimacy)
Method: Cluster analysis
Results: Two clusters emerged: I Marital rejection; II
Social distance
Conclusions: Ten items were selected to represent these
two scales for further research.
Trlin, A. D., & Johnston, R. J. Dimensionality of attitudes
towards immigrants: A New Zealand example. Australian
Journal of Psychology, 1973, 25, 183-189.
Purpose: To identify dimensions underlying attitudes toward
immigrants in New Zealand
Subjects: 317 randomly-sampled registered voters (11
Auckland electorates)
Variables: 6-point Bogardus Social Distance Scale score
for each of 14 birthplace groups
Method: Principal components condensation with varimax
rotation
Results: Three factors emerged: I White, non-English
speaking; II Non-white, or non-European; III White,
English speaking (UK, USA).
Conclusions: Separate analyses for native- and foreign-
born subsamples yielded similar dimensions.
Vroegh, K. Masculinity and femininity as perceived by
Hawaiians. Perceptual and Motor Skills, 1972, 35,
119-125.
Purpose: To identify dimensions underlying Hawaiians' per-
ception of masculinity-femininity (Mf)
Subjects: 109 lower-income American-Hawaiians, aged 15-65
years
Variables: Scores on 53 Mf items
Method: Elementary factor condensation without rotation
Results: Six factors emerged: I Dependable-frivolous; II
Competent-incompetent; III Emotionally unstable-emotion-
ally mature; IV Impulsive-calm; V Passive-active; VI
Heterosexual-not heterosexual.
Wilson, G. D., & Lee, H. S. Social attitude patterns in
Korea. Journal of Social Psychology, 1974, 94, 27-30.
Purpose: To identify dimensions of social attitudes in
Korea
Subjects: 356 Koreans, representative by occupation and
sex
Variables: Unspecified number of item scores from the
Wilson-Patterson Attitude Inventory (adapted for Korean
culture)

Method: Principal components condensation, apparently
without rotation
Results: Five factors emerged: I Serious-minded view of
life; II Religious and patriotic issues; III Authoritarian and punitive attitudes; IV Race and international
politics; V Sexual freedom.
5. Family
6. Social Change and Social Programs
B. Sexual Behavior
Abramson, P. R., & Mosher, D. L. Development of a measure
of negative attitudes toward masturbation. Journal of
Consulting and Clinical Psychology, 1975, 43, 485-490.
Purpose: To identify dimensions underlying attitudes
toward masturbation, as assessed by the Negative Attitudes Toward Masturbation Inventory (NATMI)
Subjects: 96 male and 102 female college students
Variables: 30 NATMI item scores
Method: Unspecified condensation with varimax rotation
Results: Three factors emerged: I Positive attitudes
toward masturbation; II False beliefs about the harmful
nature of masturbation; III Personally experienced nega-
tive affects associated with masturbation.
Burdsal, C., Greenberg, G., Bell, M., & Reynolds, S. A
factor-analytic examination of sexual behaviors and
attitudes and marihuana use.
Purpose: To identify dimensions of self-reported sexual
attitudes and behavior, family relationships, and
marihuana use.
Subjects: 358 college students
Variables: 44 questionnaire items
Method: Principal axis condensation with orthogonal varimax, oblique maxplane, graphical, and maxplane rotations
Results: Twelve factors emerged: I Liberal vs. conservative attitudes; II Age-experience; III Symbolic sexual
preoccupation; IV Romantic love vs. cynicism; V Experience-
linked drug effects; VI Affectual dependence; VII Mature
satisfaction; VIII Conservative vs. liberal sexual practices; IX High vs. low sexual activity; X Sexual revolu-
tion; XI Sex; XII Traditional vs. cynical love roles.
Byrne, D., Fisher, J. D., Lamberth, J., & Mitchell, H. E.
Evaluations of erotica: Facts or feelings? Journal of
Personality and Social Psychology, 1974, 29, 111-116.
Purpose: To identify dimensions underlying self-ratings of
affect assessed by a feelings scale (FS)
Subjects: 1) 32 married couples (one member of each couple
was college student); 2) 154 male and female college
students (data collected by W. Griffit)
Variables: 5-point scores on 11 FS dimensions
Method: Unspecified condensation with varimax rotation by
sample
Results: Two factors emerged in both analyses: I Positive
affect; II Negative affect.
Cook, R. F., Fosen, R. H., & Pacht, A. Pornography and the
sex offender: Patterns of previous exposure and arousal
effects of pornographic stimuli. Journal of Applied

<u>Psychology</u>, 1971, <u>55</u>, 503-511.
Purpose: To identify the structure of the erotic response
   to visual depictions of sexual behavior
Subjects: 66 (story-lines presentation) and 63 (random
   presentation) sex and criminal code offenders
Variables: Arousal ratings for 26 slides depicting different
   sexual behaviors
Method: Unspecified condensation with oblique rotation,
   independently for presentation groups
Results: Three story-line factors emerged: I Heavy petting
   and intercourse; II Foreplay; III Anticipation. Four
   random order factors emerged: I Female genitalia; II
   Foreplay; III Fellatio; IV Intercourse.
Hariton, E. B., & Singer, J. L. Women's fantasies during
   sexual intercourse: Normative and theoretical implica-
   tions. <u>Journal of Consulting and Clinical Psychology</u>,
   1974, <u>42</u>, 313-322.
Purpose: To identify dimensions underlying women's thought
   or perceptual reactions during coitus
Subjects: 141 suburban housewives
Variables: Scores on 18 variables describing thought or
   perceptual reaction during coitus
Method: Unspecified condensation with promax rotation
Results: Three factors emerged: I Erotic fantasy; II
   Negative sexuality; III Submissive fantasy (bipolar).
Mathews, A. M., Bancroft, J. H. J., & Slater, P. The prin-
   cipal components of sexual preference. <u>British Journal
   of Social and Clinical Psychology</u>, 1972, <u>11</u>, 35-43.
Purpose: To identify dimensions of sexual and long-term
   partner preferences
Subjects: 15 psychologists, 15 psychiatrists, 15 hospital
   porters, 15 young soldiers, 15 homosexuals, all male
Variables: Preference ratings for 50 pictures of females,
   rated for attractiveness as long-term partner and as
   short-term sexual partner.
Method: Inverse principal components condensation, independently for long- and short-term ratings and change scores
Results: The three factors emerging from the separate
   preference ratings were similar: I General attractiveness; II Sexy vs. sexless; III Unconventional. Three
   factors also emerged from the analysis of change scores,
   but only one was labeled: I Marriage vs. short-term
   sexual partner.
1. Birth Control and Abortion
   Corenblum, B., & Fischer, D. G. Factor analysis of attitudes
      toward abortion. <u>Perceptual and Motor Skills</u>, 1975, <u>40</u>,
      587-591.
   Purpose: To identify dimensions underlying attitudes toward
      abortion
   Subjects: 385 male and 315 female college students
   Variables: 31 semantic differential ratings for the concept
      abortion

Method: Principal components condensation with varimax
    rotation
Results: Four factors emerged: I Immoral-repelling; II
    Cruel-destructive; III Emotional-serious; IV Rational-
    impulsive.
Kothandapani, V. Validation of feeling, belief, and intention
    to act as three components of attitude and their contri-
    bution to prediction of contraceptive behavior. Journal
    of Personality and Social Psychology, 1971, 19, 321-333.
Purpose: To identify dimensions of attitudes toward contra-
    ceptive behavior
Subjects: 50 users and 50 nonusers of contraceptives, all
    low-income black women
Variables: Scores on 4 verbal measures each of feeling,
    belief, and intention to act related to contraceptive
    behavior
Method: Principal axes condensation with 1) varimax and
    2) oblique rotation
Results: Three factors, highly similar across solutions,
    emerged: I Feeling; II Belief; III Intention to act.
C. Attitudes and Opinions
Blake, B. F., Weigl, K., & Perloff, R. Perceptions of the
    ideal community. Journal of Applied Psychology, 1975,
    60, 612-615.
Purpose: To identify attribute dimensions of perceptions
    of the ideal community
Subjects: 2,253 adults living in small towns, 1,161 in
    medium-sized, 1,205 in large cities in Indiana
Variables: Importance ratings for 11 attributes of the
    ideal community
Method: Principal components condensation with varimax
    rotation by group
Results: Three factors emerged: I System maintenance and
    change; II Personal development, or recreation; III Rela-
    tionship.
Borus, J. F., Fiman, B. G., Stanton, M. D., & Dowd, A. F.
    The Racial Perceptions Inventory. Archives of General
    Psychiatry, 1973, 29, 270-275.
Purpose: To identify dimensions underlying the Racial
    Perceptions Inventory (RPI)
Subjects: 429 combat-trained enlisted troops
Variables: 66 RPI item scores
Method: Principal components condensation with varimax
    rotation
Results: Three factors emerged: I Attitudes toward inte-
    gration; II Perceptions of racial discrimination; III
    Backlash feelings.
Brady, D., & Rappoport, L. Policy-capturing in the field:
    The nuclear safeguards problem. Organizational Behavior
    and Human Performance, 1973, 9, 253-266.
Purpose: To identify dimensions underlying (stringency of
    control) attitudes toward nuclear safeguards for two
    groups of experts

Subjects: 45 Atomic Energy Commission (AEC) staff members
and consultants, and 44 nuclear industry (NI) personnel,
all selected by experts as being influential in policy
determination
Variables: 4-point stringency-of-control scores for 12
items relevant to any comprehensive safeguard system,
plus choice of system type
Method: Unspecified condensation with varimax rotation,
independently for AEC and NI groups
Results: Three AEC factors emerged: I Regulatory power;
II Records and transport; III Measurements. Three NI
factors emerged: I Intrusiveness; II Regulatory power;
III Accountability.
Burdsal, C. Jr. A factor analytic study of racial attitudes.
Journal of Social Psychology, 1975, 97, 255-259.
Purpose: To identify dimensions of racial attitudes
Subjects: 308 whites and 147 blacks from 1 midwestern town
Variables: 19 racial attitude item scores, age, sex, race
Method: Principal axes condensation with varimax, maxplane,
and graphical rotation
Results: Seven factors emerged: I Race; II Relaxed associa-
tion vs. tense isolation; III General racial tension; IV
Sex; V Youthful integration vs. aged segregation; VI
Genetics; VII Separationist vs. desegregationist.
Coan, R. W., Hanson, R. W., & Dobyns, Z. P. The development
of some factored scales of general beliefs. Journal of
Social Psychology, 1972, 86, 161-162.
Purpose: To identify dimensions of general beliefs
Subjects: Unspecified; data reported by Coan (book in press
in 1972)
Variables: 130 item scores from the precursor of the
General Beliefs inventory
Method: Unspecified factor analysis
Results: Seventeen (unlabeled here) factors emerged.
Conclusions: Subsequent work (unspecified here) by Hanson
(unpublished thesis, 1970) resulted in General Beliefs,
which contains 73 items composing six scales: I Con-
ventional theistic religion vs. nontheistic viewpoint;
II Future-productive vs. present-spontaneous orientation;
III Detachment vs. involvement; IV Relativism vs. absolu-
tism; V Scientism-determinism; VI Optimism vs. pessimism.
Dielman, T. E., Stiefel, G., & Cattell, R. B. A check on the
factor structure of the Opinions of Mental Illness Scale.
Journal of Clinical Psychology, 1973, 29, 92-95.
Purpose: To identify dimensions underlying the Opinions of
Mental Illness scale (OMI)
Subjects: 138 college students
Variables: Test and 1-week retest scores on 51 OMI items
Method: Principal-axes condensation with oblimax and rotoplot-
assisted graphic rotation, independently for test and re-
test data

Results: Ten test, and 12 retest, factors were extracted;
the seven factors considered good matches across admin-
istrations were reported: I Interpersonal etiology; II
Mental hygiene ideology; III Social restrictiveness; IV
Authoritarianism; V Medical model orientation; VI
Benevolence; VII (Realism or cynicism - singlet).

Eysenck, H. J. Social attitudes and social class. British
Journal of Social and Clinical Psychology, 1971, 10,
201-212.
Purpose: To identify dimensions of social attitudes
Study One:
Subjects: 2,902 persons divided into 18 categories on the
basis of sex, age group (young, middle-aged, old) and
social class (middle, skilled, unskilled)
Variables: 28 item scores from The Psychology of Politics
questionnaire
Method: Principal components condensation with promax
rotation, independently for the 18 groups
Results: Two unlabeled factors consistently emerged across
analyses.

Study Two:
Subjects: Proportional subsamples from above: 382 middle-
class, 1,235 skilled, 553 unskilled
Variables: 28 item scores, plus age and sex
Method: Principal components condensation with promax
rotation, independently for the three groups and com-
bined groups.
Results: Eight first-order factors consistently emerged:
I Authoritarian; II Religious; III Ethnocentrism; IV
Humanitarianism; V Sexual morals; VI Toughmindedness;
VII Age; VIII Sex. Two nearly orthogonal second-order
factors emerged: I Humanitarianism (v. authoritarianism);
II Religionism.

Eysenck, H. J. The structure of social attitudes. British
Journal of Social and Clinical Psychology, 1975, 14,
323-331.
Purpose: To identify dimensions underlying social attitudes
Subjects: 368 individuals randomly sampled from the urban
population
Variables: Scores on 88 social attitude questions, social
class, age, and sex
Method: Principal components condensation with oblique pro-
max rotation
Results: Ten first-order factors emerged: I Permissiveness;
II Socialism; III Racism; IV Laissez-faire; V Pacifism;
VI Capitalism; VII Religion; VIII Reactionary individualism;
IX Human nature; X Libertarianism. Three second-order fac-
tors emerged: I Conservative-radical; II Toughmindedness
vs. tender-mindedness; III Politico-economic conservatism
contrasted with socialism.
Conclusions: Separate analyses with subgroups yielded essentially the same results.

Feather, N. T., & Peay, E. R. The structures of terminal and
   instrumental values: Dimensions and clusters. <u>Australian</u>
   <u>Journal</u> <u>of</u> <u>Psychology</u>, 1975, <u>27</u>, 151-164.
   Purpose: To identify clusters of terminal and instrumental
      values
   Subjects: 548 university and 530 college of education stu-
      dents
   Variables: 18 terminal and 18 instrumental Value Survey
      scores
   Method: Cluster analysis, independently for each subject
      group and for terminal and instrumental and all 36 values
   Results: Terminal value structures appeared similar for
      the two subject groups, but instrumental value groupings
      differed; neither were sufficiently clustered to permit
      selection of a representative structure.
   Conclusions: Results are also reported for multidimensional
      scaling analyses.
Ferguson, L. W. Primary social attitudes of the 1960s and
   those of the 1930s. <u>Psychological</u> <u>Reports</u>, 1973, <u>33</u>,
   655-664.
   Purpose: To identify dimensions underlying primary social
      attitudes in the 1960s
   Subjects: 1,471 university students (in late 1960s)
   Variables: Scores on 10 Thurstone-edited social attitudes
      scales
   Method: Centroid condensation with graphic rotation
   Results: Three factors emerged: I Religionism; II Humani-
      tarianism; III Nationalism.
   Conclusions: These results were compared with 1930s' results.
Fishkin, J., Keniston, K., & MacKinnon, C. Moral reasoning
   and political ideology. <u>Journal</u> <u>of</u> <u>Personality</u> <u>and</u>
   <u>Social</u> <u>Psychology</u>, 1973, <u>27</u>, 109-119.
   Purpose: To identify dimensions of preferences for political
      slogans
   Subjects: 75 college students
   Variables: 5-point preference ratings for 31 political slogans
   Method: Principal components condensation with varimax
      rotation
   Results: Three factors emerged: I Violent radicalism; II
      Peaceful radicalism; III Conservatism.
   Conclusions: These results were considered further relative
      to moral development classification.
Goldman, R. D., Platt, B. B., & Kaplan, R. B. Dimensions of
   attitudes toward technology. <u>Journal</u> <u>of</u> <u>Applied</u> <u>Psychology</u>,
   1973, <u>57</u>, 184-187.
   Purpose: To identify dimensions of attitudes toward tech-
      nology
   Subjects: 456 college students
   Variables: 80 item scores from an attitudes-toward-mechani-
      zation questionnaire
   Method: Principal components condensation with varimax
      rotation

82

Results: Six factors of ten extracted, were interpreted:
I Global mechanism; II Mechanical curiosity; III Preference for handmade goods; IV Alienation; V Spiritual
benefits of technology; VI Human vitalism.
Gottlieb, J., & Corman, L. Public attitudes toward mentally
retarded children. American Journal of Mental Deficiency,
1975, 80, 72-80.
Purpose: To identify dimensions of public attitudes toward
retarded children
Subjects: 430 adult residents of the greater Boston area
Variables: Scores on 48 attitude items from a questionnaire
pertaining to retarded children
Method: Principal components condensation with varimax
rotation
Results: Four factors emerged: I Positive stereotype; II
Segregation in the community; III Segregation in the
classroom; IV Perceived physical and intellectual han-
dicap.
Greene, D. L., & Winter, D. G. Motives, involvements, and
leadership among Black college students. Journal of
Personality, 1971, 30, 319-332.
Purpose: To identify dimensions underlying judges' ratings
of black college students' attitudes, beliefs, and be-
haviors
Subjects: 38 black college students
Variables: 4 TAT-derived scores, judged by 3 raters
Method: Unspecified condensation with varimax rotation
Results: Three factors emerged: I Factor I; II Pragmatic;
III Idiosyncratic to one judge.
Greever, K. B., Tseng, M. S., & Friedland, B. U. Development
of the Social Interest Index. Journal of Consulting and
Clinical Psychology, 1973, 41, 454-458.
Purpose: To identify dimensions underlying the Social
Interest Index (SII)
Subjects: 83 (college students presumably)
Variables: 32 SII item scores
Method: Principal components condensation with varimax
rotation
Results: Five factors emerged: I Social interest; II Love;
III Friendship; IV Work; V Self-significance.
Hoiberg, A., & Booth, R. F. Changes in attitudes and charac-
teristics of Marine recruits during the 1960s. Psycho-
logical Reports, 1973, 32, 1079-1086.
Purpose: To identify dimensions underlying Marine recruits'
attitudes
Subjects: 704, 476, and 481 Marine recruits (tested about
Week 10 of training)
Variables: Scores on 14 attitude questionnaire items
Method: Cluster analysis
Results: One cluster emerged as highly interrelated across
samples: I Attitudes toward leadership or authority
figures in the Marine Corps.

Jones, J. M. Attitudinal valence and semantic differential potency scales. Psychological Reports, 1971, 28, 991-994.
  Purpose: To identify dimensions underlying semantic differential (SD) ratings on two issues for "for" and "against" Ss
  Subjects: 185 college students, divided into subgroups (for and against) on each issue
  Variables: 16 SD ratings on War in Vietnam and Negro-white Dating
  Method: Principal components condensation with varimax rotation, independently for "for" and "against" Ss with each issue
  Results: Four factors emerged in the for-war, six in the against-war, and six for each dating analysis; none were labeled.
Kerlinger, F. N. The structure and content of social attitude referents: A preliminary study. Educational and Psychological Measurement, 1972, 32, 613-630.
  Purpose: To identify dimensions underlying social attitude referents (objects)
  Subjects: 530 teachers and graduate education students from 3 states
  Variables: Scores on 50 Referents Attitude Scale items
  Method: Principal factors condensation with promax rotation (varimax for second-order)
  Results: Six first-order factors emerged: I Religiosity; II Educational traditionalism; III Civil rights; IV Child-centered education; V Social liberalism; VI Economic conservatism. Two second-order factors emerged: I Conservatism; II Liberalism.
Kernan, J. B., & Trebbi, G. G. Jr. Attitude dynamics as a hierarchical structure. Journal of Social Psychology, 1973, 89, 193-202.
  Purpose: To identify dimensions of attitudes
  Subjects: 57 business administration undergraduates
  Variables: Ratings of the concept MBA degree on 3 scales: 10 items designed to measure a cognitive attitude component, 10 the affective, and 10 the behavioral component (57 x 3 matrix)
  Method: Unspecified condensation with varimax rotation
  Results: Three factors emerged: I Affect; II Cognition; III Behavior.
  Conclusions: Changes along these dimensions were examined as a function of reception of "anti-MBA" messages.
Kirk, S. A. The psychiatric sick role and rejection. Journal of Nervous and Mental Disease, 1975, 161, 318-325.
  Purpose: To identify dimensions of attitudes toward (social rejection of) the psychiatric sick role
  Subjects: 806 community-college students
  Variables: 15 social distance or rejection item ratings for 2 case vignettes

84

Method: Unspecified condensation with quartimax rotation
Results: Three factors emerged: I Rejection; II Uncertainty; III Acceptance of the actor.
Conclusions: Items loading on Factor I were selected for further use in the study.

Leonard, C. V., & Flinn, D. E. Suicidal ideation and behavior in youthful nonpsychiatric populations. Journal of Consulting and Clinical Psychology, 1972, 38, 366-371.
Purpose: To identify dimensions of attitudes toward suicide and its study
Subjects: 352 college students, 128 USAF basic trainees, 123 USAF psychiatric referrals
Variables: Unspecified number of item scores from an attitudes toward suicide and the study of suicide questionnaire
Method: Unspecified factor analysis, independently for 3 groups and total
Results: Two factors emerged across analysis: I Attitude toward the study of suicide; II Attitude toward the survey.

Lerner, R. M., Pendorf, J., & Emery, A. Attitudes of adolescents and adults toward contemporary issues. Psychological Reports, 1971, 28, 139-145.
Purpose: To identify dimensions of adolescents' and adults' attitudes toward contemporary issues
Subjects: 50 college-age adolescents
Variables: Scores on 29 contemporary issues questionnaire items
Method: Principal components condensation without rotation
Results: A fair approximation to a single-factor solution (24% of total variance accounted for by Factor I, with no other factor accounting for as much as 9%) was obtained.

Mayerberg, C. K., & Bean, A. G. The structure of attitude toward quantitative concepts. Multivariate Behavioral Research, 1974, 9, 311-324.
Purpose: To identify dimensions of attitudes toward quantitative concepts
Subjects: 311 graduate students
Variables: 14 semantic differential ratings each for the concepts: algebra, statistics, mathematics, numbers, calculations, formulas, not summed across scales or concepts (i.e., 84 variables)
Method: Principal factors condensation with oblique rotation
Results: Nine factors, falling in two categories, emerged: within concepts (III Positive attitude toward statistics; VI Positive attitude toward formulas; VII Positive attitude toward algebra; and less clearly IV Positive attitude toward calculations), and across concepts (I Lucid/clear; II Important/useful; V Simple/easy; VIII Enjoyable/pleasant (only roughly correct); IX Good).

McCarrey, M. W., Peterson, L., Edwards, S., & Von Kulmiz, P. Landscape office attitudes: Reflections of perceived degree of control over transactions with the environment. Journal of Applied Psychology, 1974, 59, 401-403.

Purpose: To identify dimensions underlying attitudes toward open office landscaping

Subjects: 600 employees of a Canadian Federal Government Department

Variables: Scores on 53 Open Office Opinion Inventory items

Method: Principal components condensation with orthoblique rotation

Results: Six factors emerged: I Overall preference combined with greater job satisfaction; II More effective communication, solidarity and contact; III Recurrent greater productivity; IV Often repeated lack of auditory privacy; V Recurring lack of personal privacy and consequence; VI Lack of confidentiality communications. One second-order factor emerged: I Perceived device of control over the individual's two-way transactions with others in the landscape office environment.

Mirels, H. L., & Garrett, J. B. The Protestant Ethic as a personality variable. Journal of Consulting and Clinical Psychology, 1971, 36, 40-44.

Purpose: To identify dimensions underlying the Protestant Ethic attitude

Subjects: 1) 117 and 2) 222 (in 2 groups) male college students

Variables: Scores on 1) 30 Protestant Ethic attitude items 2) plus 16 fillers

Method: Principal components condensation 1) with varimax rotation, independently for each subject-variable set

Results: Four non-independent factors emerged in the first analysis. An unspecified number of factors emerged in the second set of analyses; however, 19 items with high loadings on the largest unrotated component were selected for further use.

Nelsen, E. A., & Uhl, N. P. A factorial study of the attitude scales on the College Student Questionnaire with students at a predominantly black university. Multivariate Behavioral Research, 1974, 9, 395-405.

Purpose: To identify dimensions underlying the College Student Questionnaire (CSQ) for students at a predominantly black university

Subjects: 772 entering freshmen at a predominantly black unversity

Variables: 66 CSQ attitude item scores

Method: Principal axes condensation with oblique rotation

Results: Six factors, of eleven extracted, were interpreted: I Family independence vs. cohesiveness; II Autonomy vs. influenceability from peers; III Non-affiliative vs. affiliative peer orientation; IV Concern for social problems; V Aesthetic interest; VI Interest in politics and world affairs.

86

Conclusions: These factors were compared with the original
CSQ attitude scales.
Perry, A., & Friedman, S. T. Dimensional analysis of atti-
tudes toward commercial flying. Journal of Applied
Psychology, 1973, 58, 388-390.
Purpose: To identify dimensions of attitudes toward commercial flying
Subjects: 355 college students
Variables: 50 item scores from an attitudes-toward-flying-
and-the-airline-industry questionnaire
Method: Principal axes condensation with varimax rotation
Results: Fourteen interpretable factors emerged: I Adver-
tising effectiveness; II Return for the money; III Image
of the industry; IV Airline differentiation; V Age and
safety; VI Role of travel agency; VII Government regu-
lations; VIII Disadvantage of flying; IX Discrimination
among passengers; X Transportation mode selection; XI
Safety; XII Airline's performance; XIII Service; XIV
Discount for young passengers.
Robertson, A., & Cochrane, R. The Wilson-Patterson Conservatism Scale: A reappraisal. British Journal of Social
and Clinical Psychology, 1973, 12, 428-430.
Purpose: To identify dimensions underlying the Wilson-
Patterson Conservatism Scale (C Scale)
Subjects: 329 male and female first- and third-year
university students in three colleges
Variables: 50 C-Scale item scores
Method: Principal components condensation with varimax
rotation, independently by sex and for total group
Results: Four factors, of 17 extracted, were interpreted
(and identical for all groups): I Religious; II
Prurient dimension of sexuality; III Racial; IV Birching.
Conclusions: The C Scale does not appear to be unidimen-
sional.
Rosenbaum, L. L., Rosenbaum, W. B., & McGinnies, E. Semantic
differential factor structure stability across subject,
concept, and time differences. Multivariate Behavioral
Research, 1971, 6, 451-469.
Purpose: To identify dimensions underlying semantic differential (SD) ratings of political concepts
Subjects: 33 Johnson supporters and 33 Goldwater supporters,
all college students
Variables: Scores on 11 SD scales for each of 6 concepts,
administered pre- and post-election day (1964)
Method: Principal components condensation with orthogonal
rotation, independently for pre- and post- within each
group and by concept
Results: Two pre- factors emerged in both groups across
concepts: I Evaluation; II Dynamism. The number of
factors within concepts varied from two to three; for
the concept Johnson, two pre- factors emerged for both
groups but three post- factors (I Evaluation; II Dynamism;
III Johnson's re-election, or Goldwater's post-election

87

status) emerged. The final analysis (across concepts and subjects) yielded two factors (I Evaluation; II Dynamism) for both times.

Schneider, J. F. Note on the factorial structure of a symbolic measure of authoritarianism. Psychological Reports, 1974, 34, 775-777.
    Purpose: To identify dimensions of a symbolic measure of authoritarianism
    Subjects: 141 newly-recruited soldiers
    Variables: 15 item scores from a symbolic measure of authoritarianism
    Method: Principal components condensation with varimax rotation
    Results: Four factors emerged: I Preference for irregular arrangement and deviation from familiar or established order; II Preference for disorderly hatched oblongs and w-shaped line drawings; III Preference for balanced figures; IV Preference for unusual division which results in a kind of imbalanced drawing.

Simon, G. C. The factorial invariance of Attitudes Toward People (ATP). Journal of Social Psychology, 1972, 86, 315-316.
    Purpose: To determine if ATP, found in Border South females, exists for a northern sample
    Subjects: 83 female and 89 male NYU students
    Variables: Scores on the Chien Anomie, Chien Anti-Police, BHI-Suspicion, Rosenberg Faith-In-People, Rotter IE Control, Christie F, Anomia, Machiavellian, Wrightsman PHN, and Agger-Goldstein Political Cynicism scales
    Method: Unspecified condensation without rotation, independently by sex
    Results: The first factor (of six extracted) was considered reliable and meaningful; it was tentatively labeled ATP for both sexes.

Stone, L. A., & Coles, G. J. Multidimensional judgment scaling of well-known political figures. Journal of Social Psychology, 1972, 87, 127-137.
    Purpose: To identify dimensions of similarity judgments of political figures
    Subjects: 6 academically-employed Ph.D. psychologists, heterogeneous as to political philosophies
    Variables: 105 similarity judgments (stimuli - 15 nationally-known politicians)
    Method: Principal components condensation with varimax rotation, independently for a 6 x 6 judge matrix, and 15 x 15 matrices for each judge and combined group
    Results: One rather large judge factor emerged, indicating radically different viewpoints about stimuli may not affect paired similarity evaluations. The number of factors extracted for individual judges ranged from two to four, but only two were interpretable (and similar across analyses): I War; II Race. Two factors emerged for the combined analysis: I Position on USA military activity, or war posture; II Race, or civil rights.

Strumpfer, D. J. W. Scales to measure autonomous and social achievement values. Psychological Reports, 1975, 36, 191-208.
Purpose: To identify dimensions of autonomous and social achievement values
Subjects: 220 college students
Variables: Scores on 44 Bendig (1964) and (rewritten) Costello (1967) items
Method: Principal components condensation, apparently without rotation
Results: Three factors emerged: I Autonomous achievement value; II Social achievement value; III Discarded.
Thistlethwaite, D. L. Impact of disruptive external events on student attitudes. Journal of Personality and Social Psychology, 1974, 30, 228-242.
Purpose: To identify dimensions underlying attitude and press scales
Subjects: 1) 1,036 males with freshman standing in 1968-69, 2) 822 males with junior standing, 3) random 20% from 1) and 2)
Variables: Scores on 33 attitude and press scales
Method: Principal axes condensation, apparently without rotation, independently for the three subject sets
Results: Eight virtually identical factors emerged across analyses: I Scientism; II Faculty rapport; III Liberalism; IV Political participation; V Estheticism; VI Gregariousness; VII Conventionalism; VIII Benevolence.
Waters, L. K., Batlis, N., & Waters, C. W. Protestant attitudes among college students. Educational and Psychological Measurement, 1975, 35, 447-450.
Purpose: To identify dimensions of protestant attitudes among college students
Subjects: 170 college students
Variables: Scores on 6 Survey of Work Values, 1 Blood Protestant Ethic Scale, 1 Mirels-Garrett Protestant Ethic Scale
Method: Principal components condensation with varimax rotation
Results: Two factors emerged: I Intrinsic (work-related); II Extrinsic (reward-related).
Wober, M. East African undergraduates' attitudes concerning the concept: Intelligence. British Journal of Social and Clinical Psychology, 1973, 12, 431-432.
Purpose: To identify dimensions of attitudes toward intelligence
Subjects: 15 male and 8 female undergraduates in Social Work
Variables: Ratings on 14 characteristics for each of 28 jobs
Method: Unspecified inverse condensation with unspecified rotation, independently for males and females
Results: Four factors emerged for both groups: I Elite; II Bonhomie; III Quickness; IV Uninterpreted.

Conclusions: Intelligence and "quickness" were not highly related.

Yellig, W. F., & Wearing, A. J. The dimensions of political ideology. Journal of Social Psychology, 1974, 93, 119-131.

Purpose: To identify dimensions of political ideology
Subjects: 63 highly educated, politically elite persons
Variables: 5-point ratings on each of 60 statements
Method: Principal components condensation (obverse); cluster analysis
Results: Four factors emerged but were unlabeled; persons with high Factor I loadings were grouped as more liberal (N=16), while persons with high Factor II loadings were classified into a second group (N=10), and 37 persons were discarded. The cluster analysis results were not specified except that subjects were again separated into two groups (N=15, 10), with 38 discarded.
Conclusions: The multidimensional scaling analysis which followed yielded one significant dimension: Left-Right.

1. Formation and Change

Booth, R. F., & Hoiberg, A. Structure and measurement of Marine recruit attitudes. Journal of Applied Psychology, 1974, 59, 236-239.

Purpose: To identify dimensions of recruits' attitudes toward the Marine Corps, across the basic training period
Subjects: 481 Marine recruits
Variables: 96 Marine Corps Opinion Questionnaire item scores, administered four times during basic
Method: Principal axes condensation with varimax rotation, by administration
Results: Five factors emerged across analyses: I Toughness of Marines; II Spirit among Marines; III Affiliation with the Marine Corps; IV Authority in the Marines; V Consideration in the Marines.

2. Influence and Behavior

Cooper, M. R., Boltwood, C. E., & Wherry, R. J. Sr. A factor analysis of air passenger reactions to skyjacking and airport security measures as related to personal characteristics and alternatives to flying. Journal of Applied Psychology, 1974, 59, 365-368.

Purpose: To identify dimensions of attitudes toward flying and response to hypothetical skyjacking situations
Subjects: 127 airline passengers
Variables: 28 item scores from a questionnaire concerning attitudes toward flying and response to hypothetical skyjacking situations.
Method: Principal factors (plus clean-up minimum residual) condensation with hierarchical varimax rotation

Results: Eight primary factors were grouped under three
subgeneral factors: A General favorable reaction to
situations (I Relatively high willingness to fly; II
Relatively high probability of taking a flight; III
Relatively favorable reaction despite highjacking; IV
Relatively favorable response despite past skyjackings);
B Preference for planes over boats or cars for long
trips (V Preference for planes over cars; VI Preference
for planes over boats); C Boats over trains for long
trips, planes or cars over trains for short trips
(VII Planes over cars for short trips; VIII Cars over
trains for short trips, planes over trains for long
trips).

Dion, K. L., & Dion, K. K. Correlates of romantic love.
Journal of Consulting and Clinical Psychology, 1973,
_41_, 51-56.
Purpose: To identify dimensions of 1) symptoms of, 2)
descriptions of, and 3) attitudes toward romantic love
Subjects: 127 male and 116 female college students
Variables: Scores on 1) 6 symptoms, 2) 23 bipolar adjec-
tives, and 3) 16 attitude items, all related to romantic
love
Method: Principal components condensation with varimax
rotation, independently for symptoms, adjectival descrip-
tions, and attitudes
Results: One (unlabeled) symptoms factor emerged. Five
adjectival description factors emerged: I Volatile;
II Circumspect; III Rational; IV Passionate; V Impetuous.
Three attitude factors emerged: I Idealistic view;
II Cynical view; III Pragmatic view.

Kerlinger, F. N. A Q validation of the structure of social
attitudes. _Educational and Psychological Measurement_,
1972, _32_, 987-995.
Purpose: To identify dimensions underlying social attitudes
Subjects: 15 known liberals, 18 known conservatives
Variables: Numbers of items per 33 piles from the Social
Attitude Q Sort and the Referents Q Sort
Method: Principal factors condensation with varimax
rotation, by Q sort
Results: Two-, three-, and four-factor solutions were obtained for each sort; for both sorts, the first factor
was a liberalism factor and subsequent ones were con-
servatism factors.

Rambo, W. W. Measurement of broad spectrum social attitudes:
Liberalism-conservatism. _Perceptual and Motor Skills_,
1972, _35_, 463-477.
Purpose: To identify dimensions underlying social attitudes
Subjects: 1) 51 college protesters and 55 college conservative political club members, 2) 51 conservative and 80
liberal black college students, 3) 143 liberal and 91
conservative middle-aged, middle-class college graduates
Variables: Scores on 123 social attitude items
Method: Principal factors condensation with varimax rotation, by sample

91

Results: Four factors emerged for blacks (#2): I Religiosity;
II Social change; III Retentionism; IV Personal discipline.
Four factors emerged for white college students (#1): I
Liberalism-conservatism; II Personal mobility; III Anti-
intellectualism; IV Change-Retentionism. Three factors
emerged for college graduates: I General; II Social
change; III Retentionism.

Schwartz, S. H., & Tessler, R. C. A test of a model for re-
ducing measured attitude-behavior discrepancies. Journal
of Personality and Social Psychology, 1972, 24, 225-236.
Purpose: To identify dimensions underlying intentions
(attitudes) related to heart, kidney, and bone marrow
transplants
Subjects: Unclear; two samples, perhaps 195 adults in a
midwestern city and 136 female clerical workers
Variables: Unclear--probably 6 intentions scores (3 transplant types, 2 recipients)
Method: Principal axes condensation with varimax rotation,
by sample
Results: Three factors emerged in both analyses: I Relative (recipient); II Stranger (recipient); III Type of
transplant (during life or after death).

Ziegler, M., & Atkinson, T. H. Information level and dimen-
sionality of liberalism-conservatism. Multivariate
Behavioral Research, 1973, 8, 195-212.
Purpose: To identify the structure of political attitudes
for high- and low-information subjects
Subjects: 195 high-information, 174 low-information college
students
Variables: Scores on 16 Congressional issues items
Method: Principal components condensation with oblimin
rotation, independently for high and low Ss
Results: Four first-order (I General liberalism; II Decentralization; III Support for or expansion of military
or police action; IV Maintenance of the status quo) and
two second-order factors (I Conservatism; II Bipolar
liberalism-conservatism) emerged for the high group.
Five first-order (I General conservatism; II General
liberalism; III Decentralization; IV Opposing dissent;
V East of amending the constitution) and two second-
order factors (I Conservatism; II Predominantly liberalism) emerged for the low group.

D. Group and Interpersonal Processes
Gorsuch, R. L. Data analysis of correlated independent
variables. Multivariate Behavioral Research, 1973,
8, 89-107.
Purpose: To identify dimensions underlying selected population variables
Subjects: 52 countries (1950s data, Cattell & Gorsuch,
1965)
Variables: Scores on 7 population variables
Method: 1) Principal components condensation with varimax
rotation; 2) image condensation with varimax and promax

rotation; 3) triangular decomposition followed by residual principal components condensation with varimax rotation

Results: In the first analysis, seven factors (essentially one through each variable) emerged. The varimax solution for the second analysis yielded two common and seven specific factors. The Schmid-Leiman promax solution yielded one higher-order (I General standard of living) and two other common factors (II Technological development; III Scientific development) as well as the seven specific factors. In the hybrid (third) analysis, two diagonal and five varimax factors were extracted.

Lynn, R., & Hampson, S. L. National differences in extraversion and neuroticism. British Journal of Social and Clinical Psychology, 1975, 14, 223-240.

Purpose: To identify dimensions underlying national prevalence rates taken from demographic and epidemiological data

Subjects: 18 advanced western nations

Variables: National prevalence rates of divorce; illegitimacy, accidents, crime, murder, suicide, alcoholism, chronic psychosis, and coronary disease, and per capita consumption rates of calories, cigarettes, and caffeine

Method: Principal components condensation with varimax and promax rotation

Results: Two virtually independent factors emerged: I Neuroticism; II Extraversion.

Conclusions: Factor scores were computed, and compared between nations.

Morris, J. D., & Guertin, W. H. Relating multidimensional sets of variables: Canonical correlation or factor analysis? Psychological Reports, 1975, 36, 859-862.

Purpose: To identify dimensions underlying a set of national and population variables for 99 countries

Subjects: 99 countries

Variables: 8 "national" and 6 "population" variables

Method: Principal axes condensation with varimax rotation, for each set of variables

Results: Three national factors emerged: I Industry; II Government size; III Exchange rate. Two population factors emerged: I Personal wealth; II Communications.

Park, T. Measuring the dynamic patterns of development: The case of Asia 1949-1968. Multivariate Behavioral Research, 1973, 8, 227-251.

Purpose: To identify dynamic patterns of development in Asia from 1949 to 1968

Subjects: 21 Asian countries

Variables: 48 national attributes, 1949-1968 annually

Method: Principal axes condensation (of super P-matrix) with varimax rotation

Results: Eleven factors emerged: I Power capability; II Trade; III Diplomatic transactions; IV Political orientation; V Christianity; VI Density; VII Wealth; VIII Religious homogeneity; IX Buddhists-Mohemmedans dichotomy; X Unnamed; XI Health.

Vincent, J. E. Scaling the universe of states on certain useful multivariate dimensions. Journal of Social Psychology, 1971, 85, 261-283.
Purpose: To identify the structure of political and social indices for a number of nations
Subjects: 129 states in the U.N.
Variables: Scores on 91 political and social indices
Method: Principal components condensation with varimax rotation
Results: Nineteen factors emerged: I Underdeveloped; II Democracy; III U.S. relations; IV Smallness; V Diffusion; VI Executive leadership; VII Turmoil; VIII Religious and linguistic heterogeneity; IX Peaceful; X Militarism; XI Mobilization; XII Internal peace; XIII Voting intensity; XIV Catholicism; XV Communist China economic relations; XVI Racial homogeneity; XVII Personalissimo; XVIII U.N. supranationalism; XIX Friendly diplomatic activity.

1. Influence and Communication

Alderfer, C. P., & Brown, L. D. Designing an "empathic questionnaire" for organizational research. Journal of Applied Psychology, 1972, 56, 456-460.
Purpose: To identify dimensions underlying items related to sarcasm in an organization
Subjects: 1) 189, 2) 203 students and faculty of a private boys' boarding school
Variables: Scores on 16 sarcasm items from 1) an event-based, and 2) an theory-based questionnaire
Method: Principal components condensation with varimax rotation, by subject-questionnaire set
Results: Four factors emerged in both analyses: I Students are sarcastic to me both face-to-face and behind my back; II Faculty members are sarcastic to me, both directly and behind my back; III Respondents are sarcastic to persons in their presence; IV Respondents are sarcastic to persons behind their backs.

Bates, J. E. Effects of a child's imitation versus nonimitation on adults' verbal and nonverbal positivity. Journal of Personality and Social Psychology, 1975, 31, 840-851.
Purpose: To identify dimensions of male undergraduate subjects' positivity toward child confederates
Subjects: 48 male college students
Variables: 19 verbal and nonverbal measures of positivity toward child confederates
Method: Principal components condensation without rotation
Results: A single-factor solution was obtained, and factor scores were used in subsequent analyses. (A two factor varimax solution yielded an uninterpretable second factor; the results of ANOVAs using those factor scores yielded the same results.)

Burdick, J. A. Cardiac activity and attitude. Journal of Personality and Social Psychology, 1972, 22, 80-86.
Purpose: To identify dimensions of speaker evaluations

Subjects: 21 male college students
Variables: 20 7-point semantic differential ratings for a
pro-French and for a pro-English speaker
Method: Principal components condensation, apparently without rotation
Results: Four factors emerged: I Evaluative; II Activity-
potency; III Physical appearance; IV Religiousness.
Butler, R. P., & Cureton, E. E. Factor analysis of small
group leadership behavior. Journal of Social Psychology,
1973, 89, 85-89.
Purpose: To identify dimensions of small group leadership
behavior
Subjects: 96 groups of four (one leader per group), all
male undergraduates
Variables: Scores on 12 Bales Categories, recorded during
a group problem-solving session of up to 30 min.
Method: Principal axes condensation with oblique promax
rotation
Results: Four factors emerged: I Task-oriented positive
leadership; II Leadership uncertainty; III Self-oriented
negative leadership; IV Rejection of leadership.
Conclusions: Only Factor III, and to a lesser extent Factor
I, correspond to Bales' logically-derived dimensions.
Daum, J. W. Internal promotion--A psychological asset or
debit? A study of the effects of leader origin.
Organizational Behavior and Human Performance, 1975,
13, 404-413.
Purpose: To identify dimensions of a satisfaction question-
naire
Subjects: 128 male college students, each of whom had
worked on a task in a 4-member group
Variables: Scores on 18 questions related to satisfaction
(group unity, etc.) plus 4 semantic differential scores
Method: Unspecified condensation with unspecified rotation
Results: Six factors emerged: I Satisfaction with group
processes; II Leader-member interaction; III Evaluation
of job parameters; IV Evaluation of capability ratings;
V Group cohesion; VI Evaluation of leader function.
Edney, J. J. Territoriality and control: A field experiment.
Journal of Personality and Social Psychology, 1975, 31,
1108-1115.
Purpose: To identify dimensions of territoriality in
dormatory rooms
Subjects: 80 pairs of male college students (60 visitor-
resident pairs; 20 visitor-visitor pairs)
Variables: Scores on 1) 16 self- and other-ratings, 2)
18 room ratings
Method: Unspecified factor analysis, independently for the
two variable sets
Results: One factor, of an unspecified number extracted,
was labeled in the self- and other-ratings analysis:
I At home. Three factors emerged for room ratings: I
Pleasant; II Stimulating; III Privacy.

Galinsky, M. J., Rosen, A., & Thomas, E. J. Distinctness of bases of social power. Psychological Reports, 1973, 33, 727-730.
Purpose: To identify dimensions of bases of social power
Subjects: 110 public assistance workers
Variables: Scores on 80 questionnaire items dealing with fellow workers' and supervisors' expertness, legitimacy, positive and negative sanctions, attraction
Method: Unspecified condensation with varimax rotation
Results: Eleven factors emerged: I Workers' positive sanctions; II Workers' legitimacy; III Supervisor's expertness; IV Supervisor's positive sanctions; V Workers' expertness; VI Supervisor's negative sanctions; VII Workers' negative sanctions 1; VIII Supervisor's legitimacy; IX Not interpretable; X Workers' negative sanctions 2; XI Not interpretable.
Hare, A. P. Four dimensions of interpersonal behavior. Psychological Reports, 1972, 30, 499-512.
Purpose: To identify dimensions of initiated behavior in small groups; data reported by Couch (unpublished dissertation, 1960)
Subjects: 12 groups of 5 college students each
Variables: Scores on 55 measures of initiated behavior
Method: Complete centroid condensation with unspecified rotation
Results: Six factors emerged: I Interpersonal dominance; II Interpersonal affect; III Social expressivity vs. task serious; IV Influence attempts; V Surface acquiescence; VI Conventional behavior.
Hrycenko, I., & Minton, H. L. Internal-external control, power position, and satisfaction in task-oriented groups. Journal of Personality and Social Psychology, 1974, 30, 871-878.
Study One
Purpose: To identify dimensions underlying 23 I-E items
Subjects: 170 male and 151 female college students
Variables: 23 I-E scale item scores
Method: Principal components condensation with varimax rotation, by sex
Results: Two factors emerged in both analyses: I Personal control; II System modifiability.
Study Two
Purpose: To identify dimensions of Ss' postexperimental satisafction with their power position in a task-oriented group
Subjects: 52 Ss from Study One (upper or lower third on Factor I)
Variables: Scores on 11 items from a postexperimental questionnaire
Method: Principal components condensation with varimax rotation
Results: Five factors emerged: I Satisfaction with task procedure chosen by the supposed group leader; II Feelings of independence in performance of their personal tasks; III Concern with personal effectiveness; IV

Interest in experiment as a function of their power
position; V General enjoyment of experiment, particu-
larly with reference to specific mechanical task requirements.
Levin, J.   Bifactor analysis of a multitrait-multimethod
matrix of leadership criteria in small groups. Journal
of Social Psychology, 1973, 89, 295-299.
Purpose: To identify an alternative solution for a multitrait-multimethod matrix of leadership criteria in
small groups (Schneider, 1970)
Subjects: 240 subjects in small group interactions
Variables: Scores for 4 traits under each of 3 methods
(self-, peer, observer ratings)
Method: Holzinger's bifactor analysis without rotation
Results: One trait (I) and three method factors (II-IV)
emerged.
Conclusions: These results were compared with Schneider's
principal components solution, which extracted no trait
factor.
Maslach, C.   Social and personal bases of individuation.
Journal of Personality and Social Psychology, 1974,
29, 411-425.
Purpose: To identify dimensions of group and personal
bases of individuation
Subjects: 40 male and 40 female college students, in 4-
person same-sex groups
Variables: 33 measures of verbal and nonverbal expressive
behaviors, test scores, manipulation checks, subjective
ratings, and timing scores from a group game
Method: Principal factors condensation with orthogonal
rotation
Results: Four factors, of 11 extracted, were interpreted:
I Individuation--singular; II Individuation--personal;
III Emotion--agitated; IV Emotion--contented.
Mehrabian, A., & Ksionzky, S.   Factors of interpersonal be-
havior and judgment in social groups. Psychological
Reports, 1971, 28, 483-492.
Purpose: To identify dimensions of interpersonal behavior
and judgment in closely-knit social groups
Subjects: 22 male and 18 female members of a student
church group
Variables: Ratings (of self and other members) on 15 scales
of a sociometric questionnaire
Method: Principal components condensation with oblique
rotation
Results: Two factors were reported: I Positive reinforcing
quality; II Negative reinforcing quality.
Norton, R., Feldman, C., & Tafoya, D.   Risk parameters across
types of secrets. Journal of Counseling Psychology, 1974,
21, 450-454.
Purpose: To identify dimensions of risk parameters for
secrets
Subjects: 190 college students

Variables: 5-point risk ratings for 49 secrets
Method: Elementary linkage cluster analysis
Results: Eight clusters emerged but were not labeled.

O'Donnell, C. R. The measurement of anxiety and evaluative components in exam and speech concepts for males. Journal of Clinical Psychology, 1973, 29, 326-327.
Purpose: To identify dimensions underlying semantic differential (SD) ratings for exam and speech concepts
Subjects: 1) 69 male college students, 2) 47 male college students
Variables: 1) 84 SD ratings (total) for the concepts "taking an exam" and "giving a speech before a group," 2) 65 SD ratings for the exam concept and 74 for the speech concept
Method: Principal axes condensation with varimax rotation, independently for each concept at each administration
Results: In all analyses, two factors (of unspecified numbers extracted) were labeled: I Anxiety; II Evaluative.

Phillips, W. R., & Lorimor, T. The impact of crisis upon the behavior of nations in the 1960's. Multivariate Behavioral Research, 1974, 9, 423-445.
Purpose: To identify pre-, during-, and post-crisis dimensions of nations' dyadic behavior
Subjects: 50 dyads of nations involved in (21) crises from July 1962 to July 1968
Variables: Scores on 2 measures of warning and defensive acts, 3 official acts of violence, 6 negative sanctions, 6 negative communications, 4 unofficial violence, and 1 non-violent demonstrations; each for 2 months pre-crisis (15 variables), 2 months during-crisis, and post-crisis (16) periods
Method: Principal axes condensation with varimax rotation, independently for each of the three periods
Results: Four pre-crisis factors emerged: I Negative communications; II Negative sanctions; III Official violence; IV Warning. Eight during-crisis factors emerged: I Negative communications; II Unofficial violent acts; III Negative sanction; IV Official violence; V Warning and defensive maneuvers; VI Unofficial embassy violence; VII Aid to rebels; VIII Boycotts and embargoes. Five post-crisis factors emerged: I Negative communications; II Negative sanction; III Official violence; IV Unofficial violence; V Aid, oral.

Pratt, R., & Rummel, R. J. Issue dimensions in the 1963 United Nations General Assembly. Multivariate Behavioral Research, 1971, 6, 251-286.
Study One
Purpose: To identify dimensions of issue conflict in the UN in 1963
Subjects: 99 UN-member countries
Variables: 3-point scores on 69 role call votes
Method: Unspecified condensation with orthogonal and oblique rotation

98

Results: Seven orthogonal factors emerged: I Racial dis-
crimination-sanctions; II UN procedures; III South
Africa; IV Cold war; V Palestine refugees; VI Economic
and social control; VII Electric voting machines. Six
oblique factors were labeled: I-V Same as above; VI
Discrimination.

Study Two (Russett, 1967 book)
Purpose: Same as above
Subjects: Unspecified (here) number of UN member countries
Variables: Unspecified (here) number of role call votes
Method: Unspecified condensation with orthogonal rotation
Results: Five factors emerged: I Cold war; II Interven-
tion in Africa; III Supranationalism; IV Palestine; V
Self determination.

Study Three
Purpose: To identify clusters of member nations
Subjects: Same as Study One
Variables: Weighted similarity ratings for voting on 10
dimensions
Method: Unspecified factor analysis; diameter cluster
analysis
Results: Five factors emerged: I Latin American; II Afro-
Asian; III Western bloc; IV Communist bloc; V African
colonial. These results were compared with a ten-level
diameter grouping.

Study Four (Russett, 1967 book)
Purpose: Same as Study Three
Subjects: Same as Study Two
Variables: Voting scores on unspecified number of role
calls
Method: Unspecified Q-type factor analysis
Results: Five clusters emerged: I Western bloc; II Communist
bloc; III Conservative Arabs; IV Brazzaville Africans;
V African colonists.

Russell, G. W. The perception and classification of collec-
tive behavior. Journal of Social Psychology, 1972, 87,
219-227.
Purpose: To identify dimensions of collective behaviors
Subjects: 42 senior sociology majors in a Collective
Behavior course
Variables: Dissimilarity ratings (4-point) for 19 instances
of collective behavior (171 judgments)
Method: Unspecified condensation with varimax rotation
Results: Four factors emerged: I Magnitude of violence;
II Amorphous-focused; III Anomie; IV Ideology.

Schriesheim, C. A., & Stogdill, R. M. Differences in factor
structure across three versions of the Ohio State
leadership scales. Personnel Psychology, 1975, 28,
189-206.
Purpose: To identify dimensions underlying three Ohio
State leadership scales--i.e., Supervisory Behavior
Description Questionnaire (SBDQ), early Leader Behavior
Description Questionnaire (LBDQ), Form XII Leader
Behavior Description Questionnaire (LBDQ-XII)--with one
sample

Subjects: 230 hourly employees of a university
Variables: 48 SBDQ, 30 LBDQ, and 20 LBDQ-XII item scores
Method: Principal axes condensation with varimax rotation,
    followed by hierarchical (multiple group condensation)
    factor analysis, by instrument
Results: Eight SBDQ varimax factors (I Personalized ap-
    proach to followers; II Insistent behavior; III Arbitrary
    resistence to change; IV Helpful and supportive vs.
    arbitrary behavior; V Flexible supportive behavior; VI
    Punitive production emphasis; VII Arbitrary production
    emphasis; VIII Punitive, arbitrary, and unyielding be-
    havior), three subgenerals (I Insistent production
    emphasis; II Consideration; III Punitive and arbitrary
    behavior), and one third-order factor (I Rater bias)
    emerged. Six LBDQ primaries (I Consideration; II En-
    couragement of standardized procedures and production;
    III Friendly interest in follower welfare; IV Supportive
    encouragement of follower contributions; V Initiating
    structure; VI Arbitrary vs. friendly and supportive
    behavior), two subgeneral (I Consideration; II Initiating
    structure), and one third-order factor (I Rater bias)
    emerged. Four LBDQ-XII primary (I Consideration; II
    Initiating structure; III Arbitrary vs. considerate
    behavior; IV Friendly vs. impersonal behavior), two sub-
    general (I Consideration; II Initiating structure), and
    one general factor (I Rater bias) emerged.
Spector, D., London, P., & Robinson, J. P. Role-playing per-
    formance as a function of incentive condition and two
    social motives. Journal of Personality and Social
    Psychology, 1972, 23, 328-332.
    Purpose: To identify dimensions of role-playing performance
    Subjects: 32 achievement- and 32 affiliation-oriented
        sixth-grade middle-class boys
    Variables: 18 role (playing) performance scores, plus
        order, motive and motive score, incentive, IQ and IQ
        range
    Method: Principal components condensation with varimax
        rotation
    Results: Two factors were interpreted: I Pantomime role-
        playing; II Verbal role-playing.
2. Social Perception and Motivation
  Bach, T. R. Adjustment differences related to pattern of
      rating of the other. Psychological Reports, 1973, 32,
      19-22.
      Purpose: To identify dimensions underlying personality
          trait ratings of "the other" in a dyadic interaction
      Subjects: 29 pairs of college students (14 similar, 15
          dissimilar levels of adjustment)
      Variables: Apparent-real discrepancy scores for 18 per-
          sonality traits
      Method: Centroid condensation with varimax rotation

Results: Five factors emerged: I Awareness of others; II
   Warm extraversion; III Thoughtfulness; IV Openness; V
   Care for others
Conclusions: Adjustment level differences were examined.
Bayes, M. A. Behavioral cues of interpersonal warmth.
   Journal of Consulting and Clinical Psychology, 1972,
   39, 333-339.
   Purpose: To identify dimensions of judged interpersonal
   warmth
   Subjects: 36 college students, rating 16 taped interviews
   Variables: Ratings of global warmth, positive content
   (self, others, surroundings), speech rate, body and
   head movement, hand movement, smile
   Method: Principal components condensation with varimax
   rotation
   Results: Two factors emerged: I Evaluation of positive
   response to others; II General activity.
Brim, J. A. Social network correlates of avowed happiness.
   Journal of Nervous and Mental Disease, 1974, 158, 432-
   439.
   Purpose: To identify dimensions of social relationship
   content
   Subjects: 1) 196 college students; 2) 92 members and 61
   non-members of a women's organization
   Variables: Scores on 1) 63 behavior items, summed over
   responses for same-sex good friend, same-sex casual
   acquaintance, same-sex non-resident close relative;
   2) 13 behavior items for S-selected number of persons
   important to her
   Method: 1) Unspecified factor analysis; 2) Principal axes
   condensation with varimax rotation, independently for
   the two sets
   Results: Five factors emerged in the 63-item analysis
   and were replicated with the 13-item subset: I Assistance; II Value similarity; III Trust; IV Concern; V
   Desired interaction.
Bryson, J. B. Factor analysis of impression formation processes. Journal of Personality and Social Psychology,
   1974, 30, 134-143.
   Purpose: To identify dimensions of impression formation
   responses
   Subjects: 85 college students
   Variables: 16 "impression" ratings (for a person described
   by an adjective)
   Method: Principal components condensation with varimax
   rotation
   Results: Two factors emerged: I General evaluation; II
   Not interpreted.
Clore, G. L., Wiggins, N. H., & Itkin, S. Judging attraction
   from nonverbal behavior: The gain phenomenon. Journal
   of Consulting and Clinical Psychology, 1975, 43, 491-497.
   Purpose: To identify dimensions of nonverbal behaviors'
   capacity to convey attraction

101

Subjects: 33 male and 33 female college students
Variables: Likability ratings for 139 nonverbal behaviors
Method: Principal components condensation with varimax
rotation, by sex
Results: Two major dimensions (I Positive, or warm; II
Negative, or cold) and at least five specific dimensions
(III Posture; IV Nervousness; V Facial expression; VI
Hand gestures; VII Eye movements) emerged.
DeMille, R., & Hirschberg, M. A. Multivariate analysis of
concepts induced from bounded predicative value-statements
Multivariate Behavioral Research, 1972, 7, 41-66.
Purpose: To identify dimensions of socially or personally
disvalued behaviors
Subjects: 257 college students
Variables: Association indices for 73 concepts
Method: Principal components condensation with varimax
rotation
Results: Eight factors were interpreted: I UNT-be untrust-
worthy; II TBL-have trouble in school; III MJD-be mal-
adjusted; IV HTH-neglect or injure one's health; V MST-
mistreat parents or intimates; VI LAW-flaunt the law or
the rules; VII XPL-exploit people; VIII DRP-drop out of
school.
Conclusions: Tests of 8 other association indices were also
reported.
Donnerstein, E., & Donnerstein, M. White rewarding behavior
as a function of the potential for black retaliation.
Journal of Personality and Social Psychology, 1972, 24,
327-333.
Purpose: To identify dimensions underlying mean reward,
mean duration, and sum high reward of white or black
"learners" by white Ss
Subjects: 40 white male college students
Variables: Mean reward, mean duration, and sum high reward scores
Method: Principal components condensation with "rigid"
rotation
Results: Two factors emerged: I Direct reward; II Indirect
aggression.
Conclusions: Factor scores were examined as a function of
race of learner and potential for retaliation.
Donnerstein, E., & Donnerstein, M. Variables in interracial
aggression: potential ingroup censure. Journal of
Personality and Social Psychology, 1973, 27, 143-150.
Purpose: To identify dimensions underlying mean shock/
reward intensity, mean shock/reward duration, and sum
of high shock/reward intensities delivered by whites
to white or black "learners"
Subjects: 1) 36, 2) 36 white male college students
Variables: Scores on mean intensity, mean duration, and
sum of high intensities of 1) shock, 2) reward administered to white or black "learner"

Method: Principal components condensation 1) without rota-
    tion, 2) with rigid orthogonal rotation
Results: Two factors emerged in both analyses: I Direct
    aggression; II Indirect aggression.
Edwards, C. N.   Interactive styles and social adaptation.
    Genetic Psychology Monographs, 1973, 87, 123-174.
Purpose: To identify dimensions of the Situational Pre-
    ference Inventory (SPI)
Subjects: 543 males and females
Variables: 66 SPI item scores
Method: Unspecified factor analysis
Results: Five factors emerged: I Instrumental; II Cooper-
    ational; III Introspection; IV Independence and autonomy;
    V Interpersonal receptivity.
Conclusions: Norms for the SIT Cooperational, Instrumental
    and Analytic scales are presented.
Erlich, H. J., & Van Tubergen, G. N.   Exploring the structure
    and salience of stereotypes. Journal of Social Psy-
    chology, 1971, 83, 113-127.
Purpose: To identify dimensions of stereotypes
Subjects: 91 business administration students
Variables: Ratings on 56 items from a Jewish checklist
    and 69 items of an atheist checklist
Method: R- and Q-type (inverse) centroid condensation with
    oblique rotation, independently for Jewish and atheist
    items
Results: The Q-solutions, said to be similar to R-solutions
    but more easily interpretable, were reported only. Three
    patterns of Jewish stereotypes, and two patterns of atheist
    stereotypes, emerged but all were left unlabelled.
Ekehammar, B., & Magnusson, D. A method to study stressful
    situations. Journal of Personality and Social Psychology,
    1973, 27, 176-179.
Purpose: To identify dimensions underlying similarity
    ratings of 20 situations
Subjects: 5 male and 5 female 14- to 15-year-olds
Variables: Similarity judgments for 20 ego-threatening,
    pain-threatening, or neutral school situations
Method: Principal components condensation with varimax
    rotation
Results: Five factors emerged: I Ego threat; II Positive;
    III Social; IV Active; V Pain threat.
Ekehammar, B., Schalling, D., & Magnusson, D. Dimensions of
    stressful situations: A comparison between a response
    analytical and a stimulus analytical approach. Multi-
    variate Behavioral Research, 1975, 10, 155-164.
Purpose: To identify dimensions underlying stressful situ-
    ations assessed from two approaches
Subjects: 1) 77 males (conscripts) aged about 20 years
    old, 2) 24 male college students aged about 20 years
Variables: Ratings for 24 stressful situations: 1) 9-
    point unpleasantness ratings; 2) ratings of experienced
    similarity (transformed mean scores)

Method: Principal components condensation with varimax
    rotation, independently for the two subject-variable
    sets
Results: Six factors emerged in each case, although the
    response analytical factors were more ambiguous. The
    stimulus analytical factors could be labeled as follows:
    I Pain; II Thrill; III Boredom at work; IV Boredom in
    passive situations; V Anticipation of ego threat; VI
    Anticipation of pain.
Feldman, J. M., & Hilterman, R. J. Stereotype attribution
    revisited: The role of stimulus characteristics,
    racial attitude, and cognitive differentiation. Journal
    of Personality and Social Psychology, 1975, 31, 1177-
    1188.
Purpose: To identify dimensions of stereotype attribution
Subjects: 48 male and 48 female white college students
Variables: Scores on 29 trait ratings of eight "realistic"
    stimulus persons
Method: Principal axes condensation with varimax rotation
Results: Five factors emerged: I Occupational (middle-
    class) stereotype; II Racial (black) stereotype; III
    Conservatism; IV Upward mobility; V Successful people.
Genthner, R. W., & Taylor, S. P. Physical aggression as a
    function of racial prejudice and the race of the target.
    Journal of Personality and Social Psychology, 1973, 27,
    207-210.
Purpose: To identify dimensions underlying measures of
    hostility and prejudice
Subjects: 250 college students
Variables: Scores on 1) 6 Buss-Durhee Hostility Inventory
    (BDI) subscales, 2) plus the Kelly-Ferson-Holtzman
    Desegregation Scale
Method: Principal components condensation with varimax
    rotation, independently for the two variable sets
Results: Two factors emerged for the BDI: I Verbal ag-
    gression, assault, irritability; II Resentment, sus-
    picion, indirect hostility. Two factors emerged in
    the second analysis, with the prejudice measure loading
    on Factor I.
Golding, S. L., & Knudson, R. M. Multivariable-multimethod
    convergence in the domain of interpersonal behavior.
    Multivariate Behavioral Research, 1975, 10, 425-448.
Purpose: To identify dimensions underlying, and relation-
    ships among, self-report assessment devices, direct
    self-ratings, and peer ratings of interpersonal behavior
Subjects: 64 high school seniors
Variables: 1) Scores on 10 Personality Report Form (PRF),
    10 Interpersonal Check List (ICL), 5 Schedule of Inter-
    personal Response (SIR), 15 Rational S-R Inventory (RSR)
    scales; 2) factors extracted from each variable set in
    (1) for second-order analysis

Method: Principal components condensation with varimax
    rotation, independently for each instrument at first-
    order level and for combined factors at second-level
Results: Four PRF factors emerged: I Aggressive defen-
    siveness and social ascendency; II Affiliativeness;
    III Autonomy; IV Social desirability. Four ICL factors
    emerged: I Friendly submissiveness; II Hostile dominance;
    III Affiliativeness; IV Hostility. Three SIR factors
    emerged: I Hostile aloofness vs. sociability; II
    Ignoring of rejection and hostility; III Sufferance
    and submission. Five RSR factors emerged: I Hostile
    dominance; III Friendly dominance; III Sufferance and
    hostile submissiveness; IV Cooperativeness; V Capitu-
    lation. Six second-order components emerged: I Ag-
    gressive dominance; II Affiliation and sociability;
    III Autonomy; IV Tolerance of rejection and hostility
    (SIR and RSR exclusively); V Undesirable hostility;
    VI Sufferance and hostile submissiveness (SIR and RSR
    exclusively).
Haase, R. F., & Markey, M. J. A methodological note on the
    study of personal space. Journal of Consulting and
    Clinical Psychology, 1973, 40, 122-125.
  Purpose: To identify relationships among measures of per-
    sonal space derived from different methods
  Subjects: 28 male and 8 female college students
  Variables: Personal space measure (1 each) derived from
    photograph, felt board, live observation, and in vivo
    participation methods
  Method: Cluster analysis
  Results: Two clusters emerged: I Active participation;
    II Static task.
Hawley, P. What women think men think: Does it affect
    their career choice? Journal of Counseling Psychology,
    1971, 18, 193-199.
  Purpose: To identify dimensions of women's perceptions
    of men's attitudes toward women
  Subjects: 33 women who could be classified in homemaker,
    feminine, or androgynous career categories
  Variables: Scores on 80 "Significant men in my life think
    women should..." items
  Method: Principal axes condensation with varimax rotation
  Results: Five factors emerged: I Woman as partner; II
    Woman as ingenue; III Woman as homemaker; IV Woman as
    competitor; V Woman as knower.
Hawley, P. Perceptions of male models of femininity related
    to career choice. Journal of Counseling Psychology,
    1972, 19, 308-313.
  Purpose: To identify dimensions underlying women's per-
    ceptions of men's views of women
  Subjects: 136 female college students (teachers-in pre-
    paration, math-science, counselors)
  Variables: 35 item scores from a questionnaire

Method: Unspecified condensation with varimax rotation
Results: Five factors emerged: I Woman as partner; II Woman as ingenue; III Woman as homemaker; IV Woman as competitor; V Woman as knower.
Jones, R. A., & Ashmore, R. D. The structure of intergroup perception: Categories and dimensions in views of ethnic groups and adjectives used in stereotype research. Journal of Personality and Social Psychology, 1973, 25, 428-438.
Purpose: To identify dimensions underlying perceived similarity of ethnic groups, and adjectives often attributed to certain groups
Subjects: 1) 34, 2) 30 college students
Variables: 1) Q sort for 50 ethnic groups, 2) 8 SD ratings for each of 49 adjectives often attributed to stereotyped groups
Method: Cluster analysis, independently by subject-variable set
Results: Seven ethnic group clusters emerged: I Dominant; II Active; III Color; IV Christian; V Spanish; VI Communist; VII Culture. Five clusters emerged for adjectives: I Complex-simple; II Internal-external; III Good-bad; IV Strong-weak; V Cooperative-competitive.
Kirby, D. M., & Gardner, R. C. Ethnic stereotypes: Norms on 208 words typically used in their assessment. Canadian Journal of Psychology, 1972, 26, 140-154.
Purpose: To identify dimensions of seven scales used in rating trait-descriptive words typically used in assessment of ethnic stereotypes.
Subjects: 225 college students completing one of the following questionnaires: behavioral specificity (37), familiarity (37), imagery (25), evaluation (34), social desirability (30), activity (31), potency (31)
Variables: 7 scale scores (summed over items): Behavioral specificity, imagery, evaluation, activity, potency (all seven-point), familiarity (five-point), social desirability (nine-point)
Method: Unspecified condensation with varimax rotation
Results: Four factors were extracted: I Evaluation; II Imagery; III Activity; IV Familiarity.
Kirby, D. M., & Gardner, R. C. Ethnic stereotypes: Determinants in children and their parents. Canadian Journal of Psychology, 1973, 27, 127-143.
Purpose: To identify dimensions of ethnic stereotypes
Subjects: 157 children, aged 9 to 17 years, and 106 parents, all rating 3 ethnic groups
Variables: Scores on 22 semantic differential scales, 3 source-of-information items (5 for children), 3 social distance measures, 1 (2 for adults) E scale, (ethnic-specific) attitude scale score, age (children only)
Method: Principal axes condensation with normalized varimax rotation, independently for children and adults for each ethnic group

Results: Four French-Canadian factors emerged for children: I French Canadian attitude; II French Canadian stereotype; III French Canadian social distance; IV French Canadian evaluative stereotype. Four Canadian Indian factors emerged for children: I Canadian Indian attitude; II Canadian Indian developmental impression; III Canadian Indian information; IV Age dependent tolerance. Four English Canadian factors emerged for children: I English Canadian stereotype; II English Canadian information; III English Canadian attitude; IV Social distance.
Three French Canadian factors emerged for adults: I French Canadian attitude; II French Canadian evaluative stereotype; III French Canadian stereotype based on common sources of information. Three Canadian Indian factors emerged for adults: I Canadian Indian attitude; II Canadian Indian stereotype; III Canadian Indian social distance. Four English Canadian factors emerged for adults: I English Canadian attitude; II English Canadian information; III English Canadian ethnocentrism; IV English Canadian stereotype.

Krus, D. J., & Tellegen, A. Consciousness III: Fact or fiction? Psychological Reports, 1975, 36, 23-30.
Purpose: To identify dimensions underlying Consciousness I, II, and III Scales (adapted from The Greening of America)
Subjects: 58 friends and relatives of (night) college students (mean age=29)
Variables: 19 item scores each from Consciousness I, II, and III Scales
Method: Principal factors condensation with varimax rotation
Results: Two unlabeled factors emerged; Consciousness I and II loaded on the first, while III loaded on the second.

Lorr, M., Suziedelis, A., & Kinnane, J. F. Modes of interpersonal response to peers. Multivariate Behavioral Research, 1973, 8, 427-438.
Purpose: To identify patterns of interpersonal response to peers
Subjects: 1) unspecified; 2) 245 college students
Variables: Scores on 1) 153 odd, 153 even items from Schedule of Interpersonal Response (SIR); 2) 70 3- to 5-item SIR subtests
Method: Principal components condensation with varimax rotation, independently for odd and even items and subtests
Results: The Scree Test indicated between six and eight factors for the odd and even analyses. Five interpretable factors emerged for subtests: I Acceptance and maintenance of relations; II Sufferance and submission; III Ignoring, rejection, and hostility; IV Avoidance of involvement; V Counter-attack and resistance to control.

Magnusson, D., & Ekehammar, B. An analysis of situational dimensions: A replication. Multivariate Behavioral Research, 1973, 8, 331-339.
Purpose: To identify dimensions of individuals' perceptions of situations
Subjects: 12 college students
Variables: 5-point similarity judgments for 36 (university experience) situations: test and 16-day retest, individual and group means
Method: Principal components condensation with varimax rotation, independently for test and retest data and individual and group means; elementary linkage analysis; hierarchical cluster analysis
Results: Five highly congruent factors emerged across factor analyses: I Positive; II Negative; III Passive; IV Social; V (Active alone at home) ambiguous. Seven linkage clusters emerged: I Positive; II Negative; III Social; IV Passive; V, VII Ambiguous; VI Active. Six hierarchical clusters emerged and were similar to linkage clusters I-VI.
Magnusson, D., & Ekehammar, B. Perceptions of and reactions to stressful situations. Journal of Personality and Social Psychology, 1975, 31, 1147-1154.
Purpose: To identify dimensions underlying perceptions of and reactions to stressful situations
Subjects: 40 ninth-graders
Variables: Scores for 12 situations: 1) 10 reaction scales, 2) similarity judgments
Method: Principal components condensation with varimax rotation, independently for the two variable sets, followed by target rotation
Results: Four factors were interpreted: I Threat of punishment; II Threat of pain; III Inanimate threat; IV Ego threat.
Mehrabian, A., & Russell, J. A. The basic emotional impact of environments. Perceptual and Motor Skills, 1974, 38, 283-301.
Purpose: To identify dimensions of semantic differential (SD) ratings of emotional response to verbally-described situations
Variables: 1) 28 SD ratings for 8 situations, and 2) 23 SD ratings for 20 situations
Method: Principal components condensation with oblique rotation, independently for the two subject-variable sets
Results: Three factors emerged in both analyses: I Pleasure; II Arousal; III Dominance.
Neufeld, R. W. J. Multidimensional scaling of perceived interests of three student stereotypes. Perceptual and Motor Skills, 1972, 34, 511-514.
Purpose: To identify dimensions underlying perceived interests of three student stereotypes

Subjects: 59 college students
Variables: Similarity judgments (45) for 3 student
stereotypes and 7 areas of academic subject matter
(in Psychology course)
Method: Centroid condensation with varimax rotation
Results: Three factors emerged: I Applied vs. pure;
II Theoretical vs. experimental; III Personality
and developmental.
Phares, E. J., & Lamiell, J. T. Internal-external control,
interpersonal judgments of others in need, and attribution of responsibility. Journal of Personality, 1975,
43, 23-38.
Purpose: To identify dimensions underlying four ratings
of case histories, presented in three different manners
Subjects: 146 college students
Variables: Ratings of case histories along dimensions of
deserving help, amount of financial assistance deserved,
degree of understanding of plight, and subject's sympathy, under presentation conditions 1) victim responsibility, 2) ambiguous responsibility, and 3) personally
responsible
Method: Principal axis condensation with no rotation, for
each presentation condition independently
Results: One factor emerged in each analysis; the factor
was unlabeled but the help, money, and sympathy items
consistently showed high loadings on it.
Pollio, H. R., Edgerly, J., & Jordan, R. The comedian's
world: Some tentative mappings. Psychological Reports,
1972, 30, 387-391.
Purpose: To identify dimensions of word associations to
contemporary comedians
Subjects: 67 undergraduate and graduate students
Variables: Associative overlap scores derived from word
associations to 37 contemporary comedians
Method: Principal axes condensation with oblique rotation
Results: Eight factors emerged: I Sex of comedian; II
Color; III Night talk, or verbal facility; IV Contiguity
I; V Contiguity II; VI Hostility; VII Generation I;
VIII Generation II.
Reed, T. R. Connotative meaning of social interaction con-
cepts: An investigation of factor structure and the
effects of imagined contexts. Journal of Personality
and Social Psychology, 1972, 24, 306-312.
Purpose: To identify dimensions underlying semantic differential (SD) ratings of social interaction concepts
under differing instructions.
Subjects: 144 college students receiving standard SD
instructions (C), 146 receiving drill instructions (D),
and 149 fear instructions (F)
Variables: 20 SD ratings, for each of 5 social interaction
concepts
Method: Principal axes condensation with simultaneous
varimax rotation, by group

Results: Three factors were found to be stable across
groups: I Affective experience; II Subjective probability; III Stability.
Conclusions: Differences were examined as a function of
imagined context.
Ryan, E. B., & Carranza, M. A. Evaluative reactions of
adolescents toward speakers of standard English and
Mexican American accented English. Journal of Personality and Social Psychology, 1975, 31, 855-863.
Purpose: To identify dimensions underlying personality
adjective ratings by adolescents of speakers of standard English and Mexican American accented English
Subjects: 21 bilingual Mexican-American, 21 Anglo, 21
black female high school students
Variables: 15 adjective scale ratings for 12 test speakers
Method: Principal factors condensation with varimax
rotation
Results: Three factors emerged: I Designated status; II
Solidarity; III Activity/potency.
Schiff, W., & Saxe, E. Person perceptions of deaf and hear-
ing observers viewing filmed interactions. Perceptual
and Motor Skills, 1972, 35, 219-234.
Purpose: To identify dimensions underlying deaf and hear-
ing observers' person perceptions
Subjects: 97 deaf and 54 hearing college students
Variables: 17 characteristics ratings for each of two
(filmed) interactants
Method: Principal components condensation with varimax
rotation, by interactant within deaf and hearing groups
Results: Three Actor 1 factors emerged for hearing sub-
jects (I Likability; II Unfriendliness; III Traditional
leadership), and four for deaf subjects (I Lethargy;
II Unfriendly-aloofness; III Modest-artistic; IV Sincere-
liberal). Similar patterns were found for Actor 2.
Sutton, A. J. The use of quadratic determinant analysis for
the measurement of profile distance in social perception.
British Journal of Psychology, 1971, 62, 253-260.
Purpose: To identify dimensions of students' role perceptions
Subjects: 291 technology college students
Variables: 35 bipolar adjective scale ratings for each
of 12 student role concepts
Method: Maximum likelihood factor condensation, apparently
without rotation
Results: Eight unlabeled factors emerged.
Conclusions: Fourteen scales loading highly on the seven
main factors were selected for the major (quadratic
determinant) analysis reported.
Taylor, H. F. Semantic differential factor scores as mea-
sures of attitude and perceived attitude. Journal of
Social Psychology, 1971, 83, 229-234.
Purpose: To identify dimensions of self-attitude and role-
player attitude as measured by semantic differential
(SD) ratings

Subjects: 40 male Yale sophomores
Variables: 12 SD ratings (related to admission of women
    to Yale) each for subject and for role-playing partner
Method: Principal axes condensation with varimax rotation,
    independently for attitudes and perceived attitudes
Results: Two similar factors emerged in both analyses:
    I Attitude; II Not labeled.
Weissman, H. N., Seldman, M., & Ritter, K. Changes in aware-
    ness of impact upon others as a function of encounter
    and marathon group experiences. Psychological Reports,
    1971, 28, 651-661.
Purpose: To identify dimensions underlying self-, estimate-,
    and peer-nomination forms (4-dimensional)
Subjects: 8 men and 2 women in a small E group, plus 12
    individuals from Weissman and Seldman (1969 unpublished
    data)
Variables: Self-, estimate-, and peer-ratings on 4 peer
    nomination form scales
Method: Unspecified condensation with varimax rotation
Results: Four factors emerged: I Unlabeled; II Trust vs.
    mistrust; III Openness vs. closedness to experience;
    IV Independence vs. dependence, plus sensitivity vs.
    insensitivity.
Zak, I. Dimensions of Jewish-American identity. Psycholo-
    gical Reports, 1973, 33, 891-900.
Purpose: To identify dimensions of Jewish-American
    identity
Subjects: 4 groups totaling 1006 Jewish-American college students
Variables: 20 Jewish-American Identity Scale item scores
Method: Principal axes condensation with varimax and
    oblique rotation, by subsample and for total
Results: Two factors emerged across groups: I Jewish
    identity; II American identity.
E. Communication
Antes, J. R., & Stone, L. A. Multidimensional scaling of
    pictorial informativeness. Perceptual and Motor Skills,
    1975, 40, 887-893.
Purpose: To identify dimensions of pictorial informative-
    ness
Subjects: 1) 10 college students, and 2) 7 using similar
    judgments
Variables: Informational similarity ratings for 32 areas
    of a picture
Method: 1) Inverse unspecified condensation, apparently
    without rotation; 2) unspecified condensation with
    varimax rotation
Results: Three judge factors emerged: I 7 judges (used
    in second analysis); II 2 judges; III 1 judge. Five
    factors emerged in the second analysis: I Presence vs
    absence of information; II Left vs right; III Meaning-
    fulness; IV Inner vs outer; V Foreground vs background.
Helson, R. Heroic and tender modes in women authors of
    fantasy. Journal of Personality, 1973, 41, 493-512.

Purpose: To identify clusters of 1) books and 2) writers
of fantasy books
Subjects: 1) 91 fantasies written for 8 12-year-olds, 2)
55 authors of those books
Variables: 1) Analysts' ratings on 5 stylistic and 10
"need" dimensions; 2) 56 work style items from the
Writers Q Sort
Method: Tyron cluster analysis, independently for books
and authors
Results: Three book clusters emerged: I Heroic; II Tender;
III Comic. Three clusters, of six extracted, were labeled
for authors: I Emotional involvement in work; II Asser-
tive ego; III Practical reality vs. receptive intuition.
Helson, R. The heroic, the comic, and the tender: Patterns
of literary fantasy and their authors. Journal of
Personality, 1973, 41, 163-184.
Purpose: To identify clusters of 1) formal dimensions or
"needs" of books, and 2) authors of those works
Subjects: 1) 91 works of fantasy written for 8 12-year-
olds, 2) the 27 male and 28 female authors
Variables: 1) Analysts' ratings on 5 formal dimensions
(importance ratings) and 10 "needs"; 2) 56 work style
items from the Writers Q Sort
Method: Tyron cluster analysis, independently for the
two S-variable sets
Results: Three book clusters emerged: I Heroic; II Tender;
III Comic. Six author clusters emerged: I Literary
ambition; II Emotional involvement in work; III Language
and elaboration; IV Restive autonomy and intuitiveness;
V Assertive ego; VI Not labeled.
Laffal, J., Monahan, J., & Richman, P. Communication of
meaning in glossolalia. Journal of Social Psychology,
1974, 92, 277-291.
Purpose: To identify dimensions of content profiles of
audience response to glossolalic speech
Subjects: 35 community college students
Variables: Scores on 114 computer-generated content categories for 10 stimuli
Method: Unspecified condensation
Results: Four factors emerged: I Death, mood description,
beginning; II Following, weakness, anger; III Leading,
fullness, emptiness, strength; IV Ending and beginning.
Rossenberg, S., & Jones, R. A method for investigating and
representing a person's implicit theory of personality:
Theodore Dreiser's view of people. Journal of Person-
ality and Social Psychology, 1972, 22, 372-386.
Purpose: To identify dimensions underlying Theodore Dreiser's
implicit personality theory
Study One:
Subjects: A Gallery of Women
Variables: (Two types of) co-occurrence indices for 99
most frequently appearing physical/psychological traits

Method: Cluster analysis
Results: Three clusters, of an unspecified number extracted,
were named: I Hard-soft; II Male-female; III Conform-
not conform.
Study Two:
Subjects: 144 college students
Variables: Scores on 14 trait properties for 99 most fre-
quently occurring physical/psychological traits appearing
in A Gallery of Women
Method: Principal axes condensation with varimax rotation
Results: Four factors emerged: I Evaluation; II Activity;
III Conform-does not conform, decided-undecided; IV Sex,
frequency of occurrence.
Conclusions: Multidimensional scaling analyses are also
reported.
   Thomas, M., & Seeman, J.   Personality integration and cognitive
processes. Journal of Personality and Social Psychology,
1972, 24, 154-161.
Purpose: To identify dimensions of ratings of pictures
portraying emotional states
Subjects: 18 high- and 15 low-self-concept (Tennessee
Self-Concept Scale) college freshmen
Variables: 35 affective adjective ratings for each of 19
pictures portraying emotional states
Method: Principal axes condensation with varimax and
promax rotation, by subject
Results: Analyses of variance were used to compare number
of factors 1) determined by the Scree test, and 2) with
eigenvalues greater than .950, between groups and by
sex for both first- and second-order solutions.
1. Language
   Gifford, R. K.   Information properties of descriptive words.
Journal of Personality and Social Psychology, 1975, 31,
727-734.
Purpose: To identify dimensions of information conveyed
by descriptive words
Subjects: 1) 30, 2) 25 female college students
Variables: 15 information ratings for each of 1) 100 personal traits words, or 2) 89 life condition words
Method: Unspecified condensation with varimax rotation,
independently for each subject-variable set
Results: Five personal traits factors emerged: I Evaluative; II How stable or dispositional the information
conveyed by the word was; III Familiarity; IV Behavior
influence; V Objectivity.   Five life conditions factors
emerged: I Centrality; II Evaluative; III Familiarity;
IV Stability; V Subjective-objective.
   Klein, R. H., & Iker, H. P.   The lack of differentiation
between male and female in Schreber's autobiography.
Journal of Abnormal Psychology, 1974, 83, 234-239.
Purpose: To identify context dimensions for "male" and
"female" in Schreber's Memoirs and K. Lorenz' On
Aggression

113

Subjects: 303 one-page segment widths of <u>Memoirs</u> and <u>On Aggression</u>
Variables: 215 most frequently appearing words: context dimensions
Method: Principal components condensation with varimax rotation; cluster analysis
Results: An unspecified number of factors and clusters emerged; male and female loaded on different factors (clusters) for the Lorenz text but on the same factor for the Schreber text.

Mabry, E. A. Dimensions of profanity. <u>Psychological Reports</u>, 1974, <u>35</u>, 387-391.
Purpose: To identify dimensions of profane language use by college students
Subjects: 283 college students
Variables: 5-point frequency-of-use ratings for 48 "sexual vernacular" terms
Method: Principal axes condensation with varimax rotation
Results: Five factors emerged: I Abrasive; II Technical; III Abrasive-expletive; IV Latent; V Euphemistic.

Osborne, J. W. Extraversion, neuroticism and word-arousal. <u>British Journal of Psychology</u>, 1973, <u>64</u>, 559-562.
Purpose: To identify arousal dimensions for word ratings
Subjects: 47 educational psychology students
Variables: 15-point ratings for 4 low- and 4 high-arousal words
Method: Principal components condensation with varimax (and procrustes) rotation
Results: Two factors emerged: I Low-arousal; II High-arousal.
Conclusions: Factor scores were related to extraversion and neuroticism.

2. Psycholinguistics

Farr, S. D. Component scores for reduced rank, transformed solutions. <u>American Educational Research Journal</u>, 1971, <u>8</u>, 93-103.
Purpose: To identify dimensions of semantic differential (SD) ratings of the concept "studying" (data collected by Kubiniec, unpublished dissertation, 1969) as an example for a discussion of factor vs. component models
Subjects: 321 male college students
Variables: 15 SD ratings for the concept "studying"
Method: Principal components condensation with normal varimax and oblique rotation
Results: Five uninterpreted factors emerged with both orthogonal and oblique clusters solution.

Haynes, J. R. The effect of double standardized scoring on the semantic differential. <u>Educational and Psychological Measurement</u>, 1975, <u>35</u>, 107-114.
Purpose: To identify dimensions underlying semantic differential (SD) ratings of the concepts myself and home
Subjects: 79 female and 121 male college students

114

Variables: 12 SD ratings for concepts myself and home
(raw and double standardized scores)
Method: Principal axes condensation with varimax rotation,
by concept
Results: Analyses of raw scores yielded three factors: I
Evaluation; II Potency; III Activity. Analyses of double-
standardized scores, however, yielded five and four
(unlabeled) factors for myself and home, respectively.
Heskin, K. J., Bolton, N., & Smith, F. V. Measuring the atti-
tudes of prisoners by the semantic differential. British
Journal of Social and Clinical Psychology, 1973, 12,
73-77.
Purpose: To identify dimensions of semantic differential
(SD) ratings of concepts relevant to prisoners
Subjects: 50 male long-term prisoners
Variables: 13 evaluative SD ratings, 2 potency SD ratings,
2 activity SD ratings for 1) each of 12 concepts, and
2) summed across the 12 concepts
Method: Principal components condensation with varimax
rotation, independently by concept and summed across
concepts
Results: Structures varied both in number of factors and
factor content between concepts. Three factors emerged
in the overall analysis: I Evaluative I; II Evaluative
II; III Activity.
Conclusions: Care should be taken in selecting SD scales,
based on the obtained factor structures which differed
across concepts.
Skikiar, R., Fishbein, M., & Wiggins, N. Individual differences in semantic space: A replication and extension.
Multivariate Behavioral Research, 1974, 9, 201-209.
Purpose: To identify dimensions of similarity judgments
of semantic differential (SD) scales
Subjects: 72 college students
Variables: Similarity judgments of 15 bipolar adjective
scales
Method: 1) Principal components condensation with varimax
rotation, for group average; 2) inverse principal components condensation for individual difference; 3) principal components condensation with varimax rotation, for
each of 9 "idealized individuals"
Results: Three group average factors emerged: I Evaluation;
II Activity; III Potency. Three factors also emerged
for the individual difference analysis and for eight
of nine "idealized individuals" (Activity did not emerge
for one ideal).
Szalay, L. B., & Bryson, J. A. Psychological meaning: Comparative analyses and theoretical implications. Journal
of Personality and Social Psychology, 1974, 30, 860-870.
Purpose: To identify psychological meaning dimensions underlying three types of measures (associative affinity,
similarity judgment, semantic differential)
Subjects: 60 college students

115

Variables: measures of meaning similarity for 12 concepts: index of interword associative affinity (IIAA), similarity judgment (SJ), grouping (G), judgment of relationship (JR), substitution (S), semantic differential (SD)
Method: Unspecified condensation with orthogonal rotation, independently for each type of measure
Results: Four factors, resembling to varying degrees the a priori clusters of concepts, emerged in each case: I Food; II Money; III School; IV Manners.
Tzeng, O. C. S. Reliability and validity of semantic differential E-P-A markers for an American English representative sample. Psychological Reports, 1975, 37, 292.
Purpose: To identify dimensions underlying affective meaning using indigenous American English markers
Subjects: 943 adults from Chicago total (40 Ss rating 25 concepts)
Variables: 12 semantic differential ratings for 375 heterogeneous concepts
Method: Principal components condensation with varimax rotation, independently for 25x943, two (presentation orders) 25x471 (472), malesx375, femalesx375 matrices
Results: Three factors emerged clearly in all cases: I Evaluation; II Potency; III Activity.
Wernimont, P. F., & Fitzpatrick, S. The meaning of money. Journal of Applied Psychology, 1972, 56, 218-226.
Purpose: To identify dimensions of the "meaning" of money
Subjects: 586 individuals from 11 groups of hard-core trainees, employed persons, and college students
Variables: 40 (modified) semantic differential ratings of the concept money
Method: Principal components condensation with varimax rotation
Results: Seven factors emerged: I Shameful failure; II Social acceptability; III Pooh-pooh attitude; IV Moral evil; V Comfortable security; VI Social unacceptability; VII Conservative business values.
Conclusions: Factor scores were used to examine differences between groups.
Williams, J. E., Best, D. L., Wood, F. B., & Filler, J. W. Changes in connotations of racial concepts and color names: 1963-1970. Psychological Reports, 1973, 33, 983-996.
Purpose: To identify dimensions underlying connotative ratings of color, color-person, and ethnic concepts
Subjects: 1) 116, 2) 102, 3) 86 college students in 1963; 4), 5), 6) 40 college students each in 1970
Variables: 15 semantic differential ratings each for color (1 & 4), color-person (2 & 5), and ethnic concepts (3 & 6)
Method: Principal components condensation with varimax rotation, independently for each year within each concept

Results: Two and three color factors emerged for 1963
and 1970, respectively, but were uninterpretable. Two
matched (I Non-caucasian evaluation; II Caucasian
identity) and one (III) uninterpretable factors emerged
for ethnic concepts. Two color-person factors (I White
person identity plus color-person; II Non-white person
identity) emerged for 1963, but four (I Color-person;
II Non-caucasian evaluation; III White person identity;
IV Black and brown person) emerged for 1970.

F. Aesthetics

Berlyne, D. E. Dimensions of perception of exotic and pre-
Renaissance paintings. Canadian Journal of Psychology,
1975, 29, 151-173.

Purpose: To identify dimensions underlying stylistic
ratings of two sets of paintings

Subjects: 1) 10 college students interested in art,
rating 20 exotic and pre-Renaissance paintings, and
2) 6 college students interested in art, rating 20
Western post-Renaissance paintings

Variables: 12 stylistic ratings

Method: Principal components condensation with varimax
rotation, independently for the two subject-stimuli
sets

Results: Three Western post-Renaissance factors emerged:
I Realism; II Subjectivism; III Classicism. Three
factors emerged for exotic and pre-Renaissance paintings:
I Stylization; II Fantasy; III Decorativeness.

Conclusions: These results are discussed in conjunction
with those of four others reported here.

Eysenck, H. J. Preference judgments for polygons, designs,
and drawings. Perceptual and Motor Skills, 1972, 34,
396-398.

Purpose: To identify dimensions underlying design
preferences and selected personality measures

Subjects: 484 adults

Variables: Preference judgments for 10 pairs of polygons,
10 pairs of designs, 10 pairs of drawings, scores on
6 extraversion and 6 neuroticism items, age, education,
present salary

Method: Principal components condensation with promax
rotation

Results: Six factors emerged: I E; II N; III Maitland-
Graves; IV Design; V-VI Polygons.

Eysenck, H. J., & Iwawaki, S. The determination of aesthetic
judgment by race and sex. Journal of Social Psychology,
1975, 96, 11-20.

Purpose: To identify dimensions of aesthetic judgment
across cultures

Subjects: 206 Japanese college students

Variables: Aesthetic judgments for 131 designs/devices

Method: Unspecified factor analysis

Results: Seven factors emerged: I Rectangular variant;
  II Circular variant; III Star variant; IV Interlacement
  variant; V Shading variant; VI Three-dimensional variant;
  VII Order or simplicity variant.
Conclusions: These results were compared with those for
  an English sample (Eysenck, 1971).
Firth, C. D., & Nias, D. K. B. What determines aesthetic
  preferences? Journal of General Psychology, 1974, 91,
  163-173.
Purpose: To identify dimensions of aesthetic preferences
  for designs of quantitatively defined content
Subjects: 88 education students
Variables: 7-point preference ratings for 18 computer-
  generated designs
Method: Principal components condensation with varimax
  rotation
Results: Three factors emerged and were contrasted with
  multidimensional scaling results: I Simple, low contour;
  II Complex, high contour; III High contour.
Conclusions: Contour may have been confounded with com-
  plexity in previous studies, and should be controlled
  in future ones.
Palmer, R. D. Cluster analysis of preference ratings of
  pictorial stimuli. Journal of Clinical Psychology,
  1975, 31, 437-438.
Purpose: To identify factors underlying preference ratings
  of pictorial stimuli
Subjects: 537 male inpatients
Variables: 437 picture preference ratings
Method: Cluster analysis
Results: Eighteen interpretable clusters emerged: I Sex
  (heterosexuality); II Romantic love; III Death-destruc-
  tion; IV Physical combat; V Attractive food; VI Mascu-
  line adventure; VII Active physical exertion; VIII
  Attractive animals; IX Distasteful or feared animals;
  X Studies of faces (personality revealed through faces);
  XI Pastoral scenes; XII Mythological other-worldly;
  XIII Three-dimensional perception; XIV Graphic surface
  interest; XV Attractive abstract patterns; XVI Softly-
  portrayed surfaces; XVII Archetechtonic structures;
  XVIII Unidentified objects (objects viewed from a point
  too near to permit accurate identification).
Swartz, P., Swartz, S., & Hill, K. Michelangelo's Pietàs:
  A semantic analysis. Perceptual and Motor Skills, 1974,
  38, 3-9.
Purpose: To identify dimensions underlying semantic dif-
  ferential (SD) ratings of Michelangelo's Pietàs
Subjects: 48 male and 48 female college students
Variables: 40 SD ratings for 4 Pietàs
Method: Alpha condensation with varimax rotation
Results: Seven factors emerged: I Form; II Receptiveness;
  III Activity; IV Affectivity; V Potency; VI Reality;
  VII Spirituality.

G. Smoking and Drug and Alcohol Use

Coan, R. W. Personality variables associated with cigarette
   smoking. Journal of Personality and Social Psychology,
   1973, 26, 86-104.

Purpose: To identify dimensions underlying the Smoking
   Survey (SS)

Subjects: 595 college students (of whom at least 175 were
   smokers)

Variables: Scores on 1) 23 and 2) 43 SS items

Method: Principal axes condensation with oblique rotation,
   independently for the two item sets

Results: Six factors emerged in the first analysis: I
   Addictive smoking; II Negative affect smoking; III
   Sensorimotor component; IV Pleasurable relaxation; V
   Stimulation; VI Habitual smoking. Eleven factors
   emerged from the analysis of all items: I Addiction;
   II Pleasurable relaxation; III Negative affect reduction;
   IV Smoking for distraction; V Unpleasant habit; VI
   Habitual action; VII Stimulation; VIII Dependence of
   smoking on mental state; IX Agitated state smoking;
   X Sensorimotor pleasure; XI Induction of concentration.

Colaiuta, V., & Breed, G. Development of scales to measure
   attitudes toward marijuana and marijuana users. Journal
   of Applied Psychology, 1974, 59, 398-400.

Purpose: To identify dimensions of attitudes toward
   marijuana

Subjects: 1) 155, 2) 162, 3) 137 male and female undergraduates and graduate students

Variables: Scores on 1) 21, 2) 20, 3) 34 items from a
   marijuana attitude questionnaire

Method: Basic-structure-successive factor condensation
   with varimax rotation, independently for the three
   subject-variable sets

Results: Eight unlabeled factors emerged in the first two
   analyses. Three factors were interpreted from the third
   analysis: I Effects; II Panacea; III Ambivalence.

Frith, C. D. Smoking behavior and its relation to the smoker's
   immediate experience. British Journal of Social and
   Clinical Psychology, 1971, 10, 73-78.

Purpose: To identify dimensions of smokers' immediate
   experience related to stated desire to smoke

Subjects: 59 male and 39 female cigarette smokers

Variables: 7-point ratings of desire to smoke for each of
   12 high- and 10 low-arousal situations, plus sex, age,
   and number of cigarettes smoked per day

Method: Principal components condensation without rotation

Results: Two factors were reported: I Number of cigarettes/
   day; II Low- vs. high-arousal.

Conclusions: T-tests between sexes for high- vs. low-arousal
   situations were also reported.

Martino, E. R., & Truss, C. V. Drug use and attitudes toward
   social and legal aspects of marijuana in a large metropolitan university. Journal of Counseling Psychology,
   1973, 20, 120-126.

Purpose: To identify dimensions underlying attitudes
toward marijuana
Subjects: 788 college students (unspecified numbers con-
tacted at three different sites)
Variables: Scores on 10 attitude-toward-marijuana items
Method: Unspecified condensation with varimax rotation,
by site and for total
Results: Two factors emerged across analyses: I Positive
statements; II Negative statements.
Mausner, B. An ecological view of cigarette smoking. Journal
of Abnormal Psychology, 1973, 81, 115-126.
Purpose: To identify dimensions of expectations of outcomes
of continuing or ceasing smoking.
Subjects: 164 adult smokers (PTA members)
Variables: 29 subjective-utility-of-ceasing-smoking item
scores
Method: Unspecified factor analysis
Results: Six factors emerged: I Self-concept; II Tension
reduction; III Health; IV Hedonic-esthetic; V Stimula-
tion; VI Social stimulation (other).
Naditch, M. P. Acute adverse reactions to psychoactive drugs,
drug usage, and psychopathology. Journal of Abnormal
Psychology, 1974, 83, 394-403.
Purpose: To identify dimensions underlying a questionnaire
measure of acute adverse reactions to drugs
Subjects: 483 male drug users (mean age=21.4 years)
Variables: Scores on 23 questionnaire items dealing with
acute adverse reactions to psychoactive drugs
Method: Unspecified factor analysis
Results: Apparently one factor emerged, lending support
to the intended unidimensionality of the instrument.
Naditch, M. P. Relation of motives for drug use and psycho-
pathology in the development of acute adverse reactions
to psychoactive drugs. Journal of Abnormal Psychology,
1975, 84, 374-385.
Purpose: To identify dimensions underlying reported motives
for drug use
Subjects: 483 male drug users (mean age=21.4 years)
Variables: Scores on 10 motives-for-drug-use items from a
questionnaire
Method: Unspecified condensation with varimax rotation
Results: Three factors emerged: I Therapeutic; II Pleasure
and curiosity; III Response to peer pressure.
Sinnett, E. R., Hagen, K., & Harvey, W. M. Credibility of
sources of information about drugs to heroin addicts.
Psychological Reports, 1975, 37, 1239-1242.
Purpose: To identify dimensions of heroin addicts' per-
ceptions of credibility of sources of information about
drugs
Subjects: 23 black heroin addicts
Variables: Trustworthiness ratings of 46 sources of infor-
mation about street drugs
Method: Principal axes condensation with varimax rotation

Results: Four factors emerged: I Friendship; II Legal
   and medical authority; III Benign authority; IV Drug
   experienced persons.
Sinnett, E. R., Press, A., Bates, R. A., & Harvey, W. M.
   Credibility of sources of information about drugs.
   Psychological Reports, 1975, 36, 299-309.
Purpose: To identify dimensions underlying perceived
   credibility of sources of information about drugs
Subjects: 1) 108 high school normals and drug abusers,
   2) 83 college drug class Ss, 3) 272 rural junior high
   and high school students
Variables: Credibility ratings for 45 sources of infor-
   mation about drugs
Method: Principal axes condensation with varimax rotation,
   by group
Results: Three factors emerged across analyses: I Authority;
   II Drug experience; III Friendship.
Stone, L. A., & Sinnett, E. R. Perceived dimensions for com-
   binations of drug substances. Perceptual and Motor
   Skills, 1973, 37, 227-232.
Purpose: To identify dimensions of perceived similarities
   among 11 drug substances
Subjects: 8 experienced users of street drugs
Variables: Similarity judgments for 11 drug substances
Method: Principal components condensation with varimax
   rotation
Results: Four factors emerged: I Psychedelic drugs in
   combination with other substances vs. pot in combination
   with downers; II Stimulant-depressant; III Psychedelic
   vs. other drugs; IV LSD in combination with other substances.

VIII. PERSONALITY

Coles, G. J., & Stone, L. A. An exploratory direct-estima-
tion approach to location of observers in percept
space. Perceptual and Motor Skills, 1974, 39, 539-
549.
Purpose: To identify judge dimensions of similarity judge-
ments of well-known personality theories
Subjects: 18 Ph.D. psychologist judges
Variables: Similarity judgments for 15 personality theories
Method: Principal components condensation with varimax
rotation
Results: Four factors emerged: I Psychoanalytic-nonpsy-
choanalytic; II Factor-typological vs. phenomenological-
existential, or quantitative-qualitative; III Learning-
other; IV Eclectic-other.
A. Personality Traits and Processes
Bartsch, T. W., & Nesselroade, J. R. Test of the trait-
state anxiety distinction using a manipulative, factor-
analytic design. Journal of Personality and Social
Psychology, 1973, 27, 58-64.
Purpose: To identify trait-state dimensions of anxiety
Subjects: 104 college students (stress, nonstress condi-
tions)
Variables: Scores on 14 anxiety and mood measures
Method: Principal axes condensation with procrustes
rotation
Results: Four factors emerged: I General body size; II
State anxiety; III Trait anxiety; IV Expression of
fatigue (or, general negative affect).
Bates, H. D., & Zimmerman, S. F. Toward the development of
a screening scale for assertive training. Psychological
Reports, 1971, 28, 99-107.
Purpose: To identify dimensions underlying a self-report
screening scale for assertive training
Subjects: 150 male and 150 female college freshmen
Variables: Scores on 42 Constriction Scale 1 (CS1) items
Method: Principal components condensation apparently
without rotation, by sex
Results: Thirteen unlabeled factors emerged for males,
and fourteen for females.
Conclusions: A short form (CS2) containing 23 constriction
and 6 filler items was constructed.
Beiser, M., Feldman, J. J., & Egelhoff, C. J. Assets and
affects: A study of positive mental health. Archives
of General Psychiatry, 1972, 27, 545-549.
Purpose: To identify dimensions underlying judged per-
sonality assets
Subjects: 123 Nova Scotians aged 18-55+, heterogeneous
as to SES, etc.
Variables: Scores on 7 judged personality assets
Method: Unspecified condensation with unspecified rotation
Results: Two factors emerged: I Interpersonal reactivity;
II Role-related planning abilities.

Bond, A., & Lader, M. The use of analogue scales in rating subjective feelings. British Journal of Medical Psychology, 1974, 47, 211-218.
  Purpose: To identify dimensions underlying analogue scales (previously used to measure drug effects) with normal subjects
  Subjects: 500 technical college, university, and hospital personnel
  Variables: Scores on 16 100 mm bipolar mood rating scales
  Method: Principal components condensation with orthogonal rotation
  Results: Three factors emerged: I Alertness; II Contentedness; III Calmness.
  Conclusions: Drug effects were examined via use of these three factor-analytically derived scales.
Bottenberg, E. H. Phenomenological and operational characterization of factor-analytically derived dimensions of emotion. Psychological Reports, 1975, 37, 1253-1254.
  Purpose: To identify dimensions of emotion
  Subjects: 92 male and 58 female college students
  Variables: 1) 50 adjective scale ratings for 30 emotions; 2) 3 factor scores from 1 plus 79 well-known personality variables; 3) unspecified number of variables defining each cluster from (2)
  Method: 1) Centroid condensation with varimax rotation, by sex and for total group; 2) cluster analysis; 3) centroid condensation with varimax rotation, within each cluster
  Results: Three factors emerged across solutions for the first analyses: I Pleasantness vs. unpleasantness; II Activation; III Depth. Two meaningful clusters emerged: I Individual tendency to-be-pleased vs. tendency to-be-displeased (A); II Individual degree of emotional activation (B). Three Cluster A factors emerged: I Extraversion; II Ego strength; III Field independent cognitive style. Three cluster B factors emerged: I Emotional excitability and impulsivity; II Cognitive simplicity-concreteness; III Not reported.
Chapman, J. L. Development and validation of a scale to measure empathy. Journal of Counseling Psychology, 1971, 18, 281-282.
  Purpose: To identify dimensions of empathy
  Subjects: 148 individuals from National Defense Education Act Summer Institutes
  Variables: 172 empathy (adjective) item scores
  Method: Unspecified factor analysis
  Results: Five unlabeled factors emerged.
Coan, R. W. Measurable components of openness to experience. Journal of Consulting and Clinical Psychology, 1972, 39, 346.
  Purpose: To identify dimensions underlying openness to experience

Subjects: 1) 383 college students, 2) 219 Ss
Variables: Scores on 1) 114, 2) 149 openness-to-experience
    items
Method: Unspecified condensation with oblique rotation,
    by S-variable set
Results: Sixteen (unlabeled but reportedly identifiable
    as openness to particular kinds of experience) factors
    emerged in the first analysis; seven large-variance
    factors appeared replicable. Seven factors emerged
    in the second analysis: I Aesthetic sensitivity vs.
    insensitivity; II Unusual perceptions and associations;
    III Openness to theoretical or hypothetical ideas; IV
    Constructive utilization of fantasy and dreams; V Open-
    ness to unconventional views of reality vs. adherence
    to mundane, material reality; VI Indulgence in fantasy
    vs. avoidance of fantasy; VII Deliberate and systematic
    thought.
Conclusions: These seven factors are represented as scales
    in the 83-item Experience Inventory.
Coan, R. W., Fairchild, M. T., & Dobyns, Z. P. Dimensions
    of experienced control. Journal of Social Psychology,
    1973, 91, 53-60.
Purpose: To identify dimensions of experienced control
Subjects: 1) 525 college students; 2) 196 college stu-
    dents
Variables: Scores on 1) 130 true-false items from the
    original Personal Opinion Survey (POS-O); 2) 149 items
    of the revised Personal Opinion Survey (POS-R)
Method: Unspecified condensation with oblique rotation,
    independently for the two subject-variable sets
Results: Eighteen factors were extracted from the POS-O
    but were not reported here. Seven POS-R factors
    emerged: I Achievement through conscientious effort;
    II Personal confidence in ability to achieve mastery;
    III Capacity of mankind to control its destiny vs.
    supernatural power or fate; IV Successful planning and
    organization; V Self-control of internal processes;
    VI Control over large-scale social and political
    events; VII Control in immediate social interaction.
Coffman, R. N., & Levy, B. I. The dimensions implicit in
    psychological masculinity-femininity. Educational
    and Psychological Measurement, 1972, 32, 975-985.
Purpose: To identify dimensions underlying psychological
    masculinity-femininity (Mf)
Subjects: 100 males and 100 females associated with an
    urban university
Variables: Scores on 12 Mf categories (scales)
Method: Unspecified condensation with unspecified rota-
    tion
Results: Six factors emerged: I Fastidiousness; II Atti-
    tudes toward affective life; III Need for social close-
    ness; IV Interest in vocations and avocations; V Puni-
    tiveness vs. mercy; VI Overt attitudes toward hetero-
    sexuality.

Farr, S. D., & Kubiniec, C. M. Stable and dynamic components of self-report self-concept. *Multivariate Behavioral Research*, 1972, 7, 147-163.
  Purpose: To identify dimensions of self-report self-concept
  Subjects: 324 male, 260 female college freshmen
  Variables: 15 semantic differential ratings each for concepts my past, my future, my real self, my ideal self
  Method: Principal components condensation with varimax rotation, by sex
  Results: Four dynamic-evaluative factors (one for each concept) appeared in both analyses, although males produced two additional factors (past, future); five stable-descriptive common to sexes also emerged, although females produced two additional factors of this type; the remaining components (three male, four female, one singlet each) tended toward the stable-descriptive pattern.
Gaudry, E., Vagg, P., & Spielberger, C. D. Validation of the state-trait distinction in anxiety research. *Multivariate Behavioral Research*, 1975, 10, 331-341.
  Purpose: To identify dimensions of anxiety
  Subjects: 345 tenth-grade girls, 255 graduate students, all Australian; tested under 3 anxiety conditions
  Variables: Scores on 20 A-trait and 60 A-state STAI items, Otis, age, TASC and mathematic test
  Method: Principal axes condensation with varimax rotation, by sample
  Results: Six factors emerged in each case: I Trait anxiety; II-IV State anxiety factors (corresponding to conditions); V Reversed-item; VI Ability.
Gorman, B. S., & Wessman, A. E. The relationship of cognitive styles and moods. *Journal of Clinical Psychology*, 1974, 30, 18-25.
  Purpose: To identify dimensions of cognitive styles
  Subjects: 20 male and 47 female community college summer school students
  Variables: Scores on 33 cognitive style and control measures, plus sex
  Method: Principal axes condensation with varimax rotation
  Results: Eight factors emerged: I Subjective hope and confidence; II Sensation-seeking and openness; III External vs. internal locus of control; IV Defensiveness vs. admissions of affective involvement; V Satisfaction with the present; VI Narrow vs. broad conceptual bandwidth; VII Field articulation (field independence); VIII Rigid conventionality.
  Conclusions: Factors were examined in relation to mood characteristics.
Hakel, M. D. Normative personality factors recovered from ratings of personality descriptors: The beholder's eye. *Personnel Psychology*, 1974, 27, 409-421.

Purpose: To identify dimensions of personality ratings
Subjects: 149 male and 141 female college students, each rating a person of his/her choice
Variables: Rating scale scores for each of Norman's (1966) personality descriptors
Method: Principal factors condensation with varimax rotation
Results: Five factors emerged: I Extroversion; II Agreeableness; III Conscientiousness; IV Emotional stability; V Culture.
Conclusions: A multidimensional scaling study is also reported.
Herrenkohl, R. C. Factor-analytic and criterion study of achievement orientation. Journal of Educational Psychology, 1972, 63, 314-326.
Purpose: To identify dimensions underlying a self-description assessment of achievement orientation
Subjects: 1) 690 college students, 2) 5,102 high school and college students
Variables: 160 achievement orientation item scores
Method: Principal components condensation with varimax rotation, by sample (odds, evens within second sample)
Results: Ten factors emerged: I Test anxiety; II Threat of failure; III Parental expectations; IV Unwillingness to risk failing; V Dislike of persons who do better than oneself; VI Concern about failing in primary roles; VII Desire to excel; VIII Sensitivity to others' knowing one's failure; IX Exerting effort to do well; X Valuation of competition.
Jackson, D. N. Multimethod factor analysis: A reformulation. Multivariate Behavioral Research, 1975, 10, 259-276.
Purpose: To identify dimensions of personality (as an illustration of multimethod factor analysis)
Subjects: 1) 240 male, 240 females (Jackson & Singer, 1967); 2) unspecified (Jackson & Carlson, 1973)
Variables: Judgments on 5 items each for 4 personality traits; 11 DPI scale scores plus corresponding self- and roommate ratings
Method: Principal components condensation with varimax rotation, within traits and then across
Results: Four trait and five method factors emerged in the first analysis; 11 trait and 3 method factors emerged in the second analysis.
Jackson, D. N., Hourany, L., & Vidmar, N. J. A four-dimensional interpretation of risk taking. Journal of Personality, 1972, 40, 483-501.
Purpose: To identify dimensions of risk taking
Subjects: 47 male, 46 female college students and 49 YWCA social group housewives
Variables: Scores on 5 personality (2 PRF, 3 JPI) scales plus types of risk-taking scores in each of 4 domains

Method: Principal components condensation with varimax (and clustran) rotation
Results: Four first-order (I Monetary; II Physical; III Social; IV Ethical) and one second-order factor (I Generalized risk taking) emerged.

Jones, R. A., & Rosenberg, S. Structural representations of naturalistic descriptions of personality. Multivariate Behavioral Research, 1974, 9, 217-230.
Purpose: To identify dimensions underlying naturalistic descriptions of personality
Subjects: 50 male, 50 female college students, each describing 5 persons s/he knew well
Variables: Trait co-occurrence measure for 99 most frequently occurring trait categories
Method: Hierarchical cluster analysis
Results: Six cluster levels were interpreted in terms of Dominant-submissive, Active-passive, Intellectual good-bad, Hard-soft, Impulsive-inhibited, and Good-bad.

Jones, R. A., Sensenig, J., & Haley, J. V. Self-descriptions: Configurations of content and order effects. Journal of Personality and Social Psychology, 1974, 30, 36-45.
Purpose: To identify dimensions of college students' self-descriptions
Subjects: 150 male, 150 female college students
Variables: Co-occurrence scores for 97 self-descriptive trait categories
Method: Diameter method hierarchical cluster analysis
Results: Six clusters were labeled: I Frequency; II Introverted-extraverted; III Intellectual good-bad; IV Hard-soft; V Male-female; VI Social good-bad.

Judd, L. R., & Smith, C. B. Discrepancy score validity in self- and ideal self-concept measurement. Journal of Conseling Psychology, 1974, 21, 156-158.
Purpose: To identify dimensions underlying semantic differential (SD) ratings of self- and ideal self-concept
Subjects: 445 college (?) students
Variables: 13 SD ratings each for self- and ideal self-concept as a speaker
Method: Principal components condensation with varimax rotation, separately for the two concepts
Results: Two self- and three ideal self-concept factors emerged but were not labeled.

Lanyon, B. J. Empirical construction and validation of a sentence completion test for hostility, anxiety, and dependency. Journal of Consulting and Clinical Psychology, 1972, 39, 420-428.
Purpose: To identify dimensions underlying items of sentence-completion tests (preliminary form) for hostility, anxiety, and dependency
Subjects: 205 checklists for boys and girls in Grades 7-9
Variables: 25 sentence-completion item scores each for 1) hostility (H), 2) anxiety (A), and 3) dependency (D)

127

Method: Principal components condensation without rotation,
    independently for three sets of 25 items and for 75 items
Results: One major factor emerged for the H, A, and D
    analyses. Four interpretable factors emerged for the
    combined items: I Hostility; II Anxiety (with some
    dependency); III Dependency; IV Anxiety.
Lund, T. An alternative content method for multidimensional
    scaling. Multivariate Behavioral Research, 1975, 10,
    181-191.
Purpose: To identify intensity, "standard emotion," dis-
    similarity, and constant-sum dimensions of judgments
    of emotions
Subjects: 1) 13 male and 2 female, 2) 11 male and 4 fe-
    male, 3) 20 male and 5 female, 4) 8 male and 4 female
    psychology students, most with some knowledge of
    scaling/factor analysis
Variables: 1) Intensity, 2) "standard emotion," 3) dis-
    similarity, 4) constant-sum judgments for 9 emotions
Method: Principal axes condensation with varimax rotation,
    independently for the four subject-variable sets and
    alternative content method for each
Results: The resultant two- (I Happiness-sadness; II
    Agitation) and three-factor (I Sadness; II Happiness;
    III Agitation) solutions are presented.
Magnusson, D., & Ekehammar, B. Anxiety profiles based on
    both situational and response factors. Multivariate
    Behavioral Research, 1975, 10, 27-43.
Purpose: To identify dimensions underlying an anxiety
    inventory
Subjects: 58 male, 58 female 15- to 17-year-olds
Variables: Scores (experienced intensity) for 18 modes
    of response (R) in 17 situations (S)
Method: Unspecified condensation with varimax rotation,
    independently for R and S
Results: Two R factors, of three extracted, were inter-
    preted: I Psychic anxiety; II Somatic anxiety. Three
    S factors emerged: I Threat of punishment; II Antici-
    pation fear; III Inanimate threat.
Marshall, N. J. Dimensions of privacy preferences. Multi-
    variate Behavioral Research, 1974, 9, 255-271.
Purpose: To identify dimensions of privacy preferences
Subjects: 1) 198 college students, 2) 149 students and
    101 of their parents
Variables: Scores on 1) 56 Privacy Preference Scale (PPS)
    items, 2) 86 PPS items
Method: Principal axes condensation with varimax rotation,
    by S-variable set
Results: Five factors emerged in the first analysis: I
    Having a room or a retreat of one's own; II Anonymity
    as a means of gaining privacy; III Self-disclosure;
    IV Solitude; V Preferred study setting. Six factors
    emerged in the second set of analyses: I Non-involve-
    ment with neighbors; II Seclusion of the home; III

Privacy with intimates; IV Anonymity; V Solitude; VI
Reserve.
Meadows, C. M. The phenomenology of joy, an empirical
investigation. Psychological Reports, 1975, 37, 39-54.
Purpose: To identify dimensions underlying 1) the Joy
Scale, and 2) the Mood Adjective Check List (MACL)
administered under joy condition
Subjects: 333 college students
Variables: Scores on 1) 61 Joy Scale items, 2) unspeci-
fied number of MACL items
Method: Principal axes condensation with promax rotation,
for each variable set
Results: Fourteen Joy Scale factors emerged: I Affilia-
tion; II Individuation; III Excitement; IV Serenity;
V Passiveness; VI Activeness; VII Perception of beauty;
VIII Self-potency; IX Positive world; X Time: immediacy;
XI Time: brevity; XII Rapid time; XIII Productive ef-
fects; XIV Ecstasy. Nine MACL factors emerged: I
Concentration; II Aggression; III Egotism; IV Vigor;
V Surgency-social affection; VI Skepticism-anxiety;
VII Fatigue-sadness; VIII Joy; IX Ecstasy.
Meddis, R. Bipolar factors in mood adjective checklists.
British Journal of Social and Clinical Psychology,
1972, 11, 178-184.
Purpose: To identify dimensions of mood
Subjects: 154 individuals, presumably college students;
half using each type of scale
Variables: 1) Nowlis (1965) or control scale scores for
38 MACL items selected to correspond to Thayer's (1965)
adjectives; 2) Nowlis or control scale scores for sub-
set of 38 MACL items selected to represent the Nowlis-
Green (1957) dimensions
Method: 1) Principal components condensation and 2) principal axes condensation, both with varimax rotation,
independently for type of rating scale
Results: Three control (I General activation - deactivation
sleep; II Tension - quiescence; III Activation) and five
Nowlis scale factors (I General activation; II Deactivation sleep; III General deactivation; IV High activation;
V Uninterpreted) emerged in the first set of analyses.
Four factors which tended to be bipolar only with the
control rating scales, emerged: I Lack of control
mixed with unpleasantness; II Deactivation; III Pleasantness; IV Positive social orientation.
Nesselroade, J. R., & Cable, D. G. "Sometimes, it's okay
to factor difference scores"--the separation of state
and trait anxiety. Multivariate Behavioral Research,
1974, 9, 273-281.
Purpose: To identify dimensions of state and trait anxiety
Subjects: 90 male, 51 female college students, tested and
retested 50 min. later
Variables: Test, retest, and difference scores for 4 state
and 4 trait anxiety measures

Method: Principal axes condensation for each type of score, with oblique rotation for test and retest scores

Results: Test and retest analyses yielded two factors (I State; II Trait) while the difference score analysis yielded one factor.

Otterbacher, J. R., & Munz, D. C. State-trait measure of experiential guilt. Journal of Consulting and Clinical Psychology, 1973, 40, 115-121.

Purpose: To identify dimensions underlying a pool of experiential guilt items

Subjects: 220 college students, in 4 groups

Variables: 9 semantic differential ratings for 83 experiential guilt items

Method: Principal components condensation with varimax rotation

Results: Factor I accounted for approximately 87% of the common factor variance.

Conclusions: Eleven items were selected for the G-Trait and G-State scales of the Perceived Guilt Index.

Pedersen, D. M. Prediction of behavioral personal space from simulated personal space. Perceptual and Motor Skills, 1973, 37, 803-813.

Purpose: To identify dimensions underlying simulated personal-space measures

Subjects: 170 male junior college students

Variables: 36 scores obtained from 3 measures of simulated personal space

Method: Principal factors condensation with varimax rotation

Results: Three factors emerged: I Pedersen measure; II Behavior personal space; III Rawls and awareness measures.

Posavac, E. J. Dimensions of trait preferences and personality type. Journal of Personality and Social Psychology, 1971, 19, 274-281.

Purpose: To identify dimensions of preferences for personality traits and social behaviors

Subjects: 135 (male) fraternity members

Variables: Ratings (desirability in friends) for 28 personality traits and social behaviors

Method: Principal components condensation with binormamin rotation

Results: Seven factors emerged: I Conventional friend; II Intellectually and politically aware friend; III Social, immoral, and bright friend; IV Interpersonally wholesome friend; V Serious and traditional friend; VI Athletic friend; VII Immoral, girl-watcher friend.

Pugh, W. M., Erickson, J., Rubin, R. T., Gunderson, E. K. E., & Rahe, R. H. Cluster analyses of life changes: II. Method and replication in Navy subpopulations. Archives of General Psychiatry, 1971, 25, 333-339.

Purpose: To identify dimensions of life changes with a second sample

130

Subjects: 2,025 USN enlisted men, in 4 subsamples
Variables: Scores on 42 Schedule of Recent Experience
life-change items
Method: Iterative intercolunmar cluster analysis, by
subsample
Results: Four major clusters emerged across subsamples:
I Personal and social; II Work; III Marital; IV
Disciplinary.
Rahe, R. H., Pugh, W. M., Erickson, J., Gunderson, E. K. E.,
& Rubin, R. T. Cluster analyses of life changes: I.
Consistency of clusters across large Navy samples.
Archives of General Psychiatry, 1971, 25, 330-332.
Purpose: To identify dimensions of life changes
Subjects: 2,678 USN enlisted men
Variables: 42 life-change item scores from the Schedule
of Recent Experience
Method: Iterative, intercolumnar cluster analysis
Results: Four major clusters emerged: I Personal and
social; II Work; III Marital; IV Disciplinary.
Ray, M. L., & Heeler, R. M. Analysis techniques for ex-
ploratory use of the multitrait-multimethod matrix.
Educational and Psychological Measurement, 1975, 35,
255-265.
Purpose: To reanalyze the Campbell-Fiske (1959) clinical
psychologist matrix
Subjects: Unspecified (here) number of clinical psycho-
logists
Variables: Ratings of 5 traits by self, staff, and team-
mate
Method: Hierarchical cluster (connectedness) analysis
Results: The results were in agreement with Campbell-
Fiske rather than Joreskog (1971) in that Assertive,
Cheerful, Broad interests, and Serious emerged as
traits but unshakable poise had minimal validity.
Reid, D. W., & Ware, E. E. Affective style and impression
formation: Reliability, validity and some inconsis-
tencies. Journal of Personality, 1972, 40, 436-450.
Purpose: To identify dimensions of affective style
Subjects: Unspecified here; Ware, unpublished study,
1966
Variables: Unspecified number (20 minimum) of semantic
differential ratings from a person perception study
Method: Unspecified factor analysis
Results: Seven factors emerged: I Rational; II Aloof;
III Unique; IV Moral; V-VII Not reported here.
Kubiniec, C. M., & Farr, S. D. Concept-scale and concept-
component interaction in the semantic differential.
Psychological Reports, 1971, 28, 531-541.
Purpose: To identify dimensions underlying semantic
differential (SD) ratings of self-concept
Subjects: 324 male and 260 female college freshmen
Variables: 15 SD ratings for concepts my past, my future,
my real self, my ideal self

Method: Principal components condensation with varimax
   rotation, by sex
Results: Fifteen factors emerged for males, but only
   eight were interpreted: I Excitable-calm and tense-
   relaxed; II Energ-lethargic and active-passive; III
   Complex-simple; IV Strong-weak; V Serious-humorous;
   VI Masculine-feminine; VII Rugged-delicate; VIII
   Severe-lenient. Fourteen factors emerged for females,
   but only seven were named: I Excitable-calm; II
   Energy-lethargic and active-passive; III Complex-
   simple; IV Strong-weak; V Serious-humorous; VI Mascu-
   line-feminine; VII Rugged-delicate.
Saklofske, D. H., & Schulz, H. W. Factor analysis of re-
   peated state hostility and guilt measures: Females.
   Psychological Reports, 1975, 37, 1152.
Purpose: To identify dimensions underlying an adjective
   checklist measure of hostility and guilt
Subjects: 67 high school females
Variables: Scores on 38 adjective checklist items, summed
   over 8 times
Method: Principal components condensation with varimax
   rotation
Results: Although none of the factors were labeled, one
   of the five to eight factors was a general factor.
Saklofske, D. H., & Schulz, H. W. Factor analysis of re-
   peated state hostility and guilt measures: Males.
   Psychological Reports, 1975, 37, 756-758.
Purpose: To identify dimensions underlying an adjective
   checklist measure of hostility and guilt
Subjects: 77 high school males
Variables: Scores on 38 adjective checklist items,
   summed over 8 times
Method: Principal factors condensation
Results: Although none of the factors were labeled, one
   of the five to seven factors was a general factor.
Schlenker, B. R. Self-presentation: Managing the impression
   of consistency when reality interferes with self-
   enhancement. Journal of Personality and Social Psy-
   chology, 1975, 32, 1030-1037.
Purpose: To identify dimensions of self-presentation
Subjects: 60 male, 60 female college students (in 4-
   person, same-sex groups)
Variables: 35 personal attribute item scores
Method: Principal axes condensation with varimax rotation
Results: Two factors emerged: I Competence; II Interper-
   sonal relations.
Sheehan, P. W. Measurement of a construct of the student-
   activist personality. Journal of Consulting and
   Clinical Psychology, 1971, 36, 297.
Purpose: To identify dimensions underlying a set of items
   designed to describe the student-activist personality
Subjects: 125 male college students

Variables: Scores on 90 items designed to measure 6
student-activist personality dimensions
Method: Principal axes condensation with varimax rotation
Results: Three factors emerged: I Ideology; II Authority;
III Conflict between real and ideal self.
Sherrill, D., & Salisbury, J. L. Manifest anxiety, extra-
version, and neuroticism: A factor-analytic solution.
Journal of Counseling Psychology, 1971, 18, 19-21.
Purpose: To identify relationships among manifest anxiety
(MA), extraversion (E), and neuroticism (N)
Subjects: 241 college students
Variables: Scores on 20 MA, 6 E, 6 N items
Method: Image factor condensation with varimax rotation
Results: Three factors emerged: I Neurotic anxiety; II
Extraversion; III Attentiveness.
Tucker, L. R. Relations between multidimensional scaling
and three-mode factor analysis. Psychometrika, 1972,
37, 3-27.
Purpose: To identify dimensions underlying relations
judgments of personality adjectives
Subjects: 87 college students
Variables: Similarity ratings for 12 adjectives
Method: Three-mode factor condensation
Results: Two large roots and a small third one were
extracted but not labeled.
Tzeng, O. C. S. Differentiation of affective and denotative
meaning systems and their influence in personality
ratings. Journal of Personality and Social Psychology,
1975, 32, 978-988.
Purpose: To identify the factor structure of personality
impressions
Subjects: 50 urban French-speaking Belgians (police
trainees or officer applicants)
Variables: 40 semantic differential ratings for each of
40 (personality) concepts
Method: Three-mode factor condensation with varimax (and
oblimax) rotation
Results: Six unlabeled scale, four concept, and three
subject factors emerged.
Wagner, F. R., & Morse, J. J. A measure of individual sense
of competence. Psychological Reports, 1975, 36, 451-
459.
Purpose: To identify dimensions underlying a paper-and-
pencil test designed to measure sense of competence
Subjects: 310 individuals working on a variety of jobs
in different work settings
Variables: Scores on 23 individual's-sense-of-competence
items
Method: Principal factors condensation with varimax (and
oblique) rotation
Results: Four factors emerged: I Competence thema;
II Task knowledge/problem-solving; III Influence; IV
Confidence.

Wardell, D.   Note on factor analysis of cognitive styles.
   Perceptual and Motor Skills, 1974, 38, 774.
   Purpose: To identify dimensions of cognitive style
   Subjects: 30 males, 30 females (Gardner, et al., 1959)
   Variables: 20 of the original variables
   Method: Principal components condensation with oblique
      (Case II) and with procrustes rotation, by sex
   Results: Eight factors emerged in the first analysis
      and were different by sex; only Leveling vs. sharpen-
      ing and possibly Extensiveness of scanning and field
      articulation could be identified.   The procrustes
      solution also yielded very complex results.
1. Behavior Correlates
Adams, R. C.   Perceptual correlates of the Rod-and-frame
      Test: A critical response.   Perceptual and Motor
      Skills, 1974, 38, 1044-1046.
   Purpose: To reanalyze the Adevai, et al. (1968) field
      independence data
   Subjects: (Unspecified here)
   Variables: RFT_E, EFT, LO, DAP, TPD, TLI, TL, M-Ts (2
      Adevai et al. variables excluded)
   Method: Basic-structure rank-reduction condensation
      with varimax rotation
   Results: Three unlabeled factors emerged.
Barton, K., Cattell, R. B., & Conner, D. V.   The identifi-
      cation of "state" factors through P-technique factor
      analysis.   Journal of Clinical Psychology, 1972, 28,
      459-463.
   Purpose: To identify state factors of a psychological/
      physiological battery using P-technique
   Subjects: One 36-year-old female, tested daily for 99
      days
   Variables: Scores on a 68-test battery of psychological
      and physiological measures
   Method: Unspecified (P) condensation with oblimax rota-
      tion
   Results: Eleven factors emerged: I Diurnal fatigue; II
      General adaptation syndrome; III Basophil-neutrophil
      pattern; IV Heightened emotional liveliness and un-
      control; V Memory (instrument factor); VI Urine (instru-
      ment factor); VII Untapped reserves; VIII Desurgency-
      surgency (general level of cortical inhibition associated
      with past punishment, cultural complexity and demands
      of the self-sentiment); IX Some form of anxiety (manic
      activity) concomitant with very active motor behaviors;
      X Adrenergic-cyclothymic state; XI Parasympathic pattern.
Bond, A. D.   Factorial independence of perceptual egocen-
      trism.   Perceptual and Motor Skills, 1974, 38, 453-454.
   Purpose: To identify the relationship between perceptual
      egocentrism and field dependence
   Subjects: 34 male and 19 female kindergarteners
   Variables: Scores on Colored Progressive Matrices, Draw-
      a-person, Children's Embedded-figures, SES, age, and
      4 perceptual egocentrism measures

Method: Principal components condensation with varimax
   rotation
Results: Four factors emerged: I Decentration; II Induc-
   tive reasoning; III Spatial visualization; IV Uninterpretable.
Bone, R. N., & Eysenck, H. J. Extraversion, field-dependence
   and the Stroop test. Perceptual and Motor Skills, 1972,
   34, 873-874.
Purpose: To identify relationships among extraversion,
   field-dependence, and the Stroop test
Subjects: 97 male, 97 female college students
Variables: Scores on extraversion, neuroticism, and psychoticism; rod-and-frame test; 4 Stroop test scores
Method: Principal components condensation with promax
   rotation, by sex
Results: Three factors emerged for males: I Stroop
   (test-specific); II PEN (also test-specific); III
   Extravert speed. Four factors emerged for females:
   I Unlabeled; II PEN; III Stroop speed; IV Stroop
   interference.
Brown, S. R., & Hendrick, C. Introversion, extraversion and
   social perception. British Journal of Social and
   Clinical Psychology, 1971, 10, 313-319.
Purpose: To identify dimensions of introverts' and extro-
   verts' subjective experiences
Study One:
Subjects: 6 graduate students: 1 high introvert and 1
   high extrovert, each describing themselves; 2 middle
   introverts and 2 middle extroverts, each describing
   both the model introvert and the model extrovert
Variables: Q-sorts of 24 E-I and 24 neuroticism items
   from the Maudsley Personality Inventory (MPI)
Method: Unspecified factor analysis
Results: Two factors emerged: I Model introvert's self-
   description (only); II Extrovert descriptions
Study Two:
Subjects: The model introvert and model extrovert from
   above
Variables: Same as above, sorted for 1) actual and 2)
   potential social self, and 3) actual and 4) potential
   private self
Method: Unspecified factor analysis
Results: Two factors emerged: I Extrovert descriptions
   plus introvert's potential private and social selves;
   II Introvert actual private and social selves
Study Three:
Subjects: 10 high introverts and 10 high extroverts, all
   college seniors
Variables: Same as Study Two, with additional sort for
   "ideal leader"
Method: Principal axes condensation with varimax rotation
Results: Three factors emerged: I Prototype of typical
   extrovert; II Prototype of (non-neurotic) introvert;
   III Extraverted and non-neurotic.

Burdick, J. A., & Stewart, D. Y. Differences between "show" and "no show" volunteers in a homosexual population. Journal of Social Psychology, 1974, 92, 159-160.
Purpose: To identify dimensions underlying a set of variables collected for homosexual volunteers for an experiment
Subjects: 67 male members of a homosexual club
Variables: Scores on the EPI Extroversion, Lie, and Neuroticism scales, plus age, highest educational level, annual income, and Kinsey Scale of sexual activity score
Method: Principal components condensation
Results: Three factors were reported: I Socioeconomic scale; II N, E, and homosexual pole of Kinsey; III Social extroversion.
Douty, H. I., Moore, J. B., & Hartford, D. Body characteristics in relation to life adjustment, body-image and attitudes of college females. Perceptual and Motor Skills, 1974, 39, 499-521.
Purpose: To identify relationships between morphological, body-image characteristic, and adjustment tendency measures
Subjects: 91 female college students
Variables: 16 (apparently) morphological, 8 body-image characteristic, and 6 Bell Adjustment Inventory (BAI) scores
Method: Unspecified factor (or cluster) analysis
Results: Four clusters emerged: I Body mass-contour or body-impression; II Posture; III BAI; IV Figure impression.
Gibson, H. B. The two faces of extraversion: A study attempting validation. British Journal of Social and Clinical Psychology, 1974, 13, 91-92.
Purpose: To identify dimensions underlying extraversion, sociability, and test impulsiveness
Subjects: 24 first-year psychology students
Variables: Scores on measures of parametrically-derived sociability, test impulsivity, and EPI-derived E, N, L, E-impulsivity, and E-sociability scores
Method: Principal components condensation with orthogonal rotation
Results: Five factors emerged: I Social extraversion; II Neuroticism; III ?; IV Lie; V Cautiousness.
Hundal, P. S., & Singh, M. A factor analytical study of intellectual and non-intellectual characteristics. Multivariate Behavioral Research, 1971, 6, 503-514.
Purpose: To identify dimensions underlying a battery of personality, interest, intelligence, and academic achievement measures
Subjects: 271 Indian teacher-trainees
Variables: Scores on 18 personality, 6 interests, attitude toward profession, 2 intelligence, and 4 academic achievement measures

Method: Principal components condensation with maxplane
rotation
Results: Twelve first-order factors emerged: I Academic
achievement; II Dynamic integration vs. anxiety; III
Aesthetic interest; IV Subduedness vs. independence;
V Neuroticism; VI Facilitating test performance; VII
Extraversion vs. introversion; VIII Self control; IX
Theoretical interest; X Political interest; XI Social
interest; XII Not labeled.
Innes, J. M. The relationship of word-association commonality response set to cognitive and personality variables. British Journal of Psychology, 1972, 63, 421-
428.
Purpose: To identify dimensions common to word-association
commonality response set, and cognitive and personality
variables
Subjects: 53 male and 73 female college students
Variables: 5 scores from the Kent-Rosanoff Word Association Test, EPI E and N scores, and measures of ideational fluency, associational fluency, and verbal
ability
Method: Principal components condensation, apparently
without rotation, independently by sex
Results: Three factors emerged for males: I Word association (commonality); II Fluency (with extraversion);
III Intelligence (with originality). Three factors
emerged for females: I Word association (commonality);
II Fluency (with extraversion, intelligence) - neuroticism; III Extraversion - fluency.
Krebs, D., & Adinolfi, A. A. Physical attractiveness,
social relations, and personality style. Journal of
Personality and Social Psychology, 1975, 31, 245-253.
Purpose: To identify dimensions underlying phyiscal
attractiveness (PA), personality (P), interpersonal
orientation (IO), academic achievement (AA), and
social contact (SC)
Subjects: 60 male, 60 female college students
Variables: 31 measures of PA, P, IO, AA, and SC
Method: Principal components condensation with varimax
rotation, by sex
Results: Five factors emerged: I Affectionate sociability;
II Self-protective constraint; III Independent ambitious-
ness; IV Superordinate (male), or Defensive aggressiveness (female); V Ascetic associability (male), or Passive
withdrawal (female).
Looft, W. R., & Baranowski, M. D. An analysis of five mea-
sures of sensation seeking and preference for complexity.
Journal of General Psychology, 1971, 85, 307-313.
Purpose: To identify dimensions of sensation seeking and
preference for complexity
Subjects: 119 college students
Variables: Scores from 2 visual tasks and 3 paper-and-
pencil inventories related to need to maintain optimal
level of variability

Method: Principal components condensation with varimax
    rotation
Results: Two factors emerged: I Questionnaire method;
    II Visual tasks method.
Michael, J. J., Plass, A., & Lee, Y. B. A comparison of
    the self-report and the observed report in the mea-
    surement of the self-concept: Implications for con-
    struct validity. Educational and Psychological
    Measurement, 1972, 32, 481-483.
    Purpose: To identify the relationship between self-report
        (SR) and trained observers' recordings (OR) of self-
        concept
    Subjects: 30 sixth-graders, rated by self and 2 teachers
    Variables: 48 item scores (adapted from Coopersmith, 1967)
        each
    Method: Unspecified Q factor analysis, for OR and SR
    Results: An unspecified number of factors emerged; factor
        structure reportedly differed between SR and OR.
Nystedt, L. Predictive accuracy and utilization of cues:
    Study of the interaction between an individual's cog-
    nitive organization and ecological structure.
    Perceptual and Motor Skills, 1972, 34, 171-180.
    Purpose: To identify dimensions of individuals' cognitive
        organization and ecological structure in cue utilization
    Subjects: 19 advanced psychology students
    Variables: 1 & 3) Correlation ratings (±1.0) for a 6x6
        matrix (4 tests, 2 scholastic achievement variables);
        2) 9-point ratings of scholastic achievement for 50
        "students"
    Method: Principal components condensation with unspecified
        rotation, independently for each judge (1) and ecological
        matrix (2); 3) typal analysis for combined group
    Results: Two (unlabeled) factors emerged in each case
        (1 & 2). Four types of judges were identified.
Orvik, J. M. Social desirability for the individual, his
    group, and society. Multivariate Behavioral Research,
    1972, 7, 3-32.
    Purpose: To identify clusters of individuals on the basis
        of their social desirability (SD) endorsements of per-
        sonality items
    Subjects: 15 SDS members, 15 graduate education students,
        15 psychotherapy clients
    Variables: Ratings of 82 MMPI items for 1) individual SD,
        2) group SD, 3) general SD
    Method: Principal axes condensation, and key cluster
        analysis, for each type of rating
    Results: Three factors emerged for general ratings: I
        General; II-III Group specific. The other two analyses
        did not yield a general factor.
Platt, J. J., Eisenman, R., DeLisser, O., & Darbes, A.
    Temporal perspective as a personality dimension in
    college students: A re-evaluation. Perceptual and
    Motor Skills, 1971, 33, 103-109.

Purpose: To identify relationships between temporal perspective and personality variables
Subjects: 49-54 male and 81-121 female college students
Variables: Scores on 6 temporal perspective measures and
9 personality variables
Method: Principal components condensation with varimax
rotation, by sex
Results: Six factors emerged for females: I Authoritari-
anism and sense of personal incompetence; II Density;
III Creativity, directionality; IV Moral judgment vs.
response acquiescence; V Hostility vs. impersonal past
extension; VI Impersonal future extension, introversion.
Six factors emerged for males: I Acquiescence; II
Future density; III Creativity vs. moral judgment; IV
Temporal perspective; V Active, somewhat impulsive
orientation; VI Not labeled.
Schwirian, P. M., & Gerland, M. C. Personalities and atti-
tudes of nonmedical users of methaqualone. Archives
of General Psychiatry, 1974, 30, 525-530.
Purpose: To identify self-description personality dimen-
sions for nonmedical users of methaqualone
Subjects: 66 nonmedical users of methaqualone
Variables: Scores on 17 self-description personality
items
Method: Principal components condensation, apparently
without rotation
Results: Three factors emerged: I Introversion; II
Extroversion; III Depression.
Seymour, G. E., Gunderson, E. K. E., & Vallacher, R. R.
Clustering 34 occupational groups by personality
dimensions. Educational and Psychological Measurement,
1973, 33, 267-284.
Purpose: To identify clusters of occupational groups by
personality dimensions
Subjects: 34 occupational groups (30 of 34 with applicant
N in excess of 30)
Variables: Scores on 13 Opinion Survey, 7 Friend Descrip-
tion, and 6 FIRP-B scales
Method: Cluster analysis
Results: Five levels of clusters emerged, the top level
of which contained two clusters: I Military; II
Civilian.
Starker, S. Aspects of inner experience: Autokinesis,
daydreaming, dream recall, and cognitive style.
Perceptual and Motor Skills, 1973, 36, 663-673.
Purpose: To identify dimensions of inner experience
Subjects: 55 male college students
Variables: 2 autokinesis scores, 29 Imaginal Processes
Inventory scale scores, 5 dream diary scores, and
Embedded-figures Test score
Method: Unspecified condensation with analytic varimax
rotation
Results: Five factors emerged: I Conflictual day-
dreaming; II Positive daydreaming; III Attentional
processes; IV Dream recall; V Autokinesis.

Stefic, E. C., & Lorr, M. Analysis of defensiveness in relation to psychopathology. Journal of Consulting and Clinical Psychology, 1971, 36, 205-209.
Purpose: To identify 1) dimensions of defensiveness, and 2) relationships among measures of defensiveness, psychopathology, and extroversion-introversion
Subjects: 217 undergraduate and graduate students
Variables: Scores on 1) 110 defensiveness items, and 2) homogeneous clusters: 15 defensiveness, 16 behavior deviation, 5 extroversion, 5 impulsiveness
Method: Principal components condensation with varimax rotation, by data set
Results: Six interpretable defensiveness factors emerged: I Admission of common frailties; II Submissive over-control; III Advantage taking; IV Imperfection; V Social fear; VI Risk taking. Three interpretable factors emerged in the second analysis: I Broad psychopathology not involving admission of fear; II Social extroversion-introversion with associated anxiety and depression on the introversive end; III Hostility and Suspicion.
Taube, I., Vreeland, R. The prediction of ego functioning in college. Archives of General Psychiatry, 1972, 27, 224-229.
Purpose: To identify dimensions underlying advisor/teachers' year-end assessments of prep school students
Subjects: 271 prep school students subsequently admitted to Harvard, rated by teachers
Variables: Scores on unspecified number (55+) of content categories
Method: Principal components condensation, apparently without rotation
Results: Twelve factors emerged: I Grind; II Brilliant; III Likeable; IV Active; V Popular; VI Masculine; VII Leader; VIII Intellectual interests (verbal and scientific); IX Teacher expressed concern about student; X Imaginative, weird; XI Overconfident; XII Clear thinking.
Webb, S. C. Convergent-discriminant validity of a role oriented interest inventory. Educational and Psychological Measurement, 1973, 33, 441-451.
Purpose: To identify dimensions underlying the Inventory of Religious Activities and Interests (IRAI) and three criteria
Subjects: 1) 171, 2) 90 male theological seminary students
Variables: 10 IRAI scores plus estimates of interest in 10 curriculum areas, 10 field work activities, and 10 future work activities
Method: Multimethod factor condensation with varimax rotation, by sample
Results: Ten factors emerged in each analysis: I Counselor; II Administrator; III Teacher; IV Scholar; V Evangelist; VI Spiritual guide; VII Preacher; VIII Reformer; IX Priest; X Musician.

Wilkins, G., & Epting, F. Cognitive complexity and categor-
ization of stimulus objects being judged. Psychological
Reports, 1971, 29, 965-966.
   Purpose: To identify dimensions of cognitive complexity
   ratings for 10 stimulus objects
   Subjects: 82 unspecified subjects
   Variables: Cognitive complexity scores for 10 stimulus
   objects
   Method: Principal factors condensation with varimax
   rotation
   Results: More than two factors were required to account
   for the variance.
Williams, A. F., & Wechsler, H. Dimensions of preventive
behavior. Journal of Consulting and Clinical Psy-
chology, 1973, 40, 420-425.
   Purpose: To identify dimensions of preventive (related
   to health, accidents, property and financial loss)
   behaviors
   Subjects: 1) 161 suburban women contacted by telephone
   (age=35-54); 2) 193 males (fathers of ninth-graders)
   contacted by mail survey (age=35-67)
   Variables: Scores on 1) 22, 2) 9 preventive behaviors
   Method: Principal components condensation with varimax
   rotation, by S-variable set
   Results: Five factors in the first analysis were inter-
   pretable: I Checkup; II Exercise, limit calories,
   TB test; III Cautiousness-preparedness; IV Protection
   of property; V Risk taking. Three factors were
   interpreted in the second analysis: I Dietary be-
   havior related to heart disease; II Checkup; III
   Sleep and exercise.
B. Intelligence
Bergman, H., & Engelbrektson, K. An examination of factor
structure of Rod-and-frame Test and Embedded-Figures
Test. Perceptual and Motor Skills, 1973, 37, 939-947.
   Purpose: To identify dimensions common to Rod-and-frame
   Test (RFT), Embedded-figures Test (EFT), and reference
   tests for intellectual factors
   Subjects: 93 Swedish male college students
   Variables: Scores on EFT, RFT (3 scores), marker tests
   (2) for each of 3 focal factors, marker tests (3) for
   each of 2 cognitive factors
   Method: Restricted orthogonal maximum likelihood conden-
   sation with varimax rotation
   Results: Five factors emerged: I CMU; II CFU; III NFT;
   IV CFS; V RFT.
Bock, R. D. Word and image: Sources of the verbal and
spatial factors in mental test scores. Psychometrika,
1973, 38, 437-457.
   Purpose: To identify dimensions of genetic variation in
   Primary Mental Abilities (PMA) scores
   Subjects: 85 monozygotic, 85 dizygotic twins, all aged
   14-20 years

141

Variables: 5 PMA scores
Method: Canonical condensation with varimax rotation
Results: Three factors emerged for males: I Verbal-
    meaning; II Space; III Number. Three factors
    emerged for females: I Verbal; II Spatial; III Word-
    fluency.
Botwinick, J., & Storandt, M. Speed functions, vocabulary
    ability, and age. Perceptual and Motor Skills, 1973,
    36, 1123-1128.
Purpose: To identify the relationship between verbal
    ability and the ability to perform quickly
Subjects: 38 males and females of mean age 18.6 years,
    28 of mean age 71.0 years
Variables: Scores on WAIS Vocabulary and Digit Symbol,
    Crossing-off, Reaction Time and (1 only) age
Method: Principal components condensation with varimax
    rotation, with and without (partial corrs) age
Results: Two factors emerged in the first analysis: I
    Speed; II Cognitive. Two factors emerged in the
    second analysis: I Cognitive; II Speed.
DuJovne, B. E., & Levy, D. I. The psychometric structure
    of the Wechsler Memory Scale. Journal of Clinical
    Psychology, 1971, 27, 351-354.
Purpose: To identify dimensions underlying the Weschler
    Memory Scale (WMS)
Subjects: 1) 276 persons with no known psychiatric ill-
    ness (60% female; age 16-71); 2) 135 males and 23
    females (age 16-72) with organic, psychotic disorders,
    psychoneurotic disorders, personality and transient
    situational personality disorders as diagnoses
Variables: WAIS Full Scale, V, and PIQ scores, age,
    patient/normal dichotomy, and unspecified number (all
    but 11) of WMS item scores
Method: Principal axes condensation with varimax rota-
    tion, independently for normals and patients
Results: Three factors emerged for normals: I General
    retentiveness; II Simple learning; III Associative
    flexibility. Three factors emerged for patients: I
    Mental control; II Associative flexibility; III Cog-
    nitive dysfunction.
Fleishman, J. J., & Dusek, E. R. Reliability and learning
    factors associated with cognitive tests. Psychologi-
    cal Reports, 1971, 29, 523-530.
Purpose: To identify dimensions underlying 21 paper-and-
    pencil Kit of Reference Tests for Cognitive Factors
    (KIT)
Subjects: 90 Army enlisted men
Variables: Scores on 21 KIT tests
Method: Centroid condensation, apparently without rota-
    tion
Results: Five factors emerged: I Vocabulary ability; II
    Spatial orientation and spatial relations; III Ability
    to scan quickly and to make short quick motor movements;
    IV Arithmetic ability; V Abstract reasoning in conjunc-
    tion with good memory.

Fleishman, J. J., & Dusek, E. R. Equivalence of alternative
forms of six psychometric measures during repeated
testing. Journal of Applied Psychology, 1972, 56,
186-188.
Purpose: To identify relationships between subtests of
Repetitive Psychometric Measures (RPM)
Subjects: 80 Army enlisted men
Variables: Scores on each of 7 forms of each of 6 RPM
tests
Method: Centroid condensation with varimax rotation
Results: Nine factors emerged: I Flexibility of closure;
II Number facility; III Aiming; IV Aiming x speed of
closure; V Speed of closure; VI Perceptual speed; VII
Flexibility of closure x speed of closure; VIII Vari-
ability; IX Perceptual speed x speed of closure.
Conclusions: With the exception of Speed of Closure, all
of the RPM tests appear to measure what they are pur-
ported to.
Flores, M. B., & Evans, G. T. Some differences in cognitive
abilities between selected Canadian and Filipino stu-
dents. Multivariate Behavioral Research, 1972, 7, 175-
191.
Purpose: To identify patterns of cognitive abilities for
Canadian and Filipino students
Subjects: 117 Canadian, 94 Filipino sixth-graders, 102
Canadian, 109 Filipino eighth-graders
Variables: Scores on 18 cognitive abilities (Raven's
Progressive Matrices, SRA, PMA, ETS kit) tests
Method: Principal factors condensation with promax
rotation (and Schmid-Leiman transformation), indepen-
dently by group and for combined samples
Results: Eight factors emerged for total sample and sub-
groups: I Relational thinking; II Associative learn-
ing; III Verbal comprehension; IV Spatial facility; V
Numerical facility; VI Memory; VII Ordering; VIII
Residual.
Gough, H. G., & Olton, R. M. Field independence as related
to nonverbal measures of perceptual performance and
cognitive ability. Journal of Consulting and Clinical
Psychology, 1972, 38, 338-342.
Purpose: To identify dimensions underlying several mea-
sures of field independence
Subjects: 144 male, 165 female college students
Variables: Scores on D 48 test, Gottschaldt Hidden
Figures, Space Relations, Street Gestalt, and Per-
ceptual Acuity Test
Method: Cluster analysis with oblique rotation
Results: Two clusters emerged: I Perceptual-cognitive
ability; II Spatial relations or spatial perception.
Guilford, J. P. Executive functions and a model of behavior.
Journal of General Psychology, 1972, 86, 279-287.
Purpose: To identify the relationship between expressional
and divergent-production components of behavior

143

Subjects: 34 college students
Variables: Scores on 9 printed behavioral-divergent-
production and 4 expressive tests
Method: Principal axes condensation with varimax rotation
Results: Three factors emerged: I DBU; II Facial expres-
sions; III Verbal expressions.
Guilford, J. P., & Pandey, R. E. Abilities for divergent
production of symbolic and semantic systems. Journal
of General Psychology, 1974, 91, 209-220.
Purpose: To identify factors of a battery designed to
measure divergent production of symbolic systems (DSS)
and divergent production of semantic systems (DMS)
Subjects: 177 college males
Variables: Scores on four new and one marker test each
for DSS and DMS, and the Number Operations test
Method: Unspecified condensation with orthogonal varimax
rotation
Results: Three factors emerged: I CMS; II DMS; III DSS.
Hansen, J. B. Effects of feedback, learner control, and
cognitive abilities on state anxiety and performance
in a computer-assisted instruction task. Journal of
Educational Psychology, 1974, 66, 247-254.
Purpose: To identify relationships among four ability
measures
Subjects: 98 female college students
Variables: Total scores on the Ship Destinations Test,
Object-Number Test, First and Last Names Test, Bi-
column Number Series Test
Method: Unspecified condensation with varimax rotation
Results: Two factors emerged: I Reasoning; II Associative
memory.
Harris, M. L., & Harris, C. W. A factor analytic interpre-
tation strategy. Educational and Psychological Measurement, 1971, 31, 589-606.
Purpose: To reanalyze two of Guilford's matrices as
reported by C. W. Harris (1967)
Subjects & Variables: Data from nine Guilford studies,
matrices 08 (Creative Thinking) and 23 (Cognition
and Convergent Production)
Method: Incomplete principal components, alpha, unrestricted
maximum likelihood, and Harris R-S^2 factor analyses:
orthogonal, oblique solutions
Results: For Matrix 23, there was considerable agreement
among the derived solutions but the results did not
reproduce Guilford's. For Matrix 08, there was only
limited consistency among solutions.
Hyde, J. S., Geiringer, E. R., & Yen, W. M. On the empirical
relation between spatial ability and sex differences in
other aspects of cognitive performance. Multivariate
Behavioral Research, 1975, 10, 289-309.
Purpose: To identify relationships among measures of
spatial ability, field independence, and mental arithmetic

144

Subjects: 45 female and 35 male college students
Variables: Scores on Identical Blocks, Rod-and-Frame,
   Group Embedded Figures, Mental Arithmetic Problems,
   Vocabulary, Word Fluency, Alternate Uses, Femininity,
   and Achievement Motivation tests
Method: Principal components condensation with varimax
   rotation (for total and by sex)
Results: Three factors emerged for the total group: I
   Spatial; II Creativity/sex-typing; III Achievement
   motivation. Factor structures differed between sexes,
   but the three factors extracted in the by-sex analyses
   were not named.
Kayser, B. D. Authoritarianism, self-esteem, emotionality
   and intelligence. Perceptual and Motor Skills, 1972,
   34, 367-370.
Purpose: To identify relationships among authoritarianism,
   self-esteem, emotionality, and intelligence
Subjects: 91 college students
Variables: Scores on GPA, IQ, Sex, F Scale, and evaluation,
   activity, potency, and total scores each for Actual
   Self, Ideal Self, and My Intelligence
Method: Principal axes condensation with varimax rotation
Results: Six factors emerged: I High emotionality toward
   intelligence associated with positive self-worth and
   high E-P-A intelligence; II Dynamic self-concept; III
   Ideal activity; IV Emotionality toward actual/ideal
   selves associated with self-ratings of good-wise-
   successful; V Authoritarianism, GPA, IQ, actual-self
   potency; VI Self-esteem and actual-self evaluation.
Klingler, D. E., & Saunders, D. R. A factor analysis of the
   items for nine subtests of the WAIS. Multivariate
   Behavioral Research, 1975, 10, 131-154.
Purpose: To identify dimensions underlying WAIS items
Subjects: 916 college students
Variables: Scores on 82 items from (9 subtests of) the
   WAIS
Method: Unspecified condensation with normal equamax (and
   isopromax) rotation
Results: Fifteen factors emerged: I Cultural information;
   II Contemporary affairs; III Scientific information;
   IV Basic comprehension; V Numerical; VI Precision of
   judgment; VII Internalization; VIII Activity level; IX
   Maintenance of contact; X Maintenance of perspective;
   XI Perceptual organization; XII Planning ability and
   anticipation; XIII Social awareness; XIV (Spontaneous
   empathy); XV (Psychological mindedness).
Langevin, R. Is curiosity a unitary construct? Canadian
   Journal of Psychology, 1971, 25, 360-374.
Purpose: To identify dimensions of curiosity
Subjects: 195 sixth-graders
Variables: 7 scores from 5 curiosity measures, Otis IQ,
   2 Ravens tests scores, sex, age
Method: Unrestricted maximum likelihood condensation

Results: Four factors emerged: I Intelligence; II Breadth
of interest, curiosity; III Raven; IV Depth of interest
curiosity.
Merrill, P. F. Effects of the availability of objectives
and/or rules on the learning process. Journal of
Educational Psychology, 1974, 66, 534-539.
Purpose: To identify dimensions underlying a battery of
cognitive ability tests
Subjects: 130 college students
Variables: Scores on 6 KIT cognitive ability tests
Method: Unspecified condensation with varimax rotation
Results: Two factors emerged: I Reasoning; II Associa-
tive memory.
Messick. S., & French, J. W. Dimensions of cognitive clo-
sure. Multivariate Behavioral Research, 1975, 10, 3-16.
Purpose: To identify dimensions of cognitive closure
Subjects: 541 Naval Aviation Cadets
Variables: Scores on 34 cognitive closure measures, plus
1 dummy variable
Method: Principal axes condensation with equamax (and
promax) rotation, with Schmid-Leiman transformation
Results: Fourteen first-order factors emerged: I Ideational fluency; II Flexibility of perceptual closure/
(spatial flexibility); III Flexibility of verbal closure; IV Flexibility of semantic closure; V Speed of
verbal closure; VI Quantitative reasoning; VII Space
(specific); VIII Perceptual speed (of symbol discrimination); IX Subsample differences; X Ambiguities
doublet; XI Flexibility of grammatical closure; XII
Spontaneous flexibility; XIII Speed of perceptual
closure; XIV Speed of semantic closure. Four second-
order factors emerged: I Analytic functioning/
general reasoning; II Symbolic closure; III Semantic
closure/(verbal ability); IV Figural closure.
Mos, L., Wardell, D., & Royce, J. R. A factor analysis of
some measures of cognitive style. Multivariate
Behavioral Research, 1974, 9, 47-57.
Purpose: To identify dimensions underlying a battery of
psychological differentiation and cognitive abilities
measures
Subjects: 40 male, 40 female college students
Variables: 26 scores from 11 tests of psychological dif-
ferentiation and cognitive abilities, plus sex
Method: Alpha factor condensation with (varimax and)
promax rotation
Results: Eight factors emerged: I Element articulation;
II Form articulation; III Flexibility of closure; IV
Epistemic style; V Speed of closure; VI Sex; VII Per-
ceptual speed; VIII Conceptual differentiation in
categorization.
Ohnmacht, F. W., & Fleming, J. T. Perceptual closure and
cloze performance: A replication with older subjects.
Journal of General Psychology, 1972, 87, 225-229.

Purpose: To identify factors of perceptual closure and
cloze performance measures for college-age subjects
Subjects: 92 college students
Variables: Scores on 2 speed of closure, 2 flexibility
of closure, 2 associational fluency, 2 verbal compre-
hension, and 4 cloze tests
Method: Principal components condensation with varimax
rotation
Results: Three factors emerged: I General cloze; II
Perceptual; III Verbal ability.
Conclusions: The present results were contrasted with
those for a younger sample.

Osborne, R. T., & Suddick, D. E. Blood type gene frequency
and mental ability. Psychological Reports, 1971, 29,
1243-1249.
Purpose: To identify dimensions underlying a battery of
verbal perceptual, and spatial mental tests
Subjects: 48 sets of twins, aged 13-18 years
Variables: Scores on 8 KIT and 11 other standardized
tests of verbal, perceptual, and spatial abilities
Method: Principal components condensation with varimax
rotation
Results: Three factors emerged: I Verbal comprehension;
II Perceptual speed; III Spatial relations.

Thumin, F. J. Factor analysis and reliability of the
Mental Dexterity Test. Perceptual and Motor Skills,
1974, 38, 744-746.
Purpose: To identify factors underlying the Mental
Dexterity Test (MDT)
Subjects: 97 advanced college students
Variables: 100 MDT item scores
Method: Principal factors condensation with varimax
rotation
Results: Thirty-four factors emerged, but only the
first two were interpreted: I Complex verbal ability;
II Numerical ability
Conclusions: The MDT is an omnibus test; subtests of 30
and 20 items were selected to represent Factors I and
II, respectively.

Vernon, P. E. The distinctiveness of field independence.
Journal of Personality, 1972, 40, 366-391.
Purpose: To identify dimensions underlying creativity or
divergent thinking, field independence (FI), abilities
and achievement
Subjects: 198 male, 189 female Canadian junior high
school students
Variables: Scores on 1) 8 spatial and FI tests, 2) 22
abilities and achievement tests
Method: Principal components condensation, apparently
without rotation, for each set of measures
Results: Two factors emerged in the first analysis: I
g; II S. Three factors emerged in the second analysis:
I Verbal intelligence, vocabulary, school achievement;
II Spatial-perceptual; III g.

Vidler, D. Convergent and divergent thinking, test-anxiety, and curiosity. _Journal_ _of_ _Experimental_ _Education_, 1974, _43_, 79-85.
Purpose: To identify the relationship between two motivational (test anxiety and curiosity) and two cognitive (convergent, divergent thinking) variables
Subjects: 212 college students
Variables: Scores on 6 convergent and 6 divergent thinking tests
Method: Principal components condensation with varimax rotation
Results: Two factors emerged: I Convergent thinking (all but one test); II Divergent thinking (plus one convergent test).
Vidler, D. C., & Rawan, H. R. Further validation of a scale of academic curiosity. _Psychological_ _Reports_, 1975, _37_, 115-118.
Purpose: To identify dimensions underlying an academic curiosity scale
Subjects: 611 college students
Variables: 80 self-report item scores from an academic curiosity scale
Method: Principal components condensation with varimax rotation
Results: Five unlabeled factors emerged.
Conclusions: This scale apparently does not measure a unitary construct.
C. Creativity
Bledsoe, J. C., & Khatena, J. Factor analytic study of _Something About Myself_. _Psychological_ _Reports_, 1973, _32_, 1176-1178.
Purpose: To identify dimensions of self-reported creativity, as assessed by Something About Myself (SAM)
Subjects: 672 male and female high school and college students
Variables: 50 SAM item scores
Method: Principal components condensation with varimax rotation
Results: Six interpretable factors emerged: I Environmental sensitivity; II Initiative; III Self-strength; IV Intellectuality; V Individuality; VI Artistry.
Bledsoe, J. C., & Khatena, J. Factor analytic study of the test, What Kind of Person Are You? _Perceptual_ _and_ _Motor_ _Skills_, 1974, _39_, 143-146.
Purpose: To identify dimensions underlying self-perceptions of creativity, as measured by What Kind of Person Are You? (WKPAY)
Subjects: 645 high school and college students (unspecified numbers by sex)
Variables: 50 WKPAY item scores
Method: Unspecified condensation with varimax followed by maxplane rotation, by sex

Results: Five factors were interpreted: I Acceptance of
     authority; II Self-confidence; III Inquisitiveness; IV
     Awareness of others; V Disciplined imagination.
Jacobs, S. S., & Shin, S. H. Interrelationships among intel-
     ligence, product dimension of Guilford's model and
     multi-level measure of cognitive functioning. Psycho-
     logical Reports, 1975, 37, 903-910.
   Purpose: To identify relationships among a conventional
     intelligence measure, 6 measures of verbal creative
     behavior (product dimension), and 6 measures of cogni-
     tive functioning
   Subjects: 125 eleventh-graders
   Variables: Scores on 6 measures of verbal creative behavior,
     6 cognitive functioning measures, and intelligence
   Method: Unspecified condensation with varimax rotation
   Results: Four factors emerged: I General intelligence;
     II Simple creativity, or verbal fluency; III Complex
     creativity requiring creative operations on information;
     IV High level cognitive functioning requiring creation
     of a response.
Rossman, B. B., & Horn, J. L. Cognitive, motivational and
     temperamental indicants of creativity and intelligence.
     Journal of Educational Measurement, 1972, 9, 265-286.
   Purpose: To identify dimensions common to selected mea-
     sures of creativity and of intelligence
   Subjects: 94 engineering and 94 art students
   Variables: 33 scores from cognitive, motivational, and
     temperamental indices of creativity and intelligence
   Method: Principal axes condensation with (varimax and)
     promax rotation
   Results: Six promax factors emerged: I Fluid intelli-
     gence; II Crystallized intelligence; III Memory; IV
     Fluency; V Rule-orientation versus intuitive thinking;
     VI Self-sufficient-calculated-risk-taking.
   Conclusions: It may be useful to think of creativity
     and intelligence as outgrowths of distinct (but
     overlapping) sets of influences.
D. Personality Measurement
   Cottle, T. J. Temporal correlates of dogmatism. Journal
     of Consulting and Clinical Psychology, 1971, 36, 70-81.
   Purpose: To identify dimensions underlying 1) an abbre-
     viated Dogmatism scale (and to determine if so-called
     temporal items were differentiated from non-temporal
     ones), 2) the Time Attitude Inventory (TAI), 3) semantic
     differential (SD) ratings of past, future, present, and
     time evaluation
   Subjects: 335 men and 101 women from a naval corpsman
     training station
   Variables: Scores on 1) 22 abbreviated Dogmatism scale
     items, 2) 39 TAI items, 3) unspecified number of SD
     ratings for past, present, future, and time evaluation
     concepts

Method: 1) Unspecified condensation with oblique rotation,
2) principal axes condensation with varimax rotation,
3) unspecified condensation with orthogonal rotation
Results: Four dogmatism factors emerged: I Dogmatism; II
Self protection, or low self-esteem; III Rugged indivi-
dualism; IV Not labeled. Six TAI factors emerged: I
Temporal anxiety; II Egocentric present orientation;
III Fantasy intolerance; IV-VI Not labeled. Three fac-
tors emerged for SD ratings: I Evaluation; II Potency;
III Activity.

Lieberman, L. R., & Walters, S. M. The one-year-to-live
situation: Essay vs. inventory. Journal of Clinical
Psychology, 1972, 28, 205-209.
Purpose: To identify relationships between inventory and
essay responses to the one-year-to-live problem
Subjects: 46 male and 48 female college students
Variables: Scores on 85 content categories from a short
essay and an inventory concerning the one-year-to-live
problem
Method: Cluster analysis, independently for males and
females
Results: Four clusters appeared for males: I Occasion
to cut loose, give vent to impulses, and ignore the
consequences; II Uninterpretable; III Occasion to
balance things up, get right with God and other persons, and make amends; IV Uninterpretable. Four
clusters emerged for females: I Tendency to exaggerate
typical female responses; II Not reported; III Few
items; IV Many (or more) items.

Listiak, R. L., Stone, L. A., & Coles, G. J. Clinicians'
multidimensional perceptions of MMPI scales. Journal
of Clinical Psychology, 1973, 29, 29-32.
Purpose: To identify dimensions of psychologists' perceptions of MMPI scales
Subjects: 12 Ph.D. clinical psychologists employed in
clinical settings, average experience with MMPI 7.5
years
Variables: "Percent similarity" judgments for (pair-
comparisons) 12 MMPI scales
Method: Principal components condensation with unspecified
rotation; inverse principal components condensation with
unspecified rotation
Results: Four factors emerged in the first analysis: I
Activity directedness, or outer vs. inner; II Somatization, or physical vs. cognitive expression of
symptoms; III Femininity, or role identity; IV (Resembles
Welsh's (1965) A factor). Only one, potent, judge-
observer factor emerged, indicating a single judgmental
approach was used by all judges.

Messick, S., & Jackson, D. N. Judgmental dimensions of psychopathology. Journal of Consulting and Clinical
Psychology, 1972, 38, 418-427.

Purpose: To identify dimensions of psychopathology, as
  judged by college students in terms of desirability
  of MMPI items
Subjects: 150 college students
Variables: Judgments of desirability for 566 MMPI items
Method: Principal components condensation with orthogonal
  equamax rotation
Results: Twelve factors were interpreted: I Denial of
  lack of somatic control; II Impulsivity vs. religious
  preoccupation; III Femininity; IV Tolerance of deviance;
  V Oversensitivity and fearfulness; VI Not named; VII
  Socially deviant attitudes; VIII Impulse acceptance
  vs. grandiosity; IX Listless distractibility; X Worry;
  XI Not named; XII Timid cautiousness vs. masculine
  adventuresomeness.
Morf, M. E., & Jackson, D. N. An analysis of two response
  styles: True responding and item endorsement.
  Educational and Psychological Measurement, 1972, 32,
  329-353.
Purpose: To identify dimensions underlying acquiescent
  responding
Subjects: 87 male, 109 female college students
Variables: Scores on a total of 51 variables (combination
  of self-descriptive and attitude measures, worded
  positively or negatively, keyed T/F, for exhibition,
  play, succorance, understanding
Method: Principal axes condensation with (oblique and)
  orthogonal clustran rotation
Results: Eight factors emerged: I True responding; II
  Item endorsement; III Desirability; IV Adjective en-
  dorsement; V Exhibition; VI Play; VII Succorance; VIII
  Understanding.
1. Inventories
Abbott, R. D. A factor analysis of the CPI and EPI.
  Educational and Psychological Measurement, 1971, 31,
  549-553.
Purpose: To identify relationships among CPI, EPI, and
  other selected scales
Subjects: 171 female, 115 male (college) students
Variables: Scores on 53 EPI, 18 CPI, 3 social desirability,
  and Welsh's R scale
Method: Principal components condensation with varimax
  rotation
Results: Of fourteen factors extracted, six factors (I,
  II, III, VII, VIII, X) were mixed and eight were es-
  sentially EPI factors (IV, V, VI, IX, XI, XII, XIII).
Abbott, R. D. On confounding of the Repression-sensitization
  and Manifest Anxiety Scales. Psychological Reports,
  1972, 30, 392-394.
Purpose: To identify relationships among MMPI Manifest
  Anxiety (MA) Repression-sensitization (RS), and Social

151

Desirability (SD) and EPPS non-overlapping, T-F
balanced, non-pathological content SD (BSD)
Subjects: 218 college students
Variables: Scores on MA, RS, SD from MMPI, and BSD from
EPPS
Method: Principal components condensation, apparently
without rotation
Results: The results supported an SD interpretation of
scores on the four scales and did not support the
independence of MA and RS.
Abbott, R. D. Improving the validity of affective self-
report measures through constructing personality
scales unconfounded with social desirability: A
study of the Personality Research Form. Educational
and Psychological Measurement, 1975, 35, 371-377.
Purpose: To identify relationships among Personality
Research Form (PRF) scales and Edwards' SD scale
Subjects: 109 male and 109 female college students (?)
Variables: 22 PRF and Edwards' SD scale scores
Method: Principal components condensation with varimax
rotation
Results: Six factors emerged but were unlabeled.
Conclusions: Jackson's attempt to minimize the importance
of SD in the PRF has succeeded.
Abrahamson, D., Schludermann, S., & Schludermann, E. Repli-
cation of dimensions of locus of control. Journal of
Consulting and Clinical Psychology, 1973, 41, 320.
Purpose: To identify dimensions underlying Rotter's
Internal-External Locus of Control (I-E) scale with
a new sample
Subjects: 120 male, 113 female college students
Variables: 23 I-E item scores
Method: Principal components condensation with varimax
rotation, by sex
Results: Three factors emerged in both cases: I Luck
vs. personal control; II Socio-political control; III
Control over personal likability.
Allen, J. G., & Hamsher, J. H. The development and valida-
tion of a test of Emotional Styles. Journal of Consul-
ting and Clinical Psychology, 1974, 42, 663-668.
Purpose: To identify dimensions underlying the Test of
Emotional Styles (TES)
Subjects: 177 male college students
Variables: Scale scores for orientation, expressiveness,
and responsiveness for incomplete sentences, forced
choice, and true-false forms of the TES
Method: Multimethod factor analysis (principal components
condensation with varimax rotation)

152

Results: Three factors emerged: I Responsiveness; II
Expressiveness; III Orientation.
Bagley, C., & Evan-Wong, L. Neuroticism and extraversion
in responses to Coopersmith's Self-esteem Inventory.
Psychological Reports, 1975, 36, 253-254.
Purpose: To identify dimensions underlying Coopersmith's
Self-esteem Inventory (SI) and Eysenck's Personality
Inventory (EPI)
Subjects: 143 male, 131 female 14- to 15-year-olds
(British)
Variables: Scores on unspecified numbers of 1) EPI and
2) SI scales
Method: (Apparently) principal components condensation
with promax rotation, by instrument within sex
Results: Two EPI factors emerged in both cases: I
Extraversion; II Neuroticism. Two SI factors emerged
in both cases: I Self-disparagement, unhappiness at
home, general unhappiness items; II Social confidence,
extraversion items.
Barton, K., & Cattell, R. B. Marriage dimensions and per-
sonality. Journal of Personality and Social Psychology,
1972, 21, 369-375.
Purpose: To identify dimensions underlying an expanded
Marital Role Questionnaire (MRQ); Barton, Cattell,
and Kawash, unpublished 1971 manuscript
Subjects: Unspecified here
Variables: Unspecified number of original MRQ item
scores, plus unspecified number of "new" item scores
Method: Unspecified condensation with promax rotation
Results: Twelve factors emerged: I Sexual gratification;
II Togetherness and role sharing; III Home devotion;
IV Participating in community affairs; V Social-
intellectual equality; VI Marriage instability; VII
Social integration; VIII Work performance; IX Social
influence; X Spouse independence; XI Not reported;
XII Male dominance.
Berzins, J. I., Barnes, D. F., Cohen, D. I., & Ross, W. F.
Reappraisal of the A-B therapist "type" distinction
in terms of the Personality Research Form. Journal
of Consulting and Clinical Psychology, 1971, 36, 360-
369.
Purpose: To identify relationships between an A-B Scale
and selected personality scales
Subjects: 223 male college students
Variables: Scores on A-B Scale and 1) 22 Personality
Research Form (PRF) scales, 2) 21 PRF scales (Infre-
quency excluded)
Method: Principal axes condensation with varimax rotation,
by data set
Results: Five factors emerged in both analyses: I
Impulsivity; II Autonomy; III Cognitive ascendancy;
IV Defensiveness; V B status.

Blackburn, R. MMPI dimensions of sociability and impulse control. Journal of Consulting and Clinical Psychology, 1971, 37, 166.
    Purpose: To identify dimensions underlying MMPI items used in 4 extraversion and 1 neuroticism scale
    Subjects: 175 male abnormal offenders
    Variables: Scores on 116 MMPI items (Giedt-Downing Ex, Welsh Repression, Gough Impulsivity, Gough Social participation, Welsh Anxiety)
    Method: Principal components condensation with varimax rotation
    Results: Three factors emerged: I Anxiety or neuroticism vs. social poise and freedom from shyness; II Impulsivity; II General extraversion.
Bochner, A. P., & Kaminski, E. P. Modes of interpersonal behavior: A replication. Psychological Reports, 1974, 35, 1079-1083.
    Purpose: To identify dimensions underlying the Interpersonal Behavior Inventory (IBI)
    Subjects: 287 college students, rating well-liked others and selves
    Variables: 15 IBI scale scores
    Method: Principal axes condensation with varimax rotation, independently for others and selves
    Results: Three factors emerged in both cases: I Hostility-affection; II Dominance; III Submissiveness.
Bull, P. E. Structure of occupational interests in New Zealand and America on Holland's typology. Journal of Counseling Psychology, 1975, 22, 554-556.
    Purpose: To identify dimensions underlying selected (according to Holland's typology) SVIB items for New Zealand and American samples
    Subjects: 1) 150 American men and women (data reported by Campbell & Holland, 1972), 2) 147 New Zealand college students
    Variables: 6 SVIB scores (20 items each selected for Holland's types)
    Method: Principal components condensation with varimax rotation, by sample
    Results: Three factors, similar across samples not unlabeled, emerged.
Cattell, R. B. Radial parcel factoring-vs-item factoring in defining personality structure in questionnaires: Theory and experimental checks. Australian Journal of Psychology, 1974, 26, 103-119.
    Purpose: To identify dimensions of 16PF item parcels
    Subjects: 1) 780 adults, 2) 240 college students
    Variables: 46 4-item parcels (factor-analytically derived from 184 16PF items)
    Method: Centroid (?) condensation with analytic and blind rotoplot rotation
    Results: Twenty-one and 22 factors emerged for the adult and undergrad analyses, respectively; these results were

154

compared between samples and with item factors for each
sample. Five matched (all four analyses) factors were
named: I Intelligence; II Super ego; III Premsia; IV
Radicalism; V Ergic tension.
Cattell, R. B., & Bartlett, H. W. An R-dR-technique operational distinction of the states of anxiety, stress,
fear, etc. Australian Journal of Psychology, 1971, 23,
105-123.
Purpose: To identify dimensions (state and trait) of
anxiety, stress, and fear
Subjects: 63 college students
Variables: 16 scale scores from the 16PF (anxiety markers,
hyperplane variables), 7 objective anxiety test scores
(markers from IPAT Seven State Battery), 15 stress,
rigidity, and fear measures; test, with 1-month retest
Method: Principal axes condensation with blind maxplane
and rotoplot rotation, independently for R- and dR-
matrices
Results: Fifteen R factors, and 13 dR factors emerged.
Six factor patterns were found to replicate well across
analyses: I Exvia-invia; II Anxiety; III Independence;
IV Stress; V Fear and mobilization-vs-overwroughtness;
VI Fatigue or torpor.
Cattell, R. B., & Delhees, K. H. Seven missing normal per-
sonality factors in the questionnaire primaries.
Multivariate Behavioral Research, 1973, 8, 173-194.
Purpose: To identify dimensions underlying personality
as assessed by questionnaire
Subjects: 225 college students
Variables: Scores on 2 scales each for 16PF factors, 6
depression factors, and 7 "missing" normal personality
factors
Method: Principal components condensation with procrustes
followed by visual rotoplot rotation
Results: The 16PF primaries emerged, as did six (and
perhaps seven) of the "missing" primaries and a general
(collapsed) depression factor.
Cattell, R. B., Delhees, K. H., Tatro, D. F., & Nesselroade,
J. R. Personality structure checked in primary objective test factors, for a mixed normal and psychotic
sample. Multivariate Behavioral Research, 1971, 6,
187-214.
Purpose: To identify dimensions of personality for a
mixed normal and psychotic sample
Subjects: 114 adult normals, 228 psychotics
Variables: Scores on 100 variables (from 65 subtest types)
to make 18 known factors
Method: Principal components condensation with procrustes
followed by visual rotoplot rotation
Results: The 24 extracted factors were examined for congruence with those known in the homogeneous population;
17 were significant matches, 3 poor matches, 2 instrument factors, and 2 new factors.

Cattell, R. B., & Nichols, K. E. An improved definition, from 10 researchers, of second order personality factors in Q data (with cross-cultural checks). _Journal of Social Psychology_, 1972, _86_, 187-203.

Purpose: To identify second-order factors of the 16PF for a variety of samples

Subjects: Ten samples: 1000 boys and 880 girls from USA; 1800 high education and 1100 low education German students; 300 Venezuelans; 1097 17-19, 597 21-23, 1097 17-19-year-old New Zealanders; 770 male and 2234 mixed Brazilian students

Variables: 16 16PF scale scores

Method: Principal components condensation with blind oblique rotoplot rotation, independently for each sample

Results: Nine factors were generally (10 for one, 8 for one) extracted for all analyses: I Invia vs. exvia; II Anxiety; III Cortertia; IV Subduedness vs. independence; V Naturalness vs. discreteness; VI Unstable (formerly Cool realism vs. prodigal subjectivity); VII Intelligence; VIII Weak vs. strong moral upbringing; IX Common error factor.

Cattell, R. B., Schmidt, L. R., & Bjerstedt, A. Clinical diagnosis by the objective-analytic personality batteries. _Journal of Clinical Psychology_, 1972, _28_, 239-312.

Purpose: To identify (pathological) dimensions of personality using a mixed sample

Study One:

Subjects: 34 depressive (3 types) and 32 clinical comparison (3 groups) inpatients, and 48 non-psychiatric controls (N=114)

Variables: Scores on 35 objective tests, plus age, sex, hospitalization

Method: Principal components condensation with procrustes and blind rotoplot rotation

Results: Seventeen first-order factors emerged: I Assertive ego vs. unassertiveness; II Inhibition vs. confident lack of timidity; III Not identified; IV Independence vs. subduedness (tentative); V Comention vs. objectivity; VI Exuberance vs. suppressibility; VII Cortertia vs. pathemia (tentative); VIII Mobilization of energy vs. regression; IX Anxiety vs. adjustment; X Realism vs. tensinflexia (tentative); XI Not identified; XII Asthenia vs. self-assuredness; XIII Wholehearted responsiveness vs. lack of will (tentative); XIV Stolidness vs. dissofrustrance (tentative); XV Wariness vs. impulsive variability (tentative); XVI Exvia vs. invia; XVII Dismay vs. sanguine poise.

Study Two:

Subjects: Same as Study One

Variables: One-month retest scores on same variables as Study One

156

Method: Same as Study One
Results: Seventeen factors emerged: I, III, VIII, XI,
  XIV Not identified; II Inhibition vs. confident lack
  of timidity (tentative, T); IV Independence vs. sub-
  duedness (T); V Comention vs. objectivity; VI Exuberance
  vs. suppressibility; VII Cortertia vs. pathemia (T);
  IX Anxiety vs. adjustment (T); X Realism vs. tensin-
  flexia (T); XII Asthenia vs. self-assuredness (T); XIII
  Wholehearted responsiveness vs. lack of will; XV Wariness
  vs. impulsive variability (T); XVI Exvia vs. invia (T);
  XVII Dismay vs. sanguine poise.
Study Three:
Subjects: Same as above
Variables: Difference scores between Study One and Two for
  the 35 tests
Method: Same as above
Results: Sixteen factors emerged: I-V, VII, XIII, XVI
  Not identified; VI State of exuberance vs. suppressi-
  bility; VIII Good mobilization; IX Anxiety; X State of
  realism vs. tensinflexia (T); XI Possibly state of
  dismay vs. sanguine poise; XII State of wholehearted
  responsiveness vs. lack of will (T); XIV State of
  wariness vs. impulse variability (T); XV State of exvia
  vs. invia (T); XVII State of dismay vs. sanguine poise.
Study Four:
Subjects: 1) Same as above, plus 2) unspecified (here)
  subjects of Cattell, et al. (1971)
Variables: Same as Study One
Method: Same as above (for second-order solution), independently by subject group
Results: Seven factors, of the 12 second-orders extracted,
  from the previous Cattell sample were considered matches
  for the seven second-orders emerging for Study One
  subjects; however, only one was named: I General
  slowness, or low energy. Further analyses, including
  discriminant analyses, are reported.
Chansky, N. M., Covert, R., & Westler, L. Factor structure
  of the Runner Studies of Attitude Patterns. Educational
  and Psychological Measurement, 1972, 32, 1119-1123.
  Purpose: To identify dimensions underlying the Runner
  Studies of Attitude Patterns (RSAP)
  Subjects: 435 female and 358 male college freshmen
  Variables: Scores on 14 subtests, total, and social
  desirability scale from RSAP
  Method: Minimum residual condensation, apparently without
  rotation
  Results: Four factors emerged: I Control orientation; II
  Recognition orientation; III New adolescent (inner and
  outer directed values); IV Social desirability.
Chun, K., & Campbell, J. B. Dimensionality of the Rotter
  Interpersonal Trust Scale. Psychological Reports, 1974,
  35, 1059-1070.

Purpose: To identify dimensions underlying the Rotter
    Interpersonal Trust Scale (ITS)
Subjects: 187 (paid) college students
Variables: 25 ITS trust item scores
Method: Principal components condensation; cluster analysis;
    principal axes condensation with varimax and with bi-
    quartimin rotation
Results: Ten factors were identified in the factor analyses,
    of which four corresponded to four clusters extracted:
    I Political cynicism; II Interpersonal exploitation;
    III Societal hypocrisy; IV Reliable role performance.
Chun, K., & Campbell, J. B. Notes on the internal structure
    of Wrightsman's measure of trustworthiness. *Psycholo-
    gical Reports*, 1974, 37, 323-330.
Purpose: To identify dimensions underlying the Trust-
    worthiness subscale (TS) of Wrightsman's Philosophies
    of Human Nature Scale
Subjects: 116 female and 71 male college students
Variables: 14 TS item scores
Method: 1) Hierarchical cluster analysis, by sex and for
    total, 2) principal axes condensation with varimax
    rotation
Results: In all cases, one ten-item cluster (composed of
    two subclusters) and four residual items emerged. Two
    factors were interpreted: I Global morality; II
    Specific acts of honesty.
Collins, B. E. Four components of the Rotter Internal-
    External Scale: Belief in a difficult world, a just
    world, a predictable world, and a politically respon-
    sive world. *Journal of Personality and Social
    Psychology*, 1974, 29, 381-391.
Purpose: To identify dimensions underlying the Rotter
    Internal-External Scale (IE)
Subjects: 300 college students
Variables: Scores on 46 Likert-type items adapted from
    the IE scale
Method: Principal components condensation with varimax
    rotation
Results: Four factors emerged: I The difficult-easy world;
    II The just-unjust world; III The predictable-unpredic-
    table world; IV The politically responsive-unresponsive
    world.
Collins, B. E., Martin, J. C., Ashmore, R. D., & Ross, L.
    Some dimensions of the internal-external metaphor in
    theories of personality. *Journal of Personality*, 1973,
    41, 471-492.
Purpose: To identify dimensions underlying the Personal
    Behavior Inventory (PBI)
Subjects: 163 college students
Variables: 63 PBI item scores
Method: Principal components condensation with varimax
    rotation

Results: Four factors emerged: I Other direction - con-
    formity to social expectations and low self-esteem; II
    Inner direction - commitment to traditional, socially
    desirable principles and goal setting; III Lack of
    constraints - the creative role player and self-actualiz-
    ing "free spirit"; IV Trans-situational predictability
    of behaviors.
Dempewolff, J. A. Development and validation of a feminism
    scale. Psychological Reports, 1974, 34, 651-657.
    Purpose: To identify dimensions underlying the Feminism
    I (F-1) scale
    Subjects: 106 male and 119 female college students
    Variables: 80 F-1 item scores
    Method: Principal components condensation without rota-
    tion, by sex; principal components condensation with
    oblique rotation, for 10 and 6 factors
    Results: An unspecified number of factors were extracted
    in the initial analyses; factor structure was highly
    similar between sexes. Factors for the 10- and 6-
    factor solutions were not named; since the first com-
    ponent was large by comparison to subsequent ones in
    both analyses, a shortened scale (F-2) was constructed
    on the basis of high item loadings on the first factor.
Dixon, P. W., & Ahern, E. H. Factor pattern comparisons of
    EPPS scales of high school, college, and innovative
    college program students. Journal of Experimental
    Education, 1973, 42, 17-35.
    Purpose: To identify dimensions of the EPPS for three
    samples
    Subjects: 1) 167 high school seniors, 2) 137 introductory
    psychology students, 3) 39 innovative college program
    students
    Variables: Scores on 15 EPPS scales
    Method: Image analysis condensation with oblique procrustes
    rotation, independently for the three samples
    Results: Five "need" factors emerged for high school sen-
    iors: I Friendly introversion; II Ego self-direction;
    III Social control; IV Verbal aggression; V Hetero-
    sexuality. Five "need" factors emerged for psychology
    class students: I Social dominance; II Social control 1;
    III Social control 2; IV Dependency; V Freedom. Six
    "need" factors emerged for innovative college program
    students: I Order - self-analysis; II Dependent achieve-
    ment; III Dominance and aggression; IV Freedom; V Inter-
    personal relationship; VI Social freedom.
Dixon, P. W., Fukuda, N. K., & Berens, A. E. A factor analysis
    of EPPS scales, ability, and achievement measures.
    Journal of Experimental Education, 1971, 39, 31-41.
    Purpose: To identify factors common to EPPS scales and
    ability-achievement measures
    Subjects: 169 high school students
    Variables: 15 EPPS scale, SCAT V and Q, and 7 teacher
    rating-scale scores, plus graduation rank

Method: Image analysis (principal axes condensation) with
oblique varimax rotation
Results: Six factors emerged: I Intellectual introversion;
II Dependency; III Superego strength; IV Independent
orientation; V Ego strength; VI Verbal aggression.
Conclusions: Factor analysis of the EPPS, even though it
is an ipsative measure, yielded psychologically relevant
variables.
Edwards, A. L., & Abbott, R. D. Relationships among the
Edwards Personality Inventory scales, the Edwards Per-
sonality Preference Schedule, and the Personality
Research Form scales. Journal of Consulting and
Clinical Psychology, 1973, 40, 27-32.
Purpose: To identify relationships among Edwards Personality
Inventory (EPI), Edwards Personality Preference Schedule
(EPPS), and Personality Research Form (PRF) scales
Subjects: 109 male and 109 female (college?) students
Variables: 90 scale scores from the EPI, EPPS, and PRF,
plus 2 social desirability scores (Edwards, Marlowe-
Crowne) and Welsh's R
Method: Principal components condensation with varimax
rotation
Results: Eighteen factors emerged and were listed in terms
of their highest definers; all but the last two factors
were marked by at least two scales with relatively high
loadings.
Edwards, A. L., & Abbott, R. D. Relationships between the
EPI scales and the 16PF, CPI, and EPPS scales. Educa-
tional and Psychological Measurement, 1973, 33, 231-238.
Purpose: To identify relationships between EPI, 16PF, CPI,
and EPPS scales
Subjects: 115 male, 171 female (college?) students
Variables: Scores on 15 EPPS, 16 16PF, 18 CPI, 53 EPI,
Edwards' SD, Welsh's R scales, plus sex
Method: Principal components condensation with varimax
rotation
Results: Of 22 factors extracted, 10 (I-III, V-XI) were
defined by scales from at least two inventories, and
10 (IV, XII, XIV-XXI) were relatively unique to a
given inventory.
Edwards, A. L., Abbott, R. D., & Klockars, A. J. A factor
analysis of the EPPS and PRF personality inventories.
Educational and Psychological Measurement, 1972, 32,
23-29.
Purpose: To identify dimensions underlying the EPPS and
PRF
Subjects: 109 male, 109 female (college?) students
Variables: Scores on 40 scales from EPPS, PRF, Edwards'
SD, Welsh's R, and Marlowe-Crowne MC
Method: Principal components condensation with varimax
rotation
Results: For all eleven factors extracted, at least one
EPPS and one PRF scale loaded highly.

Edwards, K. J., & Whitney, D. R. Structural analysis of
    Holland's personality types using factor and configural
    analysis. Journal of Counseling Psychology, 1972, 19,
    136-145.
Purpose: To identify dimensions underlying the Self-
    Directed Search (SDS)
Subjects: 358 male, 360 female college freshmen
Variables: Scores on 6 SDS scales each for activities (A),
    competencies (C), occupations (O), self-ratings (SR)
Method: Minres condensation with varimax rotation, by sex
Results: For both sexes' analyses of A and C, four fac-
    tors emerged: I Realistic and investigative; II Social
    and enterprising; III Conventional; IV Artistic. For
    both sexes' analyses of O and SR, four factors emerged:
    I Realistic and investigative; II Social; III Enter-
    prising and conventional; IV Artistic.
Endler, N. S., & Okada, M. A multidimensional measure of
    trait anxiety: The S-R Inventory of General Trait
    Anxiousness. Journal of Consulting and Clinical
    Psychology, 1975, 43, 319-329.
Purpose: To identify dimensions underlying the S-R Inven-
    tory of General Trait Anxiousness (S-R)
Subjects: 182 male, 204 female high school normals; 150
    male, 197 female adult normals; 34 male, 91 female
    neurotic inpatients; 35 male, 10 female psychotic in-
    patients
Variables: 4 situation, 9 response mode scores from S-R
Method: Principal components condensation with varimax
    rotation, independently for situation and response
    mode data within normal and abnormal samples
Results: Two situation factors emerged in both analyses:
    I Shame (interpersonal) anxiety; II Harm (physical
    danger) anxiety. Two response mode factors emerged
    in both analyses: I Physiological-distress; II Approach.
Evans, D. R., & Stangeland, M. Development of the Reaction
    Inventory to measure anger. Psychological Reports,
    1971, 29, 412-414.
Purpose: To identify dimensions underlying the Reaction
    Inventory (RI)
Subjects: 275 individuals, primarily college students
Variables: Scores on 76 RI items
Method: Principal axes condensation with varimax rotation
Results: Ten factors emerged: I Minor chance annoyances;
    II Destructive people; III Unnecessary delays; IV In-
    considerate people; V Self-opinionated people; VI
    Frustration in business; VII Criticism; VII Major
    chance annoyances; IX People being personal; X Authority.
Eysenck, H. J., & Eysenck, S. B. G. The orthogonality of
    psychoticism and neuroticism: A factorial study.
    Perceptual and Motor Skills, 1971, 33, 461-462.
Purpose: To identify relationships among psychoticism
    (P), neuroticism (N), extraversion (E), and dissimula-
    tion (L-scale)

161

Subjects: 216 male and 224 female 11- and 12-year-olds
Variables: Scores on 111 P, N, E, and L-scale items
Method: Principal components condensation with promax
rotation, by sex
Results: Four factors emerged in both sexes: I Psycho-
ticism; II Neuroticism; III Extraversion; IV L-scale.
Eysenck, S. B. G., & Eysenck, H. J. Attitudes to sex,
personality and lie scale scores. Perceptual and
Motor Skills, 1971, 33, 216-218.
Purpose: To identify relationships among sex-attitude,
psychoticism, and lie scale items
Subjects: 228 males and 263 females (participating in a
marketing research study)
Variables: Scores on 18 Lie Scale (EPI), 47 psychoticism,
12 Extraversion, 12 Introversion, and 8 sex-attitude
items
Method: Principal components condensation with promax
rotation
Results: Four factors emerged: I Psychoticism; II Ex-
traversion; III Neuroticism; IV Lie.
Conclusions: The sex-attitude items loaded on Factor I.
Fenigstein, A., Scheier, M. F., & Buss, A. H. Public and
private self-consciousness: Assessment and theory.
Journal of Consulting and Clinical Psychology, 1975,
43, 522-527.
Purpose: To identify dimensions underlying the Self-
consciousness Scale (SCS)
Subjects: 179 male, 253 female college students
Variables: Scores on 23 5-point SCS items
Method: Principal components condensation with varimax
rotation, by sex and combined
Results: Three factors emerged in all cases: I Private
self-consciousness; II Public self-consciousness; III
Social anxiety.
Forbes, A. R., Dexter, W. R., & Comrey, A. L. A cross-
cultural comparison of certain personality factors.
Multivariate Behavioral Research, 1974, 9, 383-393.
Purpose: To identify dimensions underlying the Comrey
Personality Scales (CPS) with a New Zealand sample
Subjects: 179 New Zealand college students
Variables: Scores on 40 FHIDs from the CPS, plus sex
Method: Minimum residual condensation with orthogonal
rotation
Results: Eight factors emerged: I Extraversion; II
Conformity; III Masculinity; IV Empathy; V Trust; VI
Stability; VII Activity; VIII Orderliness.
Friedman, S. T., & Manaster, G. J. Internal-external con-
trol: Studied through the use of proverbs. Psycho-
logical Reports, 1973, 33, 611-615.
Purpose: To identify dimensions underlying an Internal-
External Control Proverbs Test (I-E CPT)
Subjects: 464 college students

Variables: 25 I-E CPT item scores
Method: Principal axes condensation with varimax rotation
Results: Nine factors emerged: I Inner-directedness; II
   Conformity; III Action now; IV Chance; V Time is ripe;
   VI Veridicality; VII Pragmatism; VIII Initiative; IX
   Appropriateness.
Frost, B. P. A semantic differential analysis of the Leary
   Adjectival Check List. Journal of Clinical Psychology,
   1971, 27, 372-375.
Purpose: To identify dimensions of the Leary Adjectival
   Check List (LACL)
Subjects: 50 college freshmen
Variables: 10 semantic differential ratings each for 128
   adjectives from the LACL
Method: Principal components condensation with varimax
   rotation
Results: Two factors emerged: I Dynamism (coalescence of
   potency, activity); II Evaluation.
Conclusions: Further examination revealed that many, but
   not all, of the LACL adjectives are "correctly" placed
   within Leary's framework; revision of the LACL is
   suggested.
Gable, R. K., LaSalle, A. J., & Cook, K. E. Dimensionality
   of self-perception: Tennessee Self-concept Scale.
   Perceptual and Motor Skills, 1973, 36, 557-560.
Purpose: To identify dimensions underlying the Tennessee
   Self-concept Scale (TSCS)
Subjects: 125 college freshmen
Variables: Scores on 1) 12 scales, 2) 100 items from
   TSCS
Method: Principal components condensation with oblique
   rotation, independently for scales and items
Results: Two scale factors emerged: I Self-esteem; II
   Integration-conflict of self-concept. Twenty-three
   item factors were interpreted: I Emphasis of awareness/
   acceptance of one's physical appearance; II Awareness
   of/satisfaction with family relationships; III Self-
   criticism; IV Acceptance of personal worth/confirmation
   of proper understanding and treatment of one's family;
   V Awareness of one's physical condition; VI Awareness/
   acceptance of one's moral values; VII Awareness/attempts
   to enhance one's physical appearance; VIII Awareness of
   one's positive interpersonal relations; IX-XV, XXIII
   Awareness, acceptance, behavior with respect to physical
   characteristics; X, XI, XVIII, XIX Awareness of one's
   personal attributes; XII, XIII Confirmation, acceptance
   of religion in one's everyday life; XXI Awareness/
   acceptance/behavior with respect to the perception of
   moral attributes; XIV, XX One's perception of his
   interaction with other people; XVI, XXII Self-criticism
   (additional); XVII Extent of one's acceptance of his
   attitudes toward his family.

Gay, M. L., Hollandsworth, J. G. Jr., & Galassi, J. P. An assertiveness inventory for adults. Journal of Counseling Psychology, 1975, 22, 340-344.
  Purpose: To identify dimensions underlying the Adult Self-Expression Scale (ASES)
  Subjects: 464 unspecified individuals
  Variables: 48 ASES item scores
  Method: Unspecified condensation with varimax rotation
  Results: Fourteen factors emerged: I Interactions with authority figures; II Interactions with parents; III Interactions with the public; IV Intimate relations; V Asking favors; VI Refusing unreasonable requests; VII Expressing opinions; VIII Expressing annoyance or anger; IX Expressing positive feelings; X Standing up for one's legitimate rights; XI Taking the initiative in one's dealings with others without undue anxiety; XII-XIV Not reported.

Goldberg, J., Yinon, Y., & Cohen, A. A factor analysis of the Israeli Fear Survey Inventory. Psychological Reports, 1975, 36, 175-179.
  Purpose: To identify dimensions underlying the Israeli Fear Survey Inventory (IFSI)
  Subjects: 129 male and 215 female Israeli college students
  Variables: 97 IFSI item scores
  Method: Unspecified condensation with orthogonal rotation, by sex
  Results: Eight factors, of 29 extracted for males, were interpreted: I Seeing blood; II Social competence; III Medical intervention; IV Social criticism; V Animals and insects; VI High places; VII Danger signals; VIII Sickness. Eight factors, of 30 extracted for females, were interpreted: I Social competence; II Seeing blood; III Isolation; IV Social criticism; V Animals and insects; VI Weather and sudden noises; VII Dangerous places; VIII Medical intervention.

Golding, S. L., & Seidman, E. Analysis of multitrait-multimethod matrices: A two-step principal components procedure. Multivariate Behavioral Research, 1974, 9, 479-496.
  Purpose: To identify relationships between the Strong Vocational Interest Blank (SVIB), Personality Research Form (PRF), Employee Attitude Survey (EAS), and Social Competence Inventory (SCI)
  Subjects: 231 male college students
  Variables: Scores on 33 SVIB, 21 PRF, 8 EAS, SCI scale
  Method: 1) Principal components condensation (altered submatrices) with varimax rotation of all scores; 2) principal components condensation with varimax rotation, within each instrument followed by second-order analysis
  Results: Thirty-one unlabeled components emerged in the first (multimethod) analysis. Eight SVIB (I Business management and sales interests; II Aesthetic interests; III Service interest; IV Working with concepts and

things; V Management of others; VI Outdoor-nature
interests; VII Adventure interests; VIII Diffuse interests), five PRF (I Affective expression vs. control;
II Social ascendancy; III Defensiveness vs. humility;
IV Dependency vs. autonomy; V Aesthetic-intellectual),
and three EAS factors (I Quantitative ability and reasoning; II Visual-spatial ability; III Verbal comprehension) emerged. Eight second-order factors emerged:
I Social ascendance and management; II Verbal vs.
quantitative ability; III Adventuresomeness; IV Service-
oriented and dependency; V Aesthetic-intellectual; VI
Humble outdoorsman; VII Works with concepts and things
and emotionally controlled; VIII Business management
and sales interests.

Gordon, L. V. A typological assessment of "A Study of
Values" by Q-methodology. _Journal of Social Psychology_,
1972, _86_, 55-67.
  Purpose: To identify Q-types using A Study of Values (ASV)
  responses
  Subjects: Groups identified in the ASV manual, Buros
  (1965), and _Psychological Abstracts_ (1964-1967): military
  officers (3); male business students, businessmen,
  scoutmasters (9); other students and scientists (10);
  clergy or religious students (4); dental and medical
  students (4); individuals in social service occupations
  (11); education students, teachers, administrators (9);
  design students and individuals in the "arts" (5);
  medical support personnel (5)
  Variables: Mean scores on 6 ASV scales
  Method: Principal components (inverse) condensation with
  varimax rotation
  Results: Five factors emerged: I Economic-political man
  vs. social man; II Christian conservative vs. social
  man; III Theoretical man; IV Aesthetic man vs. social
  man; V Not interpreted (social service orientation).

Gough, H. G. Personality factors related to reported
severity of menstrual distress. _Journal of Abnormal
Psychology_, 1975, _84_, 59-65.
  Purpose: To identify dimensions of the Moos Menstrual
  Distress Questionnaire (MMDQ)
  Subjects: 201 women
  Variables: Scores on 8 scales each for pre-, inter-, and
  menstrual periods
  Method: Principal components condensation with varimax
  rotation
  Results: Three factors emerged: I Menstrual; II Pre-
  menstrual; III Intermenstrual distress.

Graham, J. R., Schroeder, H. E., & Lilly, R. S. Factor
analysis of items on the Social Introversion and
Masculinity-Femininity scales of the MMPI. _Journal of
Clinical Psychology_, 1971, _27_, 367-370.
  Purpose: To identify the factorial structure of the Social
  Introversion (Si) and Masculinity-Femininity (Mf) scales
  of the MMPI

Subjects: 422 individuals, aged 15-70 years (57% female; 29% psychiatric inpatients, 29% psychiatric outpatients, 42% normals)
Variables: Degree of pathology, sex, marital status, education, and age, plus 1) 70 Si item scores; 2) 60 Mf item scores
Method: Principal components condensation with varimax rotation, independently for the two variable sets
Results: Seven Si factors emerged: I Inferiority and discomfort; II Affiliation; III Social excitement; IV Demographic; V Sensitivity; VI Interpersonal trust; VII Physical-somatic concern. Seven Mf factors emerged: I Sensitivity-narcissism; II Feminine interests; III Masculine interests; IV Demographic; V Homosexual concern-passivity; VI Social extraversion; VII Exhibitionism.
Conclusions: These results appear consistent with those for other MMPI scales: neither Si nor Mf is unidimensional.

Gray, J. E. Dimensions of personality and meaning in self-ratings of personality. British Journal of Social and Clinical Psychology, 1973, 12, 319-322.
Purpose: To identify dimensions common to Eysenck Personality Inventory (EPI) and a self-concept semantic differential (SD)
Subjects: 131 nurses
Variables: EPI Neuroticism and Extraversion scores, plus 16 SD self-concept ratings
Method: Principal components condensation with varimax rotation
Results: Four factors emerged: I Evaluation; II Extraversion (and activity); III Potency; IV Neuroticism.

Greif, E. B., & Hogan, R. The theory and measurement of empathy. Journal of Counseling Psychology, 1973, 20, 280-284.
Purpose: To identify 1) dimensions underlying a CPI-derived empathy scale (ES), and 2) relationships between ES and other CPI scales
Subjects: 1) 260 male, 99 female college students; 2) 148 male college students, 79 and 183 police officers
Variables: Scores on 1) 64 ES items, 2) ES and 18 other CPI scales
Method: Minres condensation with varimax rotation, independently by groups and for total within each S-variable set.
Results: Three ES factors emerged in all analyses: I Tolerant, even-tempered disposition; II Self-possessed, outgoing, socially ascendant; III Humanistic and tolerant set of sociopolitical attitudes. Five CPI factors were identified: I Good impression and self-control; II Person orientation; III Capacity for independent thought and action, or flexibility; IV Communality, socialization, responsibility; V Femininity.

Hensley, D. R., Hensley, W. E., & Munro, H. P. Factor struc-
ture of Dean's Alienation Scale among college students.
Psychological Reports, 1975, 37, 555-561.
Purpose: To identify dimensions underlying Dean's Aliena-
tion Scale (AS) with college students
Subjects: 240 college students
Variables: 24 AS item scores
Method: Alpha factor analysis (with varimax and oblique
solutions)
Results: Eight unlabeled first-order and four second-
order factors emerged; they did not correspond to Dean's
hypothesized dimensions.
Horn, J. L., Wanberg, K. W., & Appel, M. On the internal
structure of the MMPI. Multivariate Behavioral Research,
1973, 8, 131-171.
Purpose: To identify dimensions underlying the MMPI
Subjects: 1523 male, 361 female admissions to a mental
health center
Variables: Scores on 33 non-overlapping MMPI variables
Method: Principal factors condensation with varimax rotation, by sex and total
Results: Six factors emerged (three of which were good
matches across samples): I Anxiety vs. ego strength;
II Neurotic hypochondriacal concern with bodily processes
vs. stoical indifference; III Sensitive (feminine) withdrawal vs. aggressive, outgoing sociability; IV Machia-
vellistic authoritarianism vs. ingenuous independence;
V Delicate, fundamentalistic religiosity vs. hard
skepticism; VI Eccentric, individualistic unrealism vs.
conformance with the conventional and socially desirable.
Howarth, E. A factor analysis of selected markers for objec-
tive personality factors. Multivariate Behavioral
Research, 1972, 7, 451-476.
Purpose: To identify dimensions underlying markers for
objective personality factors
Subjects: 317 female, 252 male college students
Variables: Scores on 48 markers for Cattellian factors
(U.I. 16, 17, 19-24, 32), plus age and sex
Method: 1) Principal components condensation with varimax
rotation, 2) hierarchical condensation with promax
rotation
Results: Twelve primary factors emerged (first analysis):
I Cognitive; II Extraversion 1; III Anxiety 1; IV Un-
identified; V Independence vs. concurrence; VI Anxiety 2;
VII Instrument; IX Extraversion 2; X-XI Unidentified;
XII Self-approval. Six second-order factors emerged:
I Extraversion; II Cognitive; III Anxiety 1; IV Anxiety
2; V-VI Uninterpreted.
Howarth, E., & Browne, J. A. An item factor analysis of the
Eysenck Personality Inventory. British Journal of
Social and Clinical Psychology, 1972, 11, 162-174.
Purpose: To identify dimensions underlying Extraversion (E)
and Neuroticism (N) items of the Eysenck Personality
Inventory (EPI)

167

Subjects: 666 male and 653 female college students
Variables: 57 EPI item scores, E score, N score, sex
Method: Principal axes condensation with orthogonal vari-
    max rotation, independently for males, females, and
    combined group
Results: Fifteen factors emerged in each analysis: I
    Sociability 1; II Adjustment-emotionality; III Inferi-
    ority; IV Impulsivity; V Mood swings-readjustment; VI
    Sleep; VII Superego 1; VIII Jocularity; IX Sociability 2;
    X Dominance; XI Lie scale 1; XII Social conversation
    (tentative); XIII Hypochondriac-medical; XIV Superego 2
    (tentative); XV Lie scale 2.
Conclusions: The primary (eight) factors were replicated
    with another (promax) solution; the E and N scales do
    not appear to be univocal measures of E and N.
Jackson, D. N., & Carlson, K. A. Convergent and discriminant
    validation of the Differential Personality Inventory.
    Journal of Clinical Psychology, 1973, 29, 214-219.
Purpose: To identify relationships between selected sub-
    scales of the Differential Personality Inventory (DPI),
    How Well Do You Know Your Roommate? (Roommate), and a
    self-rating
Subjects: 270 female and 100 male college students living
    in dormitories
Variables: Scores on 15 DPI scales, 12 Roommate scales,
    and 13 self-ratings
Method: Jackson's (1969) multimethod condensation with
    varimax and orthogonal graphic rotation, independently
    for each sex and combined sample
Results: The eleven factors extracted for the total
    sample were reported: I Neurotic disorganization; II
    Hostility; III Somatic complaints; IV Familial discord;
    V Depression; VI Rebelliousness; VII Cynicism; VIII
    Impulsivity; IX Health Concern; X Socially deviant
    attitudes; XI Irritability.
Jackson, D. N., & Morf, M. E. An empirical evaluation of
    factor reliability. Multivariate Behavioral Research,
    1973, 8, 439-459.
Purpose: To identify dimensions underlying parallel sets
    of personality measures
Subjects: 87 male, 109 female college students (Morf &
    Jackson, 1972)
Variables: Scores on 23 parallel sets of measures for
    (7) personality dimensions
Method: Principal components condensation with procrustes
    rotation, by set
Results: Seven factors emerged: I Item endorsement; II
    Key direction; III Tendency to respond desirably or
    undesirably; IV-VII Four content dimensions (exhibition,
    play, sentience, understanding).
Joe, V. C. Personality correlates of conservatism. Journal
    of Social Psychology, 1974, 93, 309-310.

Purpose: To identify relationships between the Conserva-
tism (C) Scale and several other measures with an
American sample
Subjects: 201 male and 215 female college students
(American)
Variables: Unspecified number (25+) of scores from the
C Scale, Jackson Personality Research Form AA (PRF),
F Scale, Crowne-Marlowe Social Desirability Scale,
Mirels-Garrett Protestant Ethic Scale, Kluckhohn-Murray
Optimism-Pessimism Scale
Method: Unspecified condensation with varimax rotation,
independently by sex
Results: Seven factors emerged in both analyses: I-V PRF
scales; VI Social desirability; VII Conservatism.
Joe, V. C., & Jahn, J. C. Factor structure of the Rotter
I-E Scale. Journal of Clinical Psychology, 1973, 29,
66-68.
Purpose: To identify dimensions underlying the Rotter I-E
Scale, using a multipoint (rather than dichotomous)
scoring system
Subjects: 120 female and 168 male college students
Variables: 23 6-point I-E item scores
Method: Principal components condensation with varimax
rotation, independently by sex
Results: Two factors emerged for both sexes: I Generalized
expectancy of reinforcement; II Expectancy of political
control.
Johnson, A. L., Metcalfe, M., & Coppen, A. An analysis of
the Marke-Nyman Temperament Scale. British Journal of
Social and Clinical Psychology, 1975, 14, 379-385.
Purpose: To identify dimensions of the Marke-Nyman
Temperament Scale (MNTS)
Subjects: 136 recovered unipolar depressed patients, and
208 controls
Variables: 60 MNTS item scores
Method: Unspecified condensation with varimax rotation
Results: Twenty factors emerged: I Confidence; II Restraint;
III Detachment; IV Reserve 1; V Sociability; VI Stamina;
VII Deliberation; VIII Staidness 1; IX Not interpreted;
X Lack of empathy; XI Difficulty of communication; XII
Daytime sleep; XIII Discretion; XIV Coldness; XV Not
interpreted; XVI Adaptability; XVII Reserve 2; XVIII
Initiative; XIX Isolation; XX Staidness 2.
Johnson, J. H., & Overall, J. E. Factor analysis of the
Psychological Screening Inventory. Journal of Consulting
and Clinical Psychology, 1973, 41, 57-60.
Purpose: To identify dimensions underlying the Psychological
Screening Inventory (PSI) with a college population
Subjects: 150 college students
Variables: 130 PSI item scores
Method: Marker variable (powered vector) condensation with
oblique rotation

169

Results: Three factors,of four extracted, were interpreted: I Introversion; II Emotional adjustment; III Social mal-adjustment.

Johnson, R. W., Flammer, D. P., & Nelson, J. G. Multiple correlations between personality factors and SVIB Occupational scales. Journal of Counseling Psychology, 1975, 22, 217-223.

Purpose: To identify dimensions underlying the California Psychological Inventory (CPI)

Subjects: 359 male college freshmen

Variables: 18 CPI scale scores

Method: Principal components condensation with varimax rotation

Results: Five factors emerged: I Positive adjustment, adjustment by social conformity, or value orientation; II Extroversion, social poise, or person orientation; III Independent thought, or independence-dependence; IV Conventionality; V Nurturance or emotional sensitivity.

Kidd, A. H., & Kidd, R. M. Relation of F-test scores to rigidity. Perceptual and Motor Skills, 1972, 34, 239-243.

Purpose: To identify relationships between the F-test and ridigity as measured by the Perceptual Rigidity test (PR)

Subjects: 100 female college students

Variables: Scores on 29 items and total F-test, plus PR score

Method: Centroid condensation with oblique rotation

Results: Two factors emerged: I Conventionality; II High defense utilization.

Kleiber, D., Veldman, D. J., & Menaker, S. L. The multi-dimensionality of locus of control. Journal of Clinical Psychology, 1973, 29, 411-416.

Purpose: To identify dimensions underlying Rotter's I-E Scale

Subjects: 219 college juniors

Variables: Scores on 23 internal and 23 external I-E Scale items in a 6-point format

Method: Principal axes condensation with varimax rotation

Results: Three factors emerged: I Disbelief in luck and chance; II System modifiability; III Individual responsibility for failure.

Conclusions: Factor scores were examined as a function of Adjective Self-Description subscales, Self-Report Inventory scales, and Dogmatism scores.

Krug, S. E., & Cattell, R. B. A test of the trait-views theory of distortion in measurement of personality by questionnaire. Educational and Psychological Measurement, 1971, 31, 721-734.

Purpose: To identify dimensions underlying trait-roles for four situations

Subjects: 159 college students

Variables: 16 trait (16PF) and 2 role scores each for 4
   situations (standard cooperative research, job seeking,
   ideal self-distortion, prospective dating)
Method: Principal axes condensation with procrustes,
   graphic, and clean-up maxplane rotation
Results: Of twenty-one factors extracted 15 16PF and 2
   role (career, dating) factors could be identified.
Lacher, M. The life styles of underachieving college stu-
   dents. Journal of Counseling Psychology, 1973, 20,
   220-226.
Purpose: To identify dimensions underlying the Omnibus
   Personality Inventory (OPI) and Opinion, Attitude, and
   Interest Survey (OAIS) for under- and overachieving
   college students
Subjects: 23 under- and 22 overachieving college students
Variables: 14 OPI, 14 OAIS, 2 SAT, 2 reading test scores,
   plus high school rank
Method: Cluster analysis (Lingoes, 1966), by group
Results: Two clusters emerged for underachievers: I
   Relatively practical, achievement-oriented, conservative,
   controlled, physical science-oriented, adjusted, non-
   intellectual, traditional; II Relatively theoretical,
   liberal, impulsive, artistic, humanities-oriented, non-
   traditional, deviant, verbal. Three clusters emerged
   for overachievers: I Relatively achievement-oriented,
   controlled, conservative, theoretical; II Higher scores
   in physical science and business interests (closer
   overall to III than I); III Relatively more verbal,
   theoretical, liberal, expressive, and impulsive than
   most other overachievers.
Lanyon, R. I., Johnson, J. H., & Overall, J. E. Factor
   structure of the Psychological Screening Inventory
   items in a normal population. Journal of Consulting
   and Clinical Psychology, 1974, 42, 219-223.
Purpose: To identify dimensions underlying the Psychological Screening Inventory (PSI)
Subjects: 400, and 400 subjects from normative sample
Variables: Scores on 129 PSI items and 5 PSI scales
Method: Principal axes condensation with varimax rotation,
   independently for two samples
Results: Five comparable factors were identified: I
   Alienation (serious psychopathology); II Expression
   dimension of extraversion-introversion; III Acting out;
   IV Protestant ethic; V Discomfort dimension of general
   maladjustment.
Lawlis, G. F. Motivational factors reflecting employment
   instability. Journal of Social Psychology, 1971, 84,
   215-223.
Purpose: To identify relationships between the 16PF,
   Motivation Analysis Test (MAT), and Response Analysis
   I (RA)
Subjects: 75 chronically unemployed individuals and 75
   employed individuals, matched on age range, intelligence,
   race, and general educational level

171

Variables: 69 scores from the 16PF, MAT, RA (nothing else
    specified)
Method: Centroid condensation with normal varimax rota-
    tion
Results: Sixteen factors of 20 were interpretable: I
    Motivation dynamic; II Introverted doubt; III Anxiety;
    IV Rebellion against parents; V Not interpreted; VI
    General fear; VII Intensified disposition toward fighting;
    VIII Not interpreted; IX General conflict with society's
    demands; X Career conflict; XI Lack of an intensified
    need for affection; XII Lowered sexual potential; XIII
    Lack of assertiveness; XIV Neurotic interaction with
    sweetheart or spouse; XV-XVI Not interpreted; XVII
    Failure to cope with threat; XVIII Not interpreted;
    XIX Passive disinterest in career; XX Limited activity
    by self-absorption.

Lazzaro, T. A., McNeil, K. A., & Beggs, D. L. A factor
    analytic study of the multidimensional properties of
    impulse control. Journal of Clinical Psychology, 1971,
    27, 495-498.
Purpose: To identify dimensions of the Self-Report Test
    of Impulse Control (STIC)
Subjects: 132 university upperclassmen, 94 university
    freshmen, 85 manpower trainees, 73 maximum security
    inmates, 58 minimum security inmates, and 33 psychiatric
    inmates
Variables: 72 STIC item scores
Method: Image analysis; principal components condensation
    with varimax rotation
Results: Four factors emerged: I Moral impulse control;
    II Control of impulses toward self; III Control of
    impulses toward others; IV Control of impulses leading
    to success.

Levin, J. Spherical model of vector extension for deter-
    mining the factor pattern of the California Psychological
    Inventory. Journal of Counseling Psychology, 1971, 18,
    579-582.
Purpose: To identify dimensions underlying the California
    Psychological Inventory (CPI); apparently reported by
    Nichols & Schnell (1963) but no reference provided
Subjects: Unspecified here (males)
Variables: 18 CPI scale scores
Method: Unspecified orthogonal factor analysis
Results: Three factors emerged: I Value orientation; II
    Person orientation; III Independence-dependence (labeled
    here via vector extension).

Lorr, M., & Knapp, R. R. Analysis of a self-actualization
    scale: The POI. Journal of Clinical Psychology, 1974,
    30, 355-357.
Purpose: To identify dimensions underlying the Personal
    Orientation Inventory (POI)
Subjects: 93 professionals, educators, counselors, and
    teachers; 123 (college?) students; 84 miscellaneous
    occupational group members

Variables: Scores on 150 POI items
Method: Principal components condensation with varimax
  rotation
Results: Fifteen factors emerged: I Self-esteem; II Ad-
  herence to principles - flexible vs. rigid; III Freedom
  of self-expression; IV Obligation to self; V Uninterpretable; VI Positive view of human nature; VII Self-
  actualization; VIII Flexible vs. perfectionistic view
  of one's performance; IX Willingness to be oneself
  regardless of mistakes or of disapproval from others
  and willingness to accept the consequences; X Acceptance
  of differences; XI (Future); XII, XV Residual or error;
  XIII Not reported; XIV Acceptance of aggression.
Lunneborg, P. W. Dimensionality of MF. _Journal of Clinical_
  Psychology, 1972, _28_, 313-317.
  Purpose: To identify dimensions of masculinity-femininity
  (MF)
  Subjects: 139 male and 384 female college students
  Variables: Item scores: 60 from MMPI Mf scale, 23 from
  MMPI Sd scale which do not overlap, 27 from CPI Fe which
  do not overlap, 19 from Gaugh MF scale, 30 from G-Z-TS
  M scale (rewritten), 62 from Heston Personal Adjustment
  Survey Mf scale (rewritten), 109 Terman-Miles Mf Exercises
  6 and 7 (rewritten), 66 Edwards Personality Inventory
  Form IA items which discriminated sexes, 53 from SVIB
  for Men M scale (rewritten) (450 total)
  Method: Principal components condensation with varimax
  rotation, independently by sex
  Results: Of nine factors emerging for females and ten
  for males, four were considered "true MF" factors (i.e.,
  occurred in both sexes): I Neuroticism (F); II Reli-
  giosity (F); III Power (M); IV Scientific interests (M).
Manners, G. E. Jr., & Steger, J. A. The stability of the
  Edwards Personal Preference Schedule and the Guilford-
  Zimmerman Temperament Survey. _Personnel Psychology_,
  1975, _28_, 501-509.
  Purpose: To identify dimensions underlying the 1) Edwards
  Personal Preference Schedule (EPPS), and 2) Guilford-
  Zimmerman Temperament Survey (GZTS) with new samples
  Subjects: 1) 325, 2) 458 male employees (from files) of
  an industrial corporation
  Variables: Scores on 1) 15 EPPS, 2) 10 GZTS scales
  Method: 1) Alpha condensation, 2) minres condensation,
  both with varimax rotation
  Results: Five EPPS factors emerged: I Affiliation; II
  Drone; III Intraception; IV Change; V Dominance. Three
  GZTS factors emerged: I Social butterfly; II Tolerance;
  III Thoughtfulness/restraint (confounded by resolution
  of Heywood cases).
  Conclusions: In both cases, the factor structure was ex-
  tremely similar to that originally reported.
Mehryar, A. H., Khajavi, F., & Hekmat, H. Comparison of
  Eysenck's PEN and Lanyon's Psychological Screening
  Inventory in a group of American students. _Journal of_
  Consulting and Clinical Psychology, 1975, _43_, 9-12.

Purpose: To identify relationships among PEN and Psychological Screening Inventory (PSI) scales
Subjects: 178 male, 297 female college students
Variables: Scores on P, E, N and Lie scales from PEN, and 5 PSI scales
Method: Principal components condensation with varimax rotation, by sex
Results: Three factors emerged in both analyses: I Psychological disturbance; II Extraversion; III Defensiveness.

Michael, W. B., Lee, Y. B., Michael, J. J., Hooke, O., & Zimmerman, W. S. A partial redefinition of the factorial structure of the Study Attitudes and Methods Survey (SAMS) Test. Educational and Psychological Measurement, 1971, 31, 545-547.
Purpose: To identify dimensions underlying old and new SAMS items
Subjects: 168 college students
Variables: Scores on 1) 167 old, 2) 67 new, 3) 144 old, 43 new SAMS items
Method: Principal components condensation with varimax rotation, independently for old, new, and selected old and new items
Results: Results are reported for the third analysis only. Eight factors emerged: I Academic drive; II Conformity; III Academic interest, or learning affect-satisfaction; IV Anxiety; V Alienation; VI Methodological and systematic approaches to study; VII Positive orientation toward teachers; VIII Manipulation, or savoir faire.
Conclusions: A revised form includes 20-25 items each for Factors I-VI and VIII.

Miller, J. K., & Farr, S. D. Bimultivariate redundancy: A comprehensive measure of interbattery relationship. Multivariate Behavioral Research, 1971, 6, 313-324.
Purpose: To identify dimensions underlying Omnibus Personality Inventory (OPI) scales and college admissions criteria
Subjects: 190 male college freshmen
Variables: 1) 14 OPI scale scores, 2) 3 college admissions criteria scores
Method: Principal components condensation without rotation, by data set
Results: Nine unnamed OPI and three admissions criteria factors emerged.

Nesselroade, J. R., & Baltes, P. B. Higher order factor convergence and divergence of two distinct personality systems: Cattell's HSPQ and Jackson's PRF. Multivariate Behavioral Research, 1975, 10, 387-407.
Purpose: To identify higher-order dimensions of the HSPQ and PRF
Subjects: 1,862 male and female seventh- through twelfth-graders
Variables: Scores on 14 HSPQ and 20 PRF scales

Method: Principal axes condensation with varimax, graphic, and oblique rotation, by instrument
Results: Seven HSPQ (I Exvia vs. invia; II Anxiety vs. good adjustment; III Cortertia vs. pathemia; IV Independence; V Emotional anxiety; VI Int.; VII Superego strength) and eight PRF factors (I Conscientiousness; II Ascendance or exhibitionistic control; III Independence; IV Aggression; V Aesthetic intellectual orientation; VI Achievement; VII Social contact; VIII Not labeled) emerged.
Conclusions: Extension analyses were also reported.

Nichols, M. P., Gordon, T. P., & Levine, M. D. Development and validation of the Life Style Questionnaire. _Journal of Social Psychology_, 1972, _86_, 121-125.
Purpose: To identify dimensions of the Life Style Questionnaire (LSA)
Subjects: 1) 93 men, 2) 141 individuals
Variables: Unspecified number of scores from the MMPI, EPPS, and LSA (58 Life style items, 7 Lie scale items)
Method: Unspecified condensation with varimax rotation, independently for each group
Results: Both analyses yielded two unlabelled factors which presumably reflect A- and B- lifestyles.

Nugent, J., Covert, R., & Chansky, N. Factor analysis of the Runner Studies of Attitudes Patterns Interview Form (1970 revision). _Educational_ and _Psychological_ _Measurement_, 1973, _33_, 491-494.
Purpose: To identify dimensions underlying the Runner Studies of Attitude Patterns, Interview Form (RSAP-IF)
Subjects: 435 female and 358 male college freshmen
Variables: 121 RSAP-IF item scores
Method: Minimum residual condensation with varimax rotation
Results: Ten factors emerged: I Af; II Pc; III Cx; IV Do; V Rt; VI T; VII Pl, Md; VIII In, Al; IX-X Not reported.

Ottomanelli, G. Second order analysis of Comrey's eleven factor solution. _Multivariate_ _Behavioral_ _Research_, 1972, _7_, 299-304.
Purpose: To identify second-order dimensions underlying the Comrey Inventory (CI)
Subjects: (Unspecified here) from Duffy, et al., 1969
Variables: 13 CI first-order factor scores (Duffy, et al. matrix)
Method: Principal components condensation with quartimax, and varimax, rotation
Results: A four-factor solution(s) was obtained: I Avoidance; II Approach; III Thoughtfulness; IV Dependence-masculinity continuum. An eight-factor solution(s) was also obtained: I Avoidance; II Approach; III Activity; IV Dependence; V Hostility; VI Masculinity; VII Thoughtfulness 1; VIII Thoughtfulness 2.

Overall, J. E., Johnson, J. H., & Lanyon, R. I. Factor
    structure and scoring of the PSI: An application of
    marker variable analysis. Multivariate Behavioral
    Research, 1974, 9, 407-422.
  Purpose: To identify dimensions underlying the Psycholo-
    gical Screening Inventory (PSI)
  Subjects: 400 males, 400 females from normative sample
  Variables: Scores on 130 PSI items 1) using 5 PSI scales
    as markers, 2) using items with highest loadings on
    factors identified in (1) as markers
  Method: Marker variable factor condensation with oblique
    rotation for (1) and (2)
  Results: Five factors emerged in both analyses: I Major
    psychopathology; II Social maladjustment; III Neuroti-
    cism; IV Introversion-extroversion; V Social desirability.
Parrott, G. Dogmatism and rigidity: A factor analysis.
    Psychological Reports, 1971, 29, 135-140.
  Purpose: To identify relationships between the Rokeach
    Dogmatism Scale (DS) and the Gough-Sanford Rigidity
    Scale (RS)
  Subjects: 1074 college students (and 2 subsamples of 300
    each)
  Variables: Scores on 40 DS and 22 RS items
  Method: Unspecified condensation with varimax rotation,
    independently for total group and subsamples, and for
    combined and DS and RS items
  Results: For RS alone, six factors emerged for total
    sample (4, 6 for subsamples) but were unnamed. For
    DS alone, nine unnamed factors emerged for total group
    (9, 10 for subsamples). For the combined analysis,
    eleven factors emerged: I Scheduling rigidity; II
    Belief in one truth; III Isolation-alienation; IV Self-
    proselytization; V Submission to ingroup authorities;
    VI Belief in one cause; VII Task conservation rigidity;
    VIII Virtuous self-denial; IX Task perseveration; X
    Adaptive rigidity; XI Outgroup repression.
  Conclusions: Three factors were primarily RS items, seven
    DS, and one mixed.
Patterson, M. L., & Strauss, M. E. An examination of the
    discriminant validity of the Social Avoidance and
    Distress scale. Journal of Consulting and Clinical
    Psychology, 1972, 39, 169.
  Purpose: To identify dimensions underlying the Social
    Avoidance and Distress scale (SAD), Interpersonal
    Anxiety scale (IA), affiliation, extraversion, and
    social desirability scales
  Subjects: 209 (unspecified) individuals
  Variables: Scores (presumably total) from SAD, IA,
    affiliation, extraversion, and social desirability
    scales
  Method: Unspecified factor analysis
  Results: Three factors emerged: I Social approach -
    avoidance; II Social anxiety; III Social desirability.

Payne, F. D. Relationships between response stability and
    item endorsement, social desirability, and ambiguity
    in the MMPI and CPI. Multivariate Behavioral Research,
    1974, 9, 127-148.
  Purpose: To identify relationships between indices of
    response stability, item endorsement, social desira-
    bility, and ambiguity indices for MMPI and CPI items
  Subjects: 95 male, 108 female college students
  Variables: Scores on 9 response stability, item endorse-
    ment, social desirability, and ambiguity indices for
    1) 547 MMPI, 2) 254 CPI, 3) 170 moderate-endorsement
    MMPI, 4) 101 moderate-endorsement CPI items
  Method: Principal components condensation with varimax
    rotation, for each set of items
  Results: Two ambiguity factors (I, II) emerged in all
    analyses; for the first two analyses, a third factor
    (III Extremeness of endorsement and of social desir-
    ability) also emerged.
Pedersen, D. M. Development of a personal space measure.
    Psychological Reports, 1973, 32, 527-535.
  Purpose: To identify dimensions underlying the Pedersen
    Personal Space Measure (PPSM)
  Subjects: 170 male junior college students
  Variables: 24 PPSM scores
  Method: Principal factors condensation with varimax
    rotation
  Results: Four factors emerged: I General simulated per-
    sonal space; II Angle of orientation; III Facing - not
    facing; IV Direct simulated personal space.
Pedhazur, E. J. Factor structure of the Dogmatism Scale.
    Psychological Reports, 1971, 28, 735-740.
  Purpose: To identify dimensions underlying the Dogmatism
    Scale (DS)
  Subjects: 309 male and 526 female teachers and graduate
    students of education
  Variables: 40 DS item scores
  Method: Principal components condensation with varimax
    and oblique rotation, by sex
  Results: Five factors emerged for both sexes and both
    solutions: I Belief in one truth; II Isolation -
    alienation; III Belief in one cause; IV Self-proselyti-
    zation; V Virtuous self-denial (males), or narrowing
    and intolerance (females).
Philip, A. E. Cross-cultural stability of second-order
    factors in the 16PF. British Journal of Social and
    Clinical Psychology, 1972, 11, 276-283.
  Purpose: To identify second-order factors of the 16PF
    for a population of British normals, for the purposes
    of comparison with previously-found factorial structures
  Subjects: 105 male and 179 female normal British adults
  Variables: 16 16PF scale scores

Method: Principal components condensation with analytic
     oblique (biquartimin criterion) rotation, independently
     by sex and for combined group
Results: Six factors emerged in all analyses: I, IV, V
     Introversion-extraversion; II Anxiety; III B; VI V.
Conclusions: These results were compared with those for
     an American and another British sample; "Anxiety vs.
     adjustment" appeared clearly in each, but more than
     one contender for the IE label.
Pishkin, V., & Thorne, F. C. A factorial study of existential
     state reactions. Journal of Clinical Psychology, 1973,
     29, 392-402.
Purpose: To identify dimensions underlying the Existential
     Study (ES)
Subjects: 193 incarcerated felons, 89 psychiatrically
     hospitalized alcoholics, 153 students of the Objectivist
     philosophy of Ayn Rand, 338 clients of an institution
     for unwed mothers, 159 college students
Variables: 200 ES item scores
Method: Principal components condensation with varimax
     rotation
Results: Five factors emerged: I Demoralization state/
     existential neurosis; II Existential confidence/morale;
     III Religious dependency defenses; IV Self-actualization
     esteem; V Concern over the human condition.
Reeves, T. G., & Shearer, R. A. Differences among campus
     groups on a measure of self-actualization. Psychological
     Reports, 1973, 32, 135-140.
Purpose: To identify dimensions underlying the Personal
     Orientation Inventory (POI) for 5 groups
Subjects: 51 black, 72 self-reported conformist white, 31
     self-reported non-conformist white, 22 self-reported
     non-conformist in attitude (but not appearance) white,
     16 "none-of-the-above" white college students
Variables: 15 POI scale scores
Method: Principal components condensation with orthogonal
     rotation
Results: Three factors emerged: I Affective phenomena,
     or how one feels in terms of self-actualization; II How
     one thinks in terms of self-actualization; III Self-
     concept.
Conclusions: Factor scores were compared between groups.
Rodrigues, A., & Comrey, A. L. Personality structure in
     Brazil and the United States. Journal of Social
     Psychology, 1974, 92, 19-26.
Purpose: To identify dimensions underlying a Portuguese
     version of the Comrey Personality Scales (CPS-P)
Subjects: 689 Brazilian college students
Variables: Scores on 40 homogeneous 4-item subgroups and
     2 validation scales from the CPS-P, plus age and sex
Method: Minimum residual condensation with orthogonal
     (tandem criteria) rotation

Results: Eight factors emerged: I Trust vs. defensiveness;
II Orderliness vs. lack of compulsion; III Social conformity vs. rebelliousness; IV Activity vs. lack of energy;
V Emotional stability vs. neuroticism; VI Extraversion
vs. introversion; VII Masculinity vs. femininity; VIII
Empathy vs. egocentrism.
Conclusions: These results were compared with those for
an American sample.
Rogers, T. B. Ratings of content as a means of assessing
personality items. Educational and Psychological Mea-
surement, 1973, 33, 845-858.
Purpose: To identify dimensions of rated content of items
from three Personality Research Form (PRF) scales
Subjects: 54 college students each in three groups
Variables: Ratings of 1) desirability (D), 2) impulsivity
(I), or 3) autonomy (A) for 20 PRF items for D, I, A,
respectively
Method: Principal components condensation with varimax
rotation, for each S-variable set
Results: Seven D factors emerged and were described as
similar to Messick's (1960). Six I factors emerged:
I Global false-keyed I; II Lack of preparation; III
Impulsive life style; IV Compulsive behavior; V D; VI
Not labeled. Six A factors emerged: I Global false-
keyed A; II Global true-keyed A; III Loneliness; IV
Interference from others; V Support from others; VI
Freedom of movement.
Rohlf, R. J. A higher-order alpha factor analysis of interest,
personality, and ability variables, including an evalua-
tion of the effect of scale interdependency. Educational
and Psychological Measurement, 1971, 31, 381-396.
Purpose: To identify dimensions common to selected interest,
personality, and ability variables
Subjects: 40 male scholarship candidates
Variables: 45 SVIB, 3 validity and 10 clinical MMPI scale,
Terman's Concept Mastery Test (CTM), and Stouffer
Mathematics Test (SMT) scores
Method: Alpha factor condensation with varimax followed
by maxplane rotation
Results: Ten first-order factors emerged but were not
named; two interest (I, II), one personality, and one
mixed-domain second-order factors emerged.
Rosen, A., & Schalling, D. On the validity of the California
Psychological Inventory Socialization scale. Journal of
Consulting and Clinical Psychology, 1974, 42, 757-765.
Purpose: To identify dimensions underlying California
Psychological Inventory (trans.) Socialization (So)
scale items
Subjects: 38 youth prison inmates (aged 18-21 years), 78
conscripts (19-21), and 70 college freshmen (20-31),
all Swedish
Variables: Scores on 18 So items each with 1) low, 2)
moderately extreme, 3) extreme distributions

179

Method: Principal components condensation with varimax
   rotation, independently for the three item sets
Results: Six, six, and five factors emerged for the three
   item sets, and the following subscales were constructed:
   I Positive interpersonal experiences; II Conformity and
   observance of convention; III Evaluation anxiety; IV
   Low self regard; V Superego strength; VI Poise vs.
   dysphoric moods and paranoid attitudes.
Ryman, D. H., Biersner, R. J., & La Rocco, J. M. Reliabilities
   and validities of the Mood Questionnaire. Psychological
   Reports, 1974, 35, 479-484.
   Purpose: To identify dimensions underlying the Mood Ques-
      tionnaire (MQ)
   Subjects: 1140 Navy recruits, tested at two time periods
   Variables: 40 MQ item scores
   Method: Principal components condensation with varimax
      rotation, by time
   Results: Five factors were stable over time: I Pleasant-
      ness; II Depression; III Anger; IV Fatigue; V Fear.
Sannito, T., Walker, R. E., Foley, J. M., & Posavac, E. J.
   A test of female sex identification: The Thorne
   femininity study. Journal of Clinical Psychology,
   1972, 28, 531-539.
   Purpose: To identify the factorial structure of the Thorne
      Femininity Scale (TFS)
   Subjects: 200 Catholic college students
   Variables: 11 TFS subtest scores (27 of the original 200
      items were deleted)
   Method: Principal-factor condensation with varimax rota-
      tion
   Results: Two factors emerged: I Delight in being feminine;
      II Enjoyment of the homemaker role.
   Conclusions: Factor scores were used in further analyses
      between seven groups of subjects.
Scherer, K. R. Judging personality from voice: A cross-
   cultural approach to an old issue in interpersonal
   perception. Journal of Personality, 1972, 40, 191-210.
   Study One:
   Purpose: To identify dimensions underlying the Personality
      Attribute rating form (PAF)
   Subjects: 28 Americans (mean age=34.3 years) and 31 Germans
      (35.4), rated by self, peers, and judged
   Variables: 35 PAF item scores
   Method: Unspecified factor analysis, independently by
      rater type within samples
   Results: Eight factors, retained as scales for Study Two,
      appeared consistently across analyses: I Dependability;
      II Task ability; III Neuroticism; IV Stability; V
      Sociability; VI Dominance; VII Likeability; VIII Ag-
      gressiveness.
   Study Two:
   Purpose: To identify relationships among PAF.scale scores
      for 4 groups of raters

Subjects: 10 Americans rating American speakers, 10 Americans rating Germans, 8 Germans rating Americans, 7 Germans rating Americans' voice samples
Variables: Eight PAF scale scores for each group of raters (32x32)
Method: Principal components condensation with varimax rotation
Results: Three factors emerged: I Nice-guy syndrome; II Leader syndrome; III Not labeled.

Schimek, J. G., & Meyer, R. M. Dimensions of alienation and pathology. Psychological Reports, 1975, 37, 727-732.
Purpose: To identify dimensions underlying the Keniston Alienation Scale (AS)
Subjects: 47 adolescent psychiatric hospital patients, 28 staff, and 78 college students
Variables: 11 AS scale scores
Method: Principal components condensation with varimax rotation (for each group as well as combined)
Results: Three factors emerged: I Interpersonal alienation; II Self-repudiation; III Cultural alienation.

Schwartz, S., & Giacoman, S. Convergent and discriminant validity of three measures of adjustment and three measures of social desirability. Journal of Consulting and Clinical Psychology, 1972, 39, 239-242.
Purpose: To identify dimensions common to adjustment and social desirability measures derived from the MMPI, Marlowe-Crowne, SDS, TAT, and Rorschach
Subjects: 19 college students, 19 psychiatric inpatients
Variables: Adjustment and social desirability scores each derived from MMPI or SDS, TAT, and Rorschach
Method: Principal components condensation
Results: Two factors emerged: I Adjustment; II Social desirability.

Scott, W. E. Jr., & Day, G. J. Personality dimensions and vocational interests among graduate business students. Journal of Counseling Psychology, 1972, 19, 30-36.
Purpose: To identify dimensions underlying the Gough Adjective Check List (ACL)
Subjects: 148 graduate business students
Variables: 24 ACL scale scores
Method: Principal components condensation with varimax rotation
Results: Three factors emerged: I General adjustment; II Self-assertiveness; III Ego control.

Seidman, E., Golding, S. L., Hogan, T. P., & LeBow, M. D. A multidimensional interpretation and comparison of three A-B scales. Journal of Consulting and Clinical Psychology, 1974, 42, 10-20.
Purpose: To identify dimensions underlying, and relationships between, the Strong Vocational Interest Blank (SVIB), Personality Research Form (PRF), and Employee Attitude Survey (EAS)

181

Subjects: 231 male college students
Variables: Scores on 1) 33 SVIB, 22 PRF, and 8 EAS scales,
2) 8 SVIB, 5 PRF, and 3 EAS factors identified in (1)
Method: Principal components condensation with varimax
rotation, independently for each instrument in (1), and
for 16 variables in (2)
Results: Eight SVIB factors emerged: I Business management and sales interests; II Aesthetic interests; III
Service interests; IV Working with concepts and things;
V Management of others interests; VI Outdoor-nature
interests; VII Adventure interests; VIII Diffuse interests. Five PRF factors emerged: I Affective ex-
pression versus control; II Social ascendency; III
Defensiveness versus humility; IV Dependency versus
autonomy; V Aesthetic-intellectual. Three EAS factors
emerged: I Quantitative ability and reasoning; II
Visual-spatial ability; III Verbal comprehension.
Eight second-order factors emerged: I Social ascendance
and management; II Verbal vs. quantitative ability; III
Adventuresomeness; IV Service-oriented and dependency;
V Aesthetic-intellectual; VI Humble outdoorsman; VII
Works with concepts and things and emotionally con-
trolled; VIII Business managements and sales interests.
Sells, S. B., Demaree, R. G., & Will, D. P. Jr. Dimensions
of personality: II. Separate factor structure in
Guilford and Cattell trait markers. Multivariate
Behavioral Research, 1971, 6, 135-185.
Purpose: To identify dimensions underlying Guilford and
Cattell personality items
Subjects: 2011 basic airmen
Variables: Scores on 1) 300 Guilford and 300 Cattell
items (T), 2) 300 Guilford items (G), 3) 300 Cattell
items (C); residual matrix for 4) Guilford items (GR),
5) Cattell items (CR)
Method: Principal factors condensation with varimax and
promax rotation, by data set
Results: Fifteen- and 18-factors solutions were obtained
for T; 12 G and 11 C factors emerged; 8 GR and 7 CR
factors were extracted. A total of 18 factors were
named: I Emotional stability; II Social extraversion;
III Artistic interest; IV Conscientiousness; V Cyclo-
thymia vs. schizothymia; VI Agreeableness vs. hostility;
VII Relaxed composure vs. suspicious excitability; VIII
Personal relations; IX General activity; X Radicalism
vs. conservatism; XI Paranoid sensitivity; XII Critical
thinking; XIII Considerateness vs. aggressive disregard
of others; XIV Dispositional anxiety; XV Serious, prudent vs. happy-go-lucky, impulsive; XVI Empathic media-
tion; XVII Stoicism, fear denial vs. weeping, fear
proneness; XVIII Tolerance of rough, uncouth behavior
vs. genteel, proper, refined.

Sherman, R. C., & Poe, C. A. Factor-analytic scales of a
    normative form of the EPPS replicated across samples
    and methodologies. Psychological Reports, 1972, 30,
    479-484.
    Purpose: To identify normative EPPS (N-EPPS) dimensions
    with a cross-validation sample
    Subjects: 314 college freshmen
    Variables: Scores on 1) 135 N-EPPS items, and 2) 15 needs
    Method: Principal axes condensation, apparently without
    rotation, independently for items and scales
    Results: Two factors (I Interpersonal orientation; II
    Assertive aggressiveness) were replicated from original
    sample in both item and scale analyses.
Silverstein, A. B., & Fisher, G. Item overlap and the "built-
    in" factor structure of the Personal Orientation Inven-
    tory. Psychological Reports, 1972, 31, 492-494.
    Purpose: To identify dimensions underlying Personal
    Orientation Inventory (POI) empirical correlations
    Subjects: Unspecified numbers of 1-4) college students
    (data from Shostrom, 1966; Klavetter & Mogar, 1967;
    LeMay & Damm, 1970 (two)), 5) psychopathic felons
    (Fisher, 1968)
    Variables: a) 12 POI empirical correlations (EC), and
    b) item overlap data (IO)
    Method: Principal factors condensation with varimax
    rotation, for EC and IO within each data set
    Results: Three factors emerged for each set of EC, and
    5 for IO; none were named but they were examined for
    congruence.
Skinner, N. St. J. F., & Howarth, E. Cross-media indepen-
    dence of questionnaire and objective-test personality
    factors. Multivariate Behavioral Research, 1973, 8,
    23-40.
    Purpose: To identify dimensions underlying 3 personality
    questionnaires and 8 objective tests
    Subjects: 79 female, 49 male college students
    Variables: Scores on 4 second-order and 16 first-order
    16PF, 2 EPI, 10 MAT, and 15 objective test measures
    Method: 1) alpha factor condensation with varimax rota-
    tion; 2) principal axes condensation with varimax or
    3) promax rotation
    Results: Nine similar factors emerged: I Extraversion;
    II Anxiety; III Alertness; IV Performance A; V Self-
    control; VI Performance B; VII Uninterpreted; VIII
    Independence; IX Uninterpreted.
Stein, K. B., Soskin, W. F., & Korchin, S. J. Dimensions
    of the Rotter Trust Scale. Psychological Reports,
    1974, 35, 999-1004.
    Purpose: To identify dimensions underlying the Rotter
    Trust Scale (TS)
    Subjects: 70 high school students
    Variables: 25 TS trust item scores
    Method: Oblique cluster analysis

Results: Three clusters emerged: I Integrity of social
role agents; II Trustworthiness of human motives; III
Dependability of people.
Stephens, J. H., Shaffer, J. W., & Zlotowitz, H. I. An
optimum A-B scale of psychotherapist effectiveness.
Journal of Nervous and Mental Disease, 1975, 160, 267-
281.
Purpose: To identify dimensions underlying Strong Voca-
tional Interest Blank (SVIB) scales and SVIB-derived
A-B scales
Subjects: 369 male medical students
Variables: 45 standard SVIB occupational interest scale
scores, plus scores on 3 A-B scales
Method: Principal axes condensation with normalized
varimax rotation
Results: Five factors emerged: I Verbal/conceptual vs.
manual/practical; II Scientific vs. sales; III Social
concern; IV Artistic vs. business-oriented; V Not
interpreted.
Stewart, D. W., & Griffith, G. M. Factor analysis of
Zuckerman's Sensation-seeking Scale. Psychological
Reports, 1975, 37, 849-850.
Purpose: To identify dimensions underlying the Sensation-
seeking Scale IV (SSS-IV)
Subjects: 156 college students
Variables: 72 SSS-IV item scores (and items within 4 sub-
scales)
Method: Principal components condensation with varimax
rotation
Results: Four factors were rotated: I Experience-seeking;
II Thrill and adventure seeking; III Boredom-suscepti-
bility; IV Not labeled (but not disinhibition).
Strahan, R., & Gerbasi, K. C. Short, homogeneous versions
of the Marlow-Crowne Social Desirability Scale.
Journal of Clinical Psychology, 1972, 28, 191-193.
Purpose: To identify dimensions of the Marlow-Crowne
Social Desirability Scale (M-C SDS), and to derive
shorter, homogeneous versions of it
Subjects: 176 male and 185 female college students
Variables: Scores on 33 M-C SDS items, plus sex and
class
Method: Principal components condensation, apparently
without rotation
Results: An unspecified number of factors (but at least
two: I General; II Uninterpreted) emerged.
Conclusions: Size of loading on Factor I was the primary
criterion for construction of two 10-item and a com-
bined 20-item scale.
Stricker, L. J. Personality Research Form: Factor struc-
ture and response style involvement. Journal of
Consulting and Clinical Psychology, 1974, 42, 529-537.
Purpose: To identify dimensions underlying the Personality
Research Form (PRF)

Subjects: 27 male and 44 female 11th-, 12-graders, and new high school grads
Variables: 20 PRF scale scores
Method: Principal axes condensation with promax rotation
Results: Six factors emerged: I Conscientiousness; II Hostility; III Ascendance; IV Dependence; V Imagination; VI Carefreeness.
Conclusions: Response style scores and sex were extracted onto these factors.
Stricker, L. J. Response styles and 16PF higher order factors. Educational and Psychological Measurement, 1974, 34, 295-313.
Purpose: To identify relationships between acquiescence (A), social desirability (SD), and defensiveness (D) response styles and 16PF factors
Subjects: 69 adolescent girls (11th or 12th grade, or just graduated)
Variables: 15 16PF scores (B excluded), 2 SD, 2 D, and 5 A measures, plus day-of-testing variable
Method: Principal axes condensation with promax rotation (for "second-order" analysis only)
Results: Four "second-order" factors emerged: I Anxiety; II Exvia; III Independence; IV (Anxiety). A single "third-order" factor emerged but was not labeled.
Stricker, L. J., Jacobs, P. I., & Kogan, N. Trait interrelations in implicit personality theories and questionnaire data. Journal of Personality and Social Psychology, 1974, 30, 198-207.
Purpose: To identify dimensions underlying MMPI PD items (unpublished study by Stricker, Jacobs, & Kogan)
Subjects: 559 female college freshmen
Variables: Scores on 46 MMPI Pd items
Method: Principal axes condensation with promax rotation
Results: Sixteen factors emerged but are not reported here.
Stroup, A. L., & Manderscheid, R. W. The California Psychological Inventory: Reappraisal of reliability. Journal of General Psychology, 1975, 92, 217-224.
Purpose: To identify dimensions underlying the California Psychological Inventory (CPI)
Subjects: 1939 midwestern college students
Variables: 18 content subscale scores from the CPI
Method: Principal components condensation with varimax rotation
Results: Four factors emerged: I Adjustment by social conformity; II Social poise, or extraversion; III Capacity for achievement; IV Super-ego strength.
Conclusions: The present results are compared with those for samples of western students and western school administrators found by previous investigators.
Taylor, J. B., Ptacek, M., Carithers, M., Griffin, C., & Coyne, L. Rating scales as measures of clinical judgment III: Judgments of the self on personality

inventory scales and direct ratings. Educational and
Psychological Measurement, 1972, 32, 543-557.
Purpose: To identify relationships among multi-item and
example-anchored personality rating scales
Subjects: 125 in- and out-patients of a state hospital
Variables: Scores on 59 instruments to measure 15 MMPI
areas
Method: Principal axes condensation with varimax rotation
Results: Nine factors emerged: I Confident cheer vs.
depressive anxiety and self-doubt; II Religious funda-
mentalism; III Conscientious self-control; IV Masculine
interests; V Family harmony; VI Social comfort; VII
Physical health; VIII Disruptive ideation; IX Methods
(multi-item) response set.
Tetenbaum, T. J. The role of student needs and teacher
orientations in student ratings of teachers. American
Educational Research Journal, 1975, 12, 417-433.
Purpose: To identify dimensions of 1) the Personality
Research Form (PRF), and 2) teacher vignette ratings
by students
Subjects: 1) Jackson's (1967) subjects, and 2) 405 graduate
education students
Variables: 1) Jackson's (1967) subscale correlation matrix
in the PRF manual; 2) scores on the College Teacher
Observation Schedule (CTOS) for 12 teacher vignettes
Method: 1) Unspecified factor analysis; 2) principal
axes condensation with varimax rotation
Results: Four PRF factors emerged: I Need for control;
II Need for intellectual striving; III Need for gre-
gariousness-dependence; IV Need for ascendancy. Four
CTOS factors emerged: I Control orientation; II
Intellectual striving orientation; III Gregariousness-
dependence orientation; IV Ascendancy orientation.
Conclusions: Twelve PRF subscales (three per factor) were
selected for use with the graduate students, and rela-
tionships between their PRF and CTOS scores were
examined.
Thorne, F. C. The Life Style Analysis. Journal of Clinical
Psychology, 1975, 31, 236-240.
Purpose: To identify factors underlying the Life Style
Analysis (LSA)
Subjects: Unspecified numbers of subjects composing two
normal and four clinical samples
Variables: 200 LSA scores
Method: Unspecified factor analysis
Results: Five "main" factors emerged: I Aggressive-
domineering life style; II Conforming life style; III
Defensive-withdrawal; IV Amoral sociopathy; V Resistive-
defiant life style.
Thorne, F. C., & Pishkin, V. A factorial study of needs in
relation to life styles. Journal of Clinical Psychology,
1975, 31, 240-248.

Purpose: To identify factors underlying the Life Style
Analysis (LSA)
Subjects: 267 incarcerated felons, 89 alcoholic inpatients,
314 college students, 336 unwed mothers, 390 chronic
schizophrenics
Variables: 200 LSA scores
Method: Principal components condensation with unspecified
rotation
Results: Five factors emerged: I Aggressive-domineering
life style; II Conforming life style; III Defensive
withdrawal; IV Amoral sociopathy; V Resistive-defiant
life style.
Conclusions: Item analyses across seven diagnostic groups
were also reported.
Trott, D. M., & Morf, M. E. A multimethod factor analysis
of the Differential Personality Inventory, Personality
Research Form, and Minnesota Multiphasic Personality
Inventory. _Journal_ _of_ _Counseling_ _Psychology_, 1972, _19_,
94-103.
Purpose: To identify dimensions common to the Differential
Personality Inventory (DPI), Personality Research Form
(PRF), and MMPI
Subjects: 151 college students seeking psychological ser-
vices
Variables: Scores on 14 DPI, 14 PRF, and 13 MMPI scales,
plus sex
Method: Principal components condensation with varimax
rotation
Results: Nineteen factors emerged: I Introversion-
extroversion; II Impulsivity vs. fearfulness and
caution; III Tendency to complain about physical
symptoms; IV Hostile, aggressive behavior; V Sex;
VI Defensiveness; VII Anti-conformity; VIII Physical
health, social reputation, and others' opinions concerns; IX Disharmony within/alienation from family;
X Pathological interpersonal sensitivity; XI Difficulty in control of hostile impulses with general,
irritable arousal; XII Depressed withdrawal vs.
affiliation; XIII Disorganization vs. order; XIV
Resistance to social pressure; XV Passivity-activity
depression; XVI Bizarreness; XVII Social and emotional
alienation; XVIII Strong achievement motivation vs.
malingering; XIX Predisposition to hysteria.
Vagg, P., Stanley, G., & Hammond, S. B. Invariance across
sex of factors derived from the Neuroticism Scale
Questionnaire. _Australian_ _Journal_ _of_ _Psychology_, 1972,
_24_, 37-44.
Purpose: To identify dimensions of the Neuroticism Scale
Questionnaire (NSQ)
Subjects: 119 female and 116 male college students
Variables: 40 item scores from the NSQ
Method: Principal axes condensation with varimax rotation,
independently by sex

Results: Seven male and eight female first-order factors
    emerged: I Unsociability (Social inhibition, females);
    II Anxiety; III Male role/tough (Non-impulsivity, for
    females); IV Vocational aspirations (Submissiveness,
    females); V Anxiety; VI High socialization; VII Group
    identification (Sober withdrawal, females); VIII Ad-
    justment. Three second-order factors emerged: I Un-
    sociability; II Anxiety or neuroticism; III Impulsive
    and tough-minded (Socialized submissiveness, females).
    When two second-order factors were extracted, good
    agreement across sex was achieved: I Introversion-
    invia; II Neuroticism-anxiety.
Viney, L. L. Multidimensionality of perceived locus of
    control: Two replications. Journal of Consulting and
    Clinical Psychology, 1974, 42, 463-464.
Purpose: To identify dimensions underlying Rotter's
    Internal-External (I-E) scale with Australian samples
Subjects: 159 14- to 19-year-old male, 137 18- to 20-
    year-old female Australians
Variables: 23 I-E scale item scores
Method: Principal components condensation with varimax
    rotation, by sample
Results: Two factors, accounting for 15% and 19% of the
    respective variances, emerged in each case: I Personal
    responsibility; II Social responsibility.
Wakefield, J. A. Jr., Bradley, P. E., Doughtie, E. B., &
    Kraft, I. A. Influence of overlapping and nonoverlapping
    items on the theoretical interrelationships of MMPI
    scales. Journal of Consulting and Clinical Psychology,
    1975, 43, 851-857.
Purpose: To identify MMPI dimensions for overlapping and
    nonoverlapping scores
Subjects: 100 adults (obtaining psychiatric services for
    their children
Variables: 3 validity and 10 clinical scale scores from
    MMPI, each scored for overlapping and nonoverlapping
    measures (26x26)
Method: Principal factors condensation with varimax
    rotation
Results: Fifteen unlabeled factors were extracted.
Wakefield, J. A. Jr., & Doughtie, E. B. The geometric re-
    lationship between Holland's personality typology and
    the Vocational Preference Inventory. Journal of
    Counseling Psychology, 1973, 20, 513-518.
Purpose: To identify dimensions underlying the Vocational
    Preference Inventory (VPI)
Subjects: 373 college students
Variables: 11 VPI scale scores
Method: Principal factors condensation with varimax
    rotation
Results: Six factors emerged: I Conventional economic;
    II Feminine, social; III Social desirability; IV
    Material world orientation; V Status; VI Artistic.

188

Wakefield, J. A. Jr., Yom, B. L., Bradley, P. E., Doughtie, E. B., Cox, J. A., & Kraft, I. A. Eysenck's personality dimensions: A model for the MMPI. British Journal of Social and Clinical Psychology, 1974, 13, 413-420.

Purpose: To identify dimensions underlying MMPI scale scores

Subjects: 205 married couples

Variables: Husband score and wife score on each of 3 validity and 10 clinical MMPI scales (i.e., 26 variables)

Method: Principal components condensation with varimax rotation

Results: Eight factors emerged: I General male personality; II Female personality; III Female introversion; IV Female neuroticism; V Male variance; VI Male introversion; VII Mf variance for both sexes; VIII Female L scale.

Conclusions: The results are discussed in relation to Eysenck's personality dimensions.

Wakefield, J. A., Yom, B. L., Doughtie, E. B., Chang, W. C., & Alston, H. L. The geometric relationship between Holland's personality typology and the Vocational Preference Inventory for blacks. Journal of Counseling Psychology, 1975, 22, 58-60.

Purpose: To identify dimensions underlying the Vocational Preference Inventory (VPI) for a black sample

Subjects: 115 black college students

Variables: 11 VPI scale scores

Method: Principal factors condensation with unspecified rotation

Results: Six factors emerged: I Realistic; II Intellectual; III Social; IV Conventional; V Enterprising; VI Artistic.

Wilcox, A. H., & Fretz, B. R. Actual-ideal discrepancies and adjustment. Journal of Counseling Psychology, 1971, 18, 166-169.

Purpose: To identify relationships between California Psychological Inventory (CPI) and concept discrepancy scores

Subjects: 43 male college students

Variables: Scores on 6 CPI scales and 3 (semantic differential) concept discrepancies

Method: Principal components condensation with varimax rotation

Results: Two factors emerged: I Adjustment; II Environmental discrepancy.

Williams, C. A study of cognitive preferences. Journal of Experimental Education, 1975, 43, 61-77.

Purpose: To identify relationships between the Cognitive Preference Test (CPT) and other selected variables

Subjects: 1) 62 liberal arts/education majors; 2) 50 vocational-technical students

Variables: 9 CPT scores, Hidden Patterns, Thing Categories, Number span, and 1) 4 ACT or 2) 6 GATB scores
Method: Principal components condensation with normal varimax rotation, independently for each sample
Results: Eight factors emerged in the first analysis: I Unlabeled (CPT); II General ability; III Science; IV Divergent thinking; V Memory span; VI Convergent thinking; VII Mathematics; VIII Preference for facts. Eight factors emerged in the second analysis: I Test format, response set, or principle application; II Numerical reasoning; III Cognitive flexibility; IV Clerical ability; V Memory span; VI Verbal reasoning; VII Fact preference; VIII Spatial perception.

Wright, L., & Wyant, K. Factor structure of self-actualization as measured from Shostrom's SAV Scale. Educational and Psychological Measurement, 1974, 34, 871-875.
Purpose: To identify dimensions underlying Shostrom's SAV Scale
Subjects: 393 college students
Variables: 26 SAV Scale item scores
Method: Principal components condensation with varimax rotation
Results: Nine factors emerged: I Identity with mankind; II Ego control of sexual and aggressive impulses; III Autonomy and independence of values; IV Spontaneity; V Realistically oriented; VI Problem centered rather than self-centered; VII Autonomy and independence of behavior; VIII Faith in others and the natural world; IX Freshness of appreciation and resistence to conformity.

Wright, T. L., & Tedeschi, R. G. Factor analysis of the Interpersonal Trust Scale. Journal of Consulting and Clinical Psychology, 1975, 43, 470-477.
Purpose: To identify dimensions underlying Rotter's Interpersonal Trust Scale (ITS)
Subjects: 1) 560 male, 679 female, 2) 381 male, 312 female, 3) 494 male, 514 female, 4) 282 male, 411 female college students
Variables: 15 ITS item scores, plus sex
Method: Principal axes condensation with varimax rotation, by sample
Results: Three factors, of four extracted, replicated across samples: I Political trust; II Paternal trust; III Trust of strangers.

Zuckerman, M. Dimensions of sensation seeking. Journal of Consulting and Clinical Psychology, 1971, 36, 45-52.
Purpose: To identify dimensions of a revised Sensation Seeking Scale (SSS)
Subjects: 160 male and 172 female college students
Variables: Scores on 50 original SS plus 63 new items
Method: Principal components condensation with varimax (and oblimin) rotation, by sex

Results: Four interpretable factors emerged for both
    sexes: I Thrill and adventure seeking; II Experience
    seeking; III Disinhibition; IV Boredom susceptibility.
2. Projective Techniques
  Campus, N. Transituational consistency as a dimension of
    personality. Journal of Personality and Social
    Psychology, 1974, 29, 593-600.
  Purpose: To identify dimensions underlying need rating
    stability across TAT cards
  Subjects: 191 college students
  Variables: Scores on 17 TAT-scored needs 1) averaged
    across 4 situations, 2) not averaged
  Method: Unspecified condensation with varimax rotation,
    for the 2 data sets
  Results: Four factors emerged in both analyses: I
    Mastery over the environment; II Conformity and sub-
    missiveness; III Overt hostility and hostility turned
    inward; IV Not labeled.
  Rimoldi, H. J. A., Insua, A. M., & Erdmann, J. B. Person-
    ality dimensions as assessed by projective and verbal
    instruments. Journal of Clinical Psychology, 1975,
    529-539.
  Purpose: To identify personality dimensions as assessed
    by selected projective and verbal tests
  Subjects: 161 college students
  Variables: 1) 21 Holtzman Inkblot Technique (HIT) scores;
    2) 19 scores from the California Psychological Inven-
    tory (CPI) and Rokeach Dogmatism Scale (RDS); 3) 21
    HIT and 19 CPI-RDS scores
  Method: Principal axes condensation with varimax and
    oblique graphic rotations
  Results: Six HIT factors emerged, and four CPI-RDS fac-
    tors emerged. Nine factors emerged from the combined
    analysis: I Good adjustment to social environment,
    self-controlled, productive, mature, tolerant, intel-
    lectually efficient, and not closed-minded (bipolar);
    II Unlabeled; III Self-assurance; IV Trustworthy per-
    son who conforms to society's norms; V Easily aroused
    emotional life and sensitivity to the most colorful
    aspect of the external world; VI Patient affection with
    control of the aggressive aspects of emotionality;
    VII Opposite of a clear, realistic and detailed ap-
    proach to the world; VIII Feelings of anxiety of a
    free-floating nature; IX Capacity for independent
    thought and action.
  Smith, P. M., & Barclay, A. G. Q analysis of the Holtzman
    Inkblot Technique. Journal of Clinical Psychology,
    1975, 31, 131-134.
  Purpose: To obtain a Q-analysis solution for three
    groups of subjects on the Holtzman Inkblot Technique
    (HIT)

Subjects: 20 normal, 20 delinquent, and 20 retarded, black male adolescent minors from socioeconomically depressed neighborhoods
Variables: Binary scores on 19 (of the 22) HIT variables, transposed to a G matrix
Method: Principal axes condensation with unspecified rotation
Results: Four unlabeled factors emerged; the normal group tended to cluster on one factor, while the delinquent and retarded groups each tended to load on two factors.

3. Rorschach Test

Ekehammar, B. A psychophysical approach to the study of individuals' perceptions of Rorschach cards. Perceptual and Motor Skills, 1971, 33, 951-965.
Purpose: To identify dimensions underlying similarity judgments of Rorschach cards
Subjects: 5 male, 5 female college students
Variables: 11-point similarity judgments for 10 Rorschach cards
Method: 1) Principal components condensation with varimax rotation, 2) elementary linkage analysis, both by sex, by individual, and for total group
Results: Three factors emerged for both sexes and group data: I Closed form; II Open form; III Color. The cluster analytic results by sex and for the group were consistent (i.e., 3-cluster solutions). In general, three factors emerged for individual data as well. Two clusters emerged in the final analysis: I Male; II Female.

Ekehammar, B. Two multidimensional methods applied to two types of stimuli. Perceptual and Motor Skills, 1972, 34, 535-542.
Study One:
Purpose: To identify dimensions underlying similarity judgments of Rorschach cards
Subjects: 5 male, 5 female college students
Variables: 11-point (correlational) similarity ratings for 10 Rorschach cards
Method: Principal components condensation with varimax rotation
Results: Two factors emerged: I Closed form; II Open form.
Conclusions: These were compared with Ekehammar (1971).
Study Two:
Subjects: 5 male, 7 female college students
Variables: 5-point (correlational) similarity ratings of 36 verbally-described situations
Method: Same as above
Results: Four factors emerged: I Positive-negative; II Active-passive; III Social; IV Uninterpreted.
Conclusions: These were compared with Magnusson (1971).

Schori, T. R., & Thomas, C. B.   The Rorschach Test: An
     image analysis.   Journal of Clinical Psychology, 1972,
     28, 195-199.
  Purpose: To identify dimensions of Rorschach responses
  Subjects: 568 medical students (97% European, 92% male)
  Variables: 35 transformed Rorschach variable (Beck, et
     al. system, 1961) scores (linear composites deleted)
  Method: Image analysis (principal axes condensation
     with varimax rotation)
  Results: Four factors emerged: I Intellectual produc-
     tivity; II Form; III Human-movement; IV Holism.

IX. CLINICAL PSYCHOLOGY

A. Personnel
    Dorr, D., Cowen, E. L., Sandler, I., & Pratt, D. M. Dimen-
        sionality of a test battery for nonprofessional mental
        health workers. Journal of Consulting and Clinical
        Psychology, 1973, 41, 181-185.
    Purpose: To identify dimensions underlying a battery of
        interest, attitude, and personality measures for non-
        professional mental health workers
    Subjects: 139 predominantly middle-class female nonpro-
        fessional mental health workers
    Variables: 15 PRF, 23 SVIB, 4 Situational Response Test,
        10 semantic differential, 1 Social Desirability, 1
        Empathy, and 3 demographic scores
    Method: Principal components condensation with varimax
        rotation
    Results: Six factors, of 20 extracted, were interpreted:
        I Cultural interest; II Semantic differential or job-
        role attitude; III Extraversion; IV Helping-person;
        V Social class; VI Interest in science.
    Lamberd, W. G., Adamson, J. D., & Burdick, J. A. A study of
        self-image experience in student psychotherapists.
        Journal of Nervous and Mental Disease, 1972, 155, 184-
        191.
    Purpose: To identify dimensions of resident psychothera-
        pists' self-concepts
    Subjects: 10 resident psychotherapists (minimum 1 yr.
        experience)
    Variables: Ratings on a 37-item psychotherapist self-
        concept checklist, administered before and after self-
        viewing on videotape
    Method: Principal components condensation, apparently
        without rotation
    Results: Three factors emerged: I Evaluative; II Activity;
        III Warmth.
B. Psychotherapy and Analysis
    Fiester, A. R., & Rudestam, K. E. A multivariate analysis
        of the early dropout process. Journal of Consulting
        and Clinical Psychology, 1975, 43, 528-535.
    Purpose: To identify dimensions of patient input, thera-
        pist input, and patient perspective therapy process
        variables which discriminated early- vs. non-dropout
        patients
    Subjects: 45 hospital center (HC), 26 state clinic (SC)
        dropouts and 75 HC, 35 SC non-dropouts accepted for
        outpatient treatment (all adults)
    Variables: Scores on 1) 13 (HC), 2) 9 (SC) patient input,
        therapist input, and patient perspective therapy pro-
        cess variables which discriminated the two groups
    Method: Principal components condensation with varimax
        rotation, by group within each setting

194

Results: Four factors emerged for HC dropouts: I Patient-therapist match; II Collaborative involvement; III Direct effective therapist; IV Intimate effective therapist. Four factors emerged for HC non-dropouts: I Collaborative involvement; II Patient satisfaction with intimate therapist; III Attacking patient anticipating didactic therapist; IV Uninterpreted. Four unlabelled, but reportedly highly similar, factors emerged in the SC analyses.

Fitzgibbons, D. J. Social class differences in patients' self-perceived treatment needs. Psychological Reports, 1972, 31, 987-997.

Purpose: To identify dimensions underlying patients' self-perceived treatment needs for high-SES Ss

Subjects: 66 male and 94 female private psychiatric hospital inpatients

Variables: Unspecified number (23?) of item scores from the Patients' Self-perceived Treatment Need Scale

Method: Principal components condensation with varimax rotation

Results: Eight factors emerged (and were compared with those for a low SES sample): I Anxiety-depression; II Psychosis; III Economic-vocational; IV Loss of control; V Alcohol; VI Religion; VII Inadequacy and physical symptoms; VIII Not interpreted.

Fitzgibbons, D. J., Cutler, R., & Cohen, J. Patients' self-perceived treatment needs and their relationship to background variables. Journal of Consulting and Clinical Psychology, 1971, 37, 253-258.

Purpose: To identify dimensions underlying patients' self-perceived treatment needs

Subjects: 118 female and 114 male psychiatric patients of caucasian, Puerto Rican, and negro backgrounds

Variables: Scores on 93 perceived-treatment-needs items

Method: Principal components condensation with varimax rotation

Results: Seven interpretable factors emerged: I Anxiety-depression; II Superego; III Psychosis; IV Physical symptoms; V Economic-vocational; VI Inadequacy; VII Marriage problems.

Garfield, S. L., Prager, R. A., & Bergin, A. E. Evaluation of outcome in psychotherapy. Journal of Consulting and Clinical Psychology, 1971, 37, 307-313.

Purpose: To identify dimensions underlying several psychotherapeutic outcome measures

Subjects: "Essentially the same sample of therapists and clients" (as the present study: 19 therapists and their 34 clients); unclear as to which of following data sources: Garfield & Bergin (1971), Prager (unpublished dissertation, 1970), Rossky & Dickey (unpublished manuscript, 1969)

Variables: 13 outcome measures

Method: Principal components condensation, apparently
    without rotation
Results: Two factors, of five extracted, were interpreted
    with surety: I Ubiquitous client self-evaluation; II
    Therapist-supervisor; III Ego strength; IV-V Mixed.
Green, B. L., Gleser, G. C., Stone, W. N., & Seifert, R. F.
    Relationships among diverse measures of psychotherapy
    outcome. Journal of Consulting and Clinical Psychology,
    1975, 43, 689-699.
Purpose: To identify dimensions underlying several mea-
    sures of psychotherapy outcome
Subjects: 50 outpatients in brief crisis-oriented psycho-
    therapy (with 7 second- and third-year residents)
Variables: 10 difference scores (5 SCL, Hamilton, 4 PEF)
    and 3 direct ratings of change
Method: Principal factors condensation with oblique rota-
    tion
Results: Three unlabeled factors emerged.
Hargreaves, W. A., Showstack, J., Flohr, R., Brady, C., &
    Harris, S. Treatment acceptance following intake as-
    signment to individual therapy, group therapy, or
    contact group. Archives of General Psychiatry, 1974,
    31, 343-349.
Purpose: To identify dimensions underlying an intake in-
    ventory (II)
Subjects: 143 applicants to an outpatient psychiatric
    clinic
Variables: 27 II items
Method: Principal components condensation with varimax
    rotation
Results: Two factors were interpreted: I Motivated; II
    Shy-upset.
Hautaluoma, J. Syndromes, antecedents, and outcomes of
    psychosis: A cluster-analytic study. Journal of
    Consulting and Clinical Psychology, 1971, 37, 332-244.
Purpose: To identify clusters of psychiatric syndromes,
    of outcomes, of antecedents, (and to identify relation-
    ships among them)
Subjects: 1,099 mental health center patients (in 3 sub-
    samples of equal N)
Variables: Scores (from records) on 1) 120 symptom items,
    2) 27 outcome items, 3) 57 antecedent items
Method: Cluster analysis
Results: Eleven syndrome clusters consistently emerged:
    I Conceptual disorganization; II Depression; III
    Severity; IV Paranoid; V Interpersonal relations; VI
    Suicidal tendencies; VII Disorientation; VIII Retarda-
    tion; IX Hostility; X Ward hostility; XI Ward retarded.
    Six replicated outcome factors emerged: I Responses to
    therapy; II Change in ward hostility; III Time in treat-
    ment; IV Change in ward thought disorders; V Acceptance
    by others and prognosis; VI Change in ward social rela-
    tionships. Twelve consistent antecedent clusters emerged:

I Retired Spanish-American; II Poor; III Dominated and overprotected; IV Intelligence and education; V Predisposition toward illness; VI Traumas; VII Slow onset of disorder; VIII Unstable veteran; IX Inconsistency of relationships; X Negative dependency; XI Married; XII Not reported.

Martin, P. J., & Sterne, A. L. Prognostic expectations and treatment outcome. Journal of Consulting and Clinical Psychology, 1975, 43, 572-576.

Purpose: To identify dimensions of the Patient Prognostic Expectancy Inventory (PPEI) and Therapist Prognostic Expectancy Inventory (TPEI)

Subjects: 150 adult inpatients (mixed diagnoses), 14 staff psychiatrists and first-year psychiatry residents

Variables: Scores on 15 PPEI (patient) or TPEI (staff) items

Method: Principal components condensation with varimax rotation, independently for patients and staff

Results: Six PPEI factors emerged: I Anticipated improvement in manic and fearful behavior; II Anxiety and somatic disturbances; III Thought disturbances; IV Social inadequacy; V Depression anxiety, and fear; VI Dependence, aggression, and unconscious control of behavior. Six TPEI factors emerged: I Anticipated reductions in impulsivity and excessive energy; II Depression, somatic disturbance, and fatigue; III Disturbances in thought; IV Psychic and somatic anxiety; V Fearfulness and unconscious control of behavior; VI Social adequacy.

Patterson, V., Levene, H., & Breger, L. Treatment and training outcomes with two time-limited therapies. Archives of General Psychiatry, 1971, 25, 161-167.

Purpose: To identify dimensions of therapist attitudes

Subjects: 8 trainees, 2 consultants, and 3 research staff of a training/treatment project

Variables: Unspecified number of scores from California Psychological Inventory, Opinions about Mental Illness Scale, Corbett Inventory of Opinions for Judging Therapeutic Improvement, and attitude and semantic differential scores for two time-limited therapies

Method: Unspecified condensation with unspecified rotation

Results: Two factors emerged: I Behavior therapy; II Idealized, graduate-school version of a psychotherapist.

Pettit, I. B., Pettit, T. F., & Welkowitz, J. Relationship between values, social class, and duration of psychotherapy. Journal of Consulting and Clinical Psychology, 1974, 42, 482-490.

Purpose: To identify dimensions underlying, and relationships between, Study of Values (SV), Ways to Live Scale (WLS), Strong Vocational Interest Blank (SVIB), and Optimal Personality Integratation Scale (OPIS)

Subjects: 1) 249 outpatients and 104 therapists, 2) 104 of the patients and 104 therapists

Variables: 1) 6 SV, 54 SVIB scale scores, 13 WLS and 20 OPIS item scores, 2) scores on the 19 factors identified in (1)

Method: Principal axes condensation with varimax rotation, independently for each instrument in (1) and for combined factors in (2)

Results: Three SV, 6 SVIB, 5 WLS, and 5 OPIS factors emerged but were not labeled. Six factors emerged for the combined analysis: I Rotary Club vs. aesthetic interest; II Authoritarian-submissive vs. independent; III Transcendentalism vs. concrete-rational; IV Neuroticism vs. dynamic mastery; V Conventional middle America vs. intellectual pursuits; VI Scientist vs. salesman.

Rice, D. G., Gurman, A. S., & Razin, A. M. Therapist sex, style, and theoretical orientation. Journal of Nervous and Mental Disease, 1974, 159, 413-421.

Purpose: To identify dimensions of therapist style

Subjects: 86 therapists

Variables: 23 item scores from a self-report style questionnaire

Method: Principal components condensation with varimax rotation

Results: Eight factors emerged: I Low activity level; II Directed focus; III Cognitive/goal emphasis; IV Traditional; V Rigid/mechanical; VI Feeling responsiveness; VII Judgmental; VIII Supportive.

Conclusions: These results were compared with those from Rice, Fey, and Kepecs (1972); differences between male and female, experienced and inexperienced therapists were examined.

Saccuzzo, D. P. What patients want from counseling and psychotherapy. Journal of Clinical Psychology, 1975, 31, 471-475.

Purpose: To identify dimensions of patients' and therapists' wants relative to counseling and psychotherapy

Subjects: 57 students,seen at a university clinic, and the (unspecified number of) therapists who saw them at intake

Variables: 14 want items from the Therapy Session Report Questionnaire

Method: Principal axes (and principal components) condensation with normal varimax rotation, for patient and therapist data independently

Results: Five factors emerged: I Self-exploration; II Catharsis; III Encouragement; IV Particular problem (patient), or Personal response (therapist); V "Other."

Shapiro, R. J. Therapist attitudes and premature termination in family and individual therapy. Journal of Nervous and Mental Disease, 1974, 159, 101-107.

Purpose: To identify dimensions of a therapist-attitude instrument

198

Subjects: 51 families referred to a clinic and evaluated
for individual or family therapy
Variables: 14 item scores from initial-evaluation questionnaire designed to measure therapist's affective reactions
to patients, degree of pathology, and treatment prognosis
(mean, if evaluated by co-therapists)
Method: Principal components condensation with varimax
rotation
Results: Three factors emerged: I Affective reaction; II
Degree of psychopathology; III Treatment prognosis.
Conclusions: Factor scores were examined relative to
family-individual therapy and premature termination.
1. Therapeutic Process
Barak, A., & LaCrosse, M. B. Multidimensional perception of
counselor behavior. Journal of Counseling Psychology,
1975, 22, 471-476.
Purpose: To identify dimensions of perceived counselor
behavior
Subjects: 202 college students (201 for Perls analysis)
Variables: 36 adjective scale ratings of films of inter-
views by Rogers, Ellis, and Perls
Method: Principal factors condensation with varimax rotation, by counselor
Results: Three similar factors emerged for Rogers and Perls:
I Expertness; II Attractiveness; III Trustworthiness.
Three factors emerged for Ellis: I Expertness and
trustworthiness; II Attractiveness (plus several other
loadings); III (Highest loading=.28 for selfless-selfish).
Berzins, J. I., Bednar, R. L., & Severy, L. J. The problem
of intersource consensus in measuring therapeutic out-
comes: New data and multivariate perspectives. Journal
of Abnormal Psychology, 1975, 84, 10-19.
Purpose: To identify 1) dimensions underlying therapeutic
outcome measures, and 2) clusters of dyads
Subjects: 79 therapist-patient dyads
Variables: Scores on 1) 15 outcome measures (residual
gains), 2) profiles on 4 factors from (1)
Method: 1) Principal components condensation with varimax
rotation, 2) correlational cluster analysis
Results: Four factors emerged in the first analysis: I
Changes in patient-experienced distress; II Changes in
observable maladjustment; III Changes in impulse expression; IV Changes in self-acceptance. Four clusters of
dyads emerged: I Consensual agreement; II Consensual
deterioration; III Patient view of lack of improvement,
psychometrist opposite; IV Flat profile.
Berzins, J. I., Ross, W. F., & Friedman, W. H. A-B therapist
distinction, patient diagnosis, and outcome of brief
psychotherapy in a college clinic. Journal of Consulting
and Clinical Psychology, 1972, 38, 231-237.
Purpose: To identify dimensions underlying psychotherapeutic outcome measures
Subjects: 391 male psychotherapy patients

Variables: 7 therapist- and 6 patient-generated post-
therapy ratings
Method: Principal axes condensation with varimax rotation
Results: Three factors emerged: I Therapist's appraisal
of own effectiveness; II Patient improvement; III
Patient-experienced rapport.
Friel, T. W., Berenson, B. G., & Mitchell, K. M.   Factor
analysis of therapeutic conditions for high and low
functioning therapists.   Journal of Clinical Psychology,
1971, 27, 291-293.
Purpose: To identify dimensions of therapeutic conditions
for high and low functioning therapists
Subjects: 13 high- and 32 low-functioning therapists (interns to 18 years of experience)
Variables: Ratings on empathy, regard, genuineness, concreteness, client self-exploration, immediacy, confrontation, and relationship to significant others, all at
base rate and post-confrontation periods of first
therapy session
Method: Unspecified condensation with varimax rotation,
independently for high- and low-functioning therapists
for each period
Results: For low-functioning therapists, three factors
emerged for base rate (I Low client self-exploration
elicited by low facilitative conditions; II Non-
immediate and non-significant others; III Low client
self-exploration as a function of more confrontation)
and three for post-confrontation (I Low levels of
facilitative conditions, immediacy, and confrontations;
II Not labeled; III Low client self-exploration).   For
high functioning therapists, three base rate factors
(I Confrontation with immediacy, empathy, genuineness,
and regard; II Client self-exploration in response to
concreteness and empathy; III Concreteness related to
reference to significant others) and three post-confrontation factors (I Immediacy with high facilitative
conditions; II Continued confrontation with added
immediacy; III Client self-exploration) emerged.
Heckel, R. V., Holmes, G. R., & Rosecrans, C. J.   A factor
analytic study of process variables in group therapy.
Journal of Clinical Psychology, 1971, 27, 146-150.
Purpose: To identify dimensions of group therapy process
Subjects: 30 male neuropsychiatric VA inpatients in group
therapy for Time 1 (Sessions 2-3), and 17 still participating for Time 2 (Sessions 12-13)
Variables: (Observer) ratings on 11 process variables
(patient verbal response), averaged for each set of 2
consecutive sessions
Method: Centroid condensation with maxplane rotation,
independently for Times 1 and 2
Results: Six factors emerged for Time 1 and were labeled
as a person exhibiting those behaviors would be: I

Egocentric participator; II Guarded, impersonal environmental commentator; III Superficial group interactor; IV Passive, noncommittal, opinion-seeker; V Therapist-directed, environmental commentator; VI Occasional opinion-seeker. Six factors emerged for Time 2: I Cohesive group builder; II Superficial environmental commentator; III Questioning-information seeker; IV Radar antenna type; V Self-oriented verbalizer; VI Occasional personal information giver.

Luborsky, L., Crabtree, L., Curtis, H., Ruff, G., & Mintz, J. The concept 'space' of transference for eight psychoanalysts. British Journal of Medical Psychology, 1975, 48, 65-70.
Purpose: To identify dimensions of transference
Subjects: 8 psychoanalysts, rating 82 segments of tape-recorded analysis sessions
Variables: Ratings on 23 concepts related to transference
Method: Principal components condensation with varimax rotation, independently for each psychoanalyst (and combined) at first-order level and for all first-order factors
Results: Four second-order factors emerged: I Expressed parental; II Infantile distortion; III Positive erotic; IV Sibling.

Menne, J. M. A comprehensive set of counselor competencies. Journal of Counseling Psychology, 1975, 22, 547-553.
Purpose: To identify dimensions of counselor-perceived counselor competencies
Subjects: 376 experienced counselors and therapists
Variables: Importance ratings for 132 competencies
Method: Principal components condensation with varimax rotation
Results: Twelve factors emerged: I Professional ethics; II Self-awareness; III Personal characteristics; IV Listening, communicating; V Testing skills; VI Counseling comprehension; VII Behavioral science; VIII Societal awareness; IX Tutoring techniques; X Professional credentials; XI Counselor training; XII Vocational guidance.
Conclusions: Differences in factor scores were examined as a function of eight background variables, including work setting.

Mintz, J., Auerbach, A. H., Luborsky, L., & Johnson, M. Patient's, therapist's, and observers' views of psychotherapy: A 'Rashomon' experience or a reasonable consensus? British Journal of Medical Psychology, 1973, 46, 83-89.
Purpose: To identify dimensions underlying those items of the Therapy Session Report (TSR) common to the therapist and patient forms
Subjects: Patient, therapist, and 2 observers for each of 48 sessions (192 ratings)

Variables: 129 TSR items common to therapist and patient
    forms
Method: Unspecified factor analysis
Results: Seven factors emerged and were retained for further
    study: I Helpful involved therapist; II Patient distress;
    III Active experiencing patient; IV Patient wants per-
    sonal response; V Affection and sex; VI Angry assertive
    patient; VII Concern with religion and self-control.
Mintz, J., & Luborsky, L. Segments versus whole sessions:
    Which is the better unit for psychotherapy process
    research. Journal of Abnormal Psychology, 1971, 78,
    180-191.
Purpose: To identify psychotherapy process dimensions
    based on segment- and session-based judgments
Subjects: 30 therapist-client pairs, rated by 3 raters
Variables: Scores on 30 psychotherapy process variables,
    rated by segment (180) or by session (2)
Method: Principal factors condensation with promax rota-
    tion, by rating base
Results: Five segment factors emerged: I Empathetic rela-
    tionship; II Therapist interventions; III Patient
    activity; IV Patient involvement; V Patient health vs.
    distress. Four session factors emerged: I Patient
    health vs. distress; II Optimal empathic relationship;
    III Active directive mode; IV Interpretive mode with
    receptive patient.
Mintz, J., Luborsky, L., & Auerbach, A. H. Dimensions of
    psychotherapy: A factor-analytic study of rating of
    psychotherapy sessions. Journal of Consulting and
    Clinical Psychology, 1971, 36, 106-120.
Purpose: To identify 1) clusters of psychotherapy vari-
    ables, and 2) dimensions underlying the clustered
    variables
Subjects: 15 experienced psychotherapists and 2 each of
    their patients, in (two) psychotherapy sessions
Variables: Scores on 1) 110 patient and therapist vari-
    ables (pooled over 3 raters, summed over two sessions),
    2) summed variables for 27 clusters defined in (1) by
    session
Method: 1) Hierarchical linkage analysis; 2) principal
    axes condensation with varimax rotation
Results: Twenty-seven clusters emerged in the first
    analysis: I Therapist unclear response; II Therapist
    hostile-defensive; III Therapist speech tempo; IV
    Therapist perceptiveness; V Therapist accepts patient;
    VI Rater likes therapist calm; VII Rater contamination;
    VIII Therapist activity; IX Patient receptive; X Ther-
    apist secure; XI Therapist empathy; XII Therapist skill;
    XIII Therapist experiential; XIV Therapist interpreta-
    tions/impact; XV Therapist intrusive; XVI Therapist
    transference; XVII Therapist creativity; XVIII Patient
    anxiety; XIX Patient guilt/shame; XX Patient depression;
    XXI Patient hostile; XXII Patient dependent; XXIII

202

Therapist clarifies; XXIV Patient effect on others;
XXV Patient health - sickness; XXVI Therapist directive;
XXVII Therapist reassurance and warmth. Four factors
were interpreted in the second analysis: I Optimal
empathic relationship; II Directive mode; III Patient
health versus distress; IV Interpretive mode.

Orlinsky, D. E., Howard, K. I., & Hill, J. A. Conjoint
therapeutic experience: Some dimensions and determinants. Multivariate Behavioral Research, 1975, 10,
463-477.

Purpose: To identify dimensions of conjoint experience
within the therapeutic dyad

Subjects: 28 patient-therapist pairs

Variables: Scores on 11 patient and 11 therapist second-
order TSR factors

Method: Unspecified condensation with varimax rotation

Results: Seven factors emerged: I Therapist agency vs.
therapist catalysis; II Patient agency vs. patient
passivity; III Productive rapport vs. unproductive
contact; IV Ambivalent nurturance-dependence; V Healing
magic vs. uncomfortable involvement; VI Sympathetic
warmth vs. conflictual erotization; VII Therapeutic
alliance vs. defensive impasse.

Rabiner, E. L., Reiser, M. F., Barr, H. L., & Gralnick, A.
Therapists' attitudes and patients' clinical status:
A study of 100 psychotherapy pairs. Archives of
General Psychology, 1971, 25, 555-569.

Purpose: To identify dimensions underlying therapist
attitudes and patient response to therapist (over time)
for 66 psychotherapy pairs

Subjects: 66 psychotherapy pairs

Variables: Scores on 146 measures (taken at one of four
random times) of therapist attitude and patient response
to therapist

Method: Centroid condensation with varimax rotation

Results: Three factors, of eight extracted, were interpreted:
I Therapist attitudinal set; II Obtrusive sick behaviors;
III Unobtrusive sick behaviors.

Raming, H. E., & Frey, D. H. A taxonomic approach to the
Gestalt theory of Perls. Journal of Counseling Psychology,
1974, 21, 179-184.

Purpose: To identify clusters of Perls' process and goal
statements

Subjects: 34 graduate counseling students

Variables: 5-point scores on 100 Perls Inventory of Counseling Goals and 131 Perls Inventory of Counseling
Process items

Method: Multiple linkage cluster analysis

Results: Three goal (I Organism and environment; II Self-
awareness; III Maturation and autonomy) and two process
clusters (I Skillful frustration of the client; II The
here-and-now) emerged.

Ryle, A., & Lipshitz, S. Towards an informed countertransference: The possible contribution of repertory grid
techniques. British Journal of Medical Psychology,
1974, 47, 219-225.
 Purpose: To identify countertransference dimensions in
 a repertory grid
 Subjects: 1) A therapist,rating 3 long-term therapy
 patients; 2) 6 nurses, rating an unspecified number of
 patients known to all of them
 Variables: 1) 27+-element, 3-construct repertory grid related to therapy session; 2) repertory grid (unspecified
 number of elements, constructs) related to patient
 description and countertransference
 Method: Principal components condensation, independently
 by grid
 Results: An unspecified number of factors were extracted;
 graphical presentation of first two components was used
 as feedback to subjects.
Saccuzzo, D. P. Naturalistic analysis of verbal behavior in
psychotherapy. Psychological Reports, 1975, 37, 911-919.
 Purpose: To identify dimensions underlying the Therapy
 Session Report (TSR)
 Subjects: 57 patients (Ps) and 19 therapists (Ts)
 Variables: 20 TSR dialogue item scores
 Method: Principal axes condensation with varimax rotation,
 independently for Ps and Ts
 Results: Seven factors emerged: I Parental family; II
 Education and career; III Fantasy; IV Domestic affairs;
 V Sex relations; VI Psychophysical reactions (P), or
 Religion (T); VII Finance (P), or Other (T).
Shaffer, W. F., Hummel, T. J., & Mastbaum, N. A. Multi-
variate analysis of variance and factor analysis of
textbook counselor leads. Journal of Clinical Psy-
chology, 1972, 28, 219-223.
 Purpose: To identify empirical dimensions of textbook
 counselor leads
 Subjects: 267 college students
 Variables: Unspecified number (less than 38) of 10-point
 helpfulness ratings of counselor leads (questionnaire
 format)
 Method: Principal-factor condensation with varimax rotation
 Results: Four factors emerged: I Rogerian statements; II
 "Bad" statements from each orientation; III Analytic;
 IV Educational-vocational counselor statements.
Weissman, H. N., Goldschmid, M. L., & Stein, D. D. Psycho-
therapeutic orientation and training: Their relation
to the practices of clinical psychologists. Journal
of Consulting and Clinical Psychology, 1971, 37, 31-37.
 Purpose: To identify dimensions underlying practices (use
 of techniques) of clinical psychologists
 Subjects: 244 APA Division 12 members

204

Variables: Scores (6-point) on extent of use of 27 therapeutic techniques
Method: Maximum likelihood condensation, apparently without rotation
Results: Nine factors emerged: I Egalitarian 1; II Normalist 1; III Dogmatist 1; IV Authoritarian 1; V Normalist 2; VI Pragmatist 1; VII Dogmatist 2; VIII Egalitarian 2; IX Pragmatist 2.

Zimmer, J. M., Hakstian, A. R., & Newby, J. F. Dimensions of counselee responses over several therapy sessions. Journal of Counseling Psychology, 1972, 19, 448-454.
Purpose: To identify dimensions of counselee response in therapy sessions
Subjects: One client responding a total of 70 times to Rogers, Ellis, and Perls, rated by 3 raters (2 graduate students, 1 faculty member in counseling)
Variables: Ratings on 30 client response characteristics
Method: Image factor condensation with varimax followed by oblique rotation
Results: Thirteen unlabeled first-order factors emerged. Seven second-order factors emerged: I Aggressive assertiveness; II Dependent help-seeking; III Insecurity; IV Hostile guardedness; V Expressiveness; VI Sincere sensitiveness; VII Ego defense.

Zimmer, J. M., & Pepyne, E. W. A descriptive and comparative study of dimensions of counselor response. Journal of Counseling Psychology, 1971, 18, 441-447.
Purpose: To identify dimensions of counselor behavior
Subjects: Rogers, Ellis, and Perls, rated on 23 responses each by two experienced counselors
Variables: Scores on 31 counselor process variables
Method: Principal components condensation with varimax rotation
Results: Six factors emerged: I Rational analyzing; II Eliciting specificity; III Confronting; IV Passive structuring; V Reconstructing; VI Interrogating.

2. Group Therapy
Dies, R. R. Group therapist self-disclosure: An evaluation by clients. Journal of Counseling Psychology, 1973, 20, 344-348.
Purpose: To identify dimensions underlying client evaluations of group therapists
Subjects: 6 male, 4 female group therapists, rated by a total of 24 clients
Variables: Scores on 20 bipolar adjective scales designed to assess therapist likability, perceived level of helpfulness, and emotional stability
Method: Unspecified factor analysis
Results: Two unlabeled factors emerged.

Dies, R. R. Group therapist self-disclosure: Development and validation of a scale. Journal of Consulting and Clinical Psychology, 1973, 41, 97-103.

Purpose: To identify dimensions underlying the Group
    Therapist Orientation Scale (GTOS)
Subjects: 87 advanced graduate students, 56 mental health
    professionals
Variables: Scores on 20 GTOS and 10 filler items
Method: Unspecified factor analysis
Results: Factor I accounted for approximately 31% of the
    total variance; 20 items had loadings in excess of .50.
Fielding, J. M. A technique for measuring outcome in group
    psychotherapy. British Journal of Medical Psychology,
    1975, 48, 189-198.
Purpose: To identify dimensions of psychotherapeutic
    change
Subjects: 8 outpatient group psychotherapy members
Variables: Scores on a Symptom Check List-derived discom-
    fort scale plus 10 patient and 2 therapist repertory
    grid variables, collected at pre-treatment and at least
    10 during-treatment sessions
Method: Principal factors condensation, apparently without
    rotation
Results: Five factors emerged: I Patient self-rated dis-
    tance; II Rating of each patient; III Transference; IV
    Self-perceived rating; V Rating of other persons.
LaRocco, J. M., Biersner, R. J., & Ryman, D. H. Mood effects
    of large group counseling among Navy recruits. Journal
    of Counseling Psychology, 1975, 22, 127-131.
Purpose: To identify dimensions underlying a modified
    Primary Affect Scale (PAS-M) which were stable across
    time
Subjects: 216 U.S. Navy recruits (who later successfully
    completed basic training)
Variables: 40 PAS-M item scores for two administrations
    (immediately post-induction, final week of basic after
    group counseling)
Method: Unspecified factor analysis, by administration
Results: Five factors (of unspecified numbers extracted)
    could be matched across time: I Pleasure; II Anger;
    III Activity; IV Fatigue; V Depression.
Long, T. J., & Bosshart, D. The Facilitator Behavior Index.
    Psychological Reports, 1974, 34, 1059-1068.
Purpose: To identify attribute dimensions for group
    facilitators
Subjects: 1) 32, 2) 51 trainees at institutes for
    facilitators of basic encounter groups
Variables: Scores on 1) 13, 2) 16 positive attribute
    scales plus negative attribute total
Method: Principal axes condensation with varimax rotation,
    by S-variable set
Results: Three factors emerged in both analyses: I Honest,
    trusting empathizer (Accurate and straightforward in
    empathy and communication, for second analysis); II
    Spontaneously expresses full range of emotions and
    feelings; III Nondirectiveness.

Mitchell, K. R., & Ng, K. T. Effects of group counseling
and behavior therapy on the academic achievement of
test-anxious students. Journal of Counseling Psychology,
1972, 19, 491-497.
   Purpose: To identify dimensions underlying measures of
   anxiety and test anxiety
   Subjects: 30 college students, tested at 5 times (treated
   as 150 Ss)
   Variables: Scores on 4 test anxiety, 1 general anxiety
   and 2 study habits measures
   Method: Principal components condensation with varimax
   rotation
   Results: Two factors emerged: I Debilitating test anxiety;
   II Study habits associated with academic difficulties vs.
   competent study habits.
Watson, J. P. Possible measures of change during group psychotherapy. British Journal of Medical Psychology, 1972,
45, 71-77.
   Purpose: To identify dimensions of repertory grids for
   therapists and patients
   Subjects: 7 therapists and 18 patients, each in one of
   four therapy groups
   Variables: Test and re-test ratings (0-100 scale) of
   each group member on 9 constructs
   Method: Principal components condensation, independently
   for each subject's repertory grid
   Results: The number of factors extracted in the 25
   analyses were not specified, nor were they labeled.
   The first factor, which was very similar across therapists, showed less change than the others; three
   components accounted for most of the variation in
   both therapists' and patients' grids.
3. Behavior and Conditioning Therapy
4. Special and Adjunctive Therapy
5. Drug Therapy
C. Hospital Care and Institutionalization
Allon, R., Graham, J. R., Lilly, R. S., & Friedman, I. Comparison of factor structures of patient and staff
responses on the Characteristics of the Treatment
Environment Scale. Journal of Clinical Psychology,
1971, 27, 385-390.
   Purpose: To identify dimensions of the Characteristics
   of the Treatment Environment Scale (CTE) for patients
   and staff
   Subjects: 1) 410 inpatients and 2) 162 staff members
   (psychiatrists, residents, nurses, aides) of an urban
   short-term psychiatric hospital
   Variables: 72 CTE item scores
   Method: Principal components condensation with varimax
   rotation, independently for patient and staff data
   Results: Five patient factors emerged: I Active staff;
   II Fostering patient autonomy; III Post-hospital
   orientation; IV Staff dominance; V Uninterpreted.

Five staff factors emerged: I Staff permissiveness; II Encouraging patient initiative; III Encouraging patient interaction; IV Delegation of responsibility to patients; V Staff-patient interaction and communication.

Bynner, J., & Romney, D. A method for overcoming the problem of concept-scale interaction in semantic differential research. British Journal of Psychology, 1972, 63, 229-234.

Purpose: To identify within- and across-concept dimensions of hospital staff attitudes toward patient types

Subjects: 209 hospital staff members (reported by Romney & Bynner, 1972)

Variables: 23 semantic differential ratings for four types of patients, 1) averaged across concepts, and 2) within each patient type

Method: Principal components condensation with orthogonal varimax rotation, independently for across-concept matrix and four within-concept matrices

Results: Five across-concepts factors emerged: I Assertiveness; II Dangerousness; III Nonconformity; IV Attractiveness; V Educability. In the within-concepts analyses, Factor I appeared for all concepts (patient types); Factor V did not emerge clearly,for any. Factor II did not emerge for "the alcoholic"; Factor III did not emerge for "the mentally ill patient"; and Factor IV did not emerge for the "mentally ill patient" and only questionably for "the alcoholic."

Coles, G. J., & Stone, L. A. Multidimensional perceptions of psychiatric patient behavior. Journal of Clinical Psychology, 1972, 28, 38-43.

Purpose: To identify dimensions of psychiatric aides' perceptions of patient behavior

Subjects: 6 psychiatric aides and 2 RNs working on a token-economy ward of a state hospital, rating male patients on the ward

Variables: Similarity judgments for 31 patients

Method: Principal components condensation with varimax rotation

Results: Six factors emerged and were labeled by the judges: I Behavioral capability-incapability; II Extent of "bugging" staff for cigarettes; III Reactivity-unreactivity to verbal instructions; IV Stubborn; V "Craziness," "disorientation," or "confusion"; VI "Fastidiousness," or "meticulous-sloppy."

Distefano, M. K. Jr., & Pryer, M. W. Work behavior dimensions of psychiatric attendants and aides. Journal of Applied Psychology, 1975, 60, 140-142.

Purpose: To identify dimensions underlying psychiatric attendants' work behavior

Subjects: 136 psychiatric attendants and aides

Variables: Scores on 80 items of a self-report job questionnaire

Method: Principal components condensation with varimax
  rotation
Results: Six factors emerged: I Supervision; II Physical
  nursing care; III Maintaining the ward milieu; IV Medical processing activities; V Recording patient behavior
  and following written plans; VI Special therapy activities.
Doherty, E. G. Labeling effects in psychiatric hospitalization: A study of diverging patterns of inpatient self-
  labeling processes. Archives of General Psychiatry,
  1975, 32, 562-568.
Purpose: To identify patterns of patient self-labeling
  and staff rating over time
Subjects: 43 inpatients
Variables: 1) patient (7-point) judgment of applicability
  of self-label "mentally ill" at 3 points in time (36
  hours or less after admission, 8th day, 29th), 2) staff
  ratings of patients on 5 rating scales at 3 points in
  time (1, 2, 4 weeks after admission)
Method: 1) Hierarchical grouping cluster analysis; 2)
  principal components condensation with normal varimax
  rotation, for each time period
Results: Three patient self-labeling patterns over time
  emerged: I Label rejectors; II Label acceptors; III
  Label deniers. An unspecified number of staff rating
  factors emerged; one (I Contact with reality and responsibility for self) emerged in each of the three
  analyses, and was chosen as a marker variable for ex-
  amination of stability.
Ellsworth, R., & Maroney, R. Characteristics of psychiatric
  programs and their effects on patients' adjustment.
  Journal of Consulting and Clinical Psychology, 1972,
  39, 436-447.
Purpose: To identify dimensions underlying the patient
  Perception of Ward Scale (POW)
Subjects: 1,141 male psychiatric VAH inpatients
Variables: Unspecified number of POW item scores
Method: Principal components condensation with varimax
  rotation
Results: Five factors emerged: I Inaccessible staff; II
  Involvement in ward management; III Satisfaction with
  ward; IV Receptive and involved staff; V Expectation
  for patient autonomy.
Fontana, A. F. Patient reputations: Manipulator, helper,
  and model. Archives of General Psychology, 1971, 25,
  88-93.
Purpose: To identify dimensions of patient reputations
Subjects: 133 short-term inpatients, rated by day nurses,
  evening nurses, and doctors
Variables: Scores on 18 Patient Reputation Scale items,
  plus a random variate
Method: Principal factors condensation with varimax rotation, independently for each group of raters

Results: Three factors matched across groups: I Manipulators; II Involved helpers; III Model patients.

Graham, J. R., Allon, R., Friedman, I., & Lilly, R. S. The Ward Evaluation Scale: A factor analytic study. Journal of Clinical Psychology, 1971, 27, 118-122.
Purpose: To identify empirical dimensions of the Ward Environment Scale (WES)
Subjects: 410 patients admitted to an urban short-term psychiatric hospital
Variables: 69 WES item scores
Method: Principal components condensation with varimax rotation
Results: Six factors emerged: I Staff interest in patients; II Cleanliness of ward; III Absence of disturbing noise on the ward; IV Staff permissiveness and sensitivity; V Patient comfort; VI Adequacy of hospital services.
Conclusions: These results were compared with the composition of the original (three) rational scales.

Graham, J. R., Friedman, I., Lilly, R. S., & Allon, R. Factor analysis of the Patient Perception of the Ward Scale. Journal of Clinical Psychology, 1971, 27, 278-284.
Purpose: To identify dimensions of the Patient Perception of the Ward Scale (PPOW) with a different population from Ellsworth's (1967)
Subjects: 410 inpatients in an urban short-term psychiatric hospital
Variables: 111 PPOW item scores
Method: Principal components condensation with varimax rotation
Results: Eight factors emerged: I Staff receptiveness; II Staff authoritarianism; III Interesting ward; IV Patient participation; V Patient-patient interaction; VI Patient responsibility; VII Staff commitment and interest; VIII Patient-staff communication.

Graham, J. R., Lilly, R. S., Allon, R., & Friedman, I. Comparison of the factor structures of staff and patient responses on the Ward Evaluation Scale. Journal of Clinical Psychology, 1971, 27, 123-128.
Purpose: To identify dimensions of the Ward Evaluation Scale (WES) for staff
Subjects: 163 psychiatric hospital staff (psychiatrists, residents, nurses, aides)
Variables: 69 WES item scores
Method: Principal components condensation with varimax rotation
Results: Five factors emerged: I Considerate staff; II Comfortable ward; III Accessible staff; IV Patient responsibility; V General dissatisfaction.
Conclusions: These results were compared with those for patients.

Gray, J. E. The NOSIE-30 ward behavior rating scale: Factor structure and sex differences. Journal of Clinical Psychology, 1972, 28, 390-393.

Purpose: To determine if the factorial structure of the NOSIE-30 can be replicated with another male sample, and to identify dimensions of the NOSIE-30 for females

Subjects: 1) 21 male head nurses and deputies rating a total of 299 male patients; 2) 21 female head nurses and deputies, rating a total of 240 female patients

Variables: 30 NOSIE-30 item scores

Method: Principal components condensation with varimax rotation, independently by sex

Results: Six factors emerged for males: I Social competence (with small loading on personal neatness items, and high negative loadings on retardation items); II Irritability; III Social interest; IV Personal neatness (with moderate loadings on social competence items); V Manifest psychosis; VI Depression. Seven factors emerged for females: I Irritability; II Social interest; III Social competence; IV Personal neatness; V Manifest psychosis; VI Depression; VII Retardation.

Conclusions: Factor structures and scores were further compared by sex.

Hogarty, G. E. The Discharge Readiness Inventory. Archives of General Psychiatry, 1972, 26, 419-426.

Purpose: To identify dimensions underlying the Discharge Readiness Inventory (DRI)

Subjects: 638, 267, 163 chronic schizophrenic inpatients

Variables: 67 DPI item scores

Method: Principal components condensation with varimax rotation, by sample

Results: Four factors replicated across samples: I Community adjustment potential; II Psychosocial adequacy; III Belligerence; IV Manifest psychopathology.

Jensema, C. J., & Shears, L. M. Atttiudes of psychiatric technician trainees. American Journal of Mental Deficiency, 1971, 76, 170-175.

Purpose: To identify dimensions underlying psychiatric technician trainees' attitudes toward self, their hospital, and mentally retarded patients

Subjects: 122 psychiatric technician trainees

Variables: 29 7-point semantic differential ratings each for the concepts Pacific State Hospital, My Ward, Mildly Mentally Retarded Person, and Severely Mentally Retarded Person

Method: Principal components condensation with unspecified rotation, independently by concept

Results: Two similar factors emerged for hospital and ward: I Evaluation of the hospital and its wards as a bureaucratic institution; II Image given to a firm, protective parent. Three factors emerged for both mildly and severely retarded persons, though they were not identical: I Degrees of incapacity to lead a normal life; II Tendency to see a humanistic, diamond-in-the-rough quality in a retarded person; III Recognition of the complexity and potential of mentally retarded individuals.

Lorei, T. W.   Hierarchical grouping of opinions about release/
    retention of psychiatric patients. Archives of General
    Psychiatry, 1971, 25, 69-73.
    Purpose: To identify dimensions underlying outcome impor-
    tance profiles (related to release of psychiatric
    patients)
    Subjects: 1,353 VAH staff members from 13 hospitals
    Variables: Importance (of avoiding) ratings for 16 possible
    outcomes of releasing psychiatric patients
    Method: Hierarchical grouping analysis, by hospital for
    first-order analysis (pooled for second-order)
    Results: A total of 78 first-order groups emerged. Six
    second-order groups emerged; dimensions differentiating
    the groups were concern about making unpopular decisions,
    about distress to patients and relatives, and about
    hospital misuse.
Morgan, D. W., Crawford, J. L., Frenkel, S. I., & Hedlund,
    J. L.   An automated patient behavior checklist. Journal
    of Applied Psychology, 1973, 58, 393-396.
    Purpose: To identify dimensions underlying (an automated)
    Patient Behavior Checklist (PBCC)
    Subjects: 103 patients in a military treatment milieu
    (rated by total of 689 staff)
    Variables: 50 item scores from the Patient Behavior Index
    Method: Unspecified factor analysis
    Results: Four factors emerged: I Acting out; II Depres-
    sion/withdrawal; III Degree of disturbance; IV Adapta-
    tion to ward.
Price, R. H., & Moos, R. H.   Toward a taxonomy of inpatient
    treatment enivronments. Journal of Abnormal Psychology,
    1975, 84, 181-188.
    Purpose: To identify an empirically based taxonomy of
    treatment programs
    Subjects: 144 inpatient treatment programs
    Variables: Scores on 10 Ward Atmosphere Scale(s)
    Method: Cluster analysis
    Results: Six clusters emerged: I Therapeutic community;
    II Relationship oriented; III Action oriented; IV In-
    sight oriented; V Control oriented; VI Disturbed
    behavior.
Romney, D., & Bynner, J.   Drug addicts as perceived by hos-
    pital staff. British Journal of Social and Clinical
    Psychology, 1972, 11, 20-34.
    Purpose: To identify dimensions of hospital staff's 1)
    attitudes toward drug addiction, and 2) perceptions of
    four types of patients
    Subjects: 209 hospital staff members (doctors, nurses,
    administrators, chasiers, etc.)
    Variables: 1) 19 attitudes-toward-drug-addiction item
    scores; 2) ratings on 23 semantic differential scales,
    averaged across perceptions of hard- and soft-drug
    addicts, alcoholic, and mentally ill patients

Method: Principal components condensation with varimax
rotation, independently for attitude- and perceptions
data
Results: Five attitudes toward addiction factors emerged:
I Intolerance of drug addicts; II Belief that drug
addiction is an illness; III Belief that drug addicts
are anti-social; IV Disapproval of doctors spending
time treating drug addicts; V Belief that drug-taking
expands consciousness. Five factors emerged for perceptions of patients: I Assertiveness; II Dangerousness; III Non-conformity; IV Attractiveness; V Educability.
Spiegel, D., & Keith-Spiegel, P. Perceptions of ward climate
by nursing personnel in a large NP hospital. Journal
of Clinical Psychology, 1971, 27, 390-393.
Purpose: To identify dimensions of ward climate perceived
by nursing personnel
Subjects: 250 nursing assistants, 61 RNs, 27 licensed
vocational nurses on 6 psychiatric units
Variables: Scores on 31 items of the Perception of the
Ward Scale (nursing form)
Method: Principal components condensation with orthogonal
rotation
Results: Five factors emerged: I Professional staff sup-
portiveness; II Communication-morale; III Ward cohesive-
ness; IV Nursing personnel influence; V Professional
staff authoritarianism.
Spiegel, D., & Younger, J. B. Ward climate and community
stay of psychiatric patients. Journal of Consulting
and Clinical Psychology, 1972, 39, 62-69.
Purpose: To identify dimensions underlying the Ward
Climate Inventory (WCI) for patients and for staff
Subjects: 254 patients, 173 staff from male wards of a
neuropsychiatric hospital
Variables: 23 WCI item scores
Method: Principal components condensation with orthogonal
rotation, independently for patients and staff
Results: Three factors emerged in both analyses: I Per-
sonal concern for patients; II Patient concern for
patients; III Ward morale.
D. Psychoanalytic Interpretation
Stone, G. C., & Gottheil, E. Factor analysis of orality
and anality in selected patient groups. Journal of
Nervous and Mental Disease, 1975, 160, 311-323.
Purpose: To identify dimensions underlying out-patient
self-reports of behavioral traits associated with
orality and anality
Subjects: 150 outpatients suffering from physical or
emotional problems
Variables: Questionnaire item scores: 1) 185 total items,
2) subsets of 40 anal trait items, 40 oral traits, 40
bowel behavior, and 40 mouth behavior items
Method: Principal components condensation with varimax
rotation, independently for all items and subsets

Results: Eight factors emerged for the 185-item set: I
Cheerful, outgoing, relaxed; II No bowel problems; III
Cautious and clean; IV Low oral interest; V Planful
leader; VI Weak personality; VII Careful, prudent; VIII
Rapid tempo, active. Seven mouth behavior factors
emerged: I Low biting, licking, sucking; II Poor
appetite; III Fast eater; IV GI discomfort; V Likes
sucking, drinking; VI Good GI system; VII Not named.
Seven bowel behavior factors emerged: I Regular bowel
function; II Eliminative problems; III Coprophilic; IV
Not retentive; V Worried about regularity; VI Coprophobic;
VII Lack of bowel concern. Seven oral trait factors
emerged: I Optimistic, easy-going; II Jealous; III
Sociable; IV Dependent; V Power orientation; VI Socially
withdrawn; VII Lack of leadership. Seven anal trait
factors emerged: I Careful, prudent, obsessive; II
Dislike of change, novelty, interruption; III Practical,
careful; IV Methodical worker; V Not named; VI Careful,
methodical; VII Slipshod and insensitive.
E. Psychodiagnosis
    Apperson, L. B., & Stinnett, P. W. Parental factors as re-
    ported by patient groups. Journal of Clinical Psychology,
    1975, 31, 419-425.
    Purpose: To identify clusters of perceived parental
    behavior for four patient groups and a control group
    Subjects: 50 subjects each classified as depressive
    reaction, alcoholic, drug addict, and control, plus
    49 schizophrenics and 8 hyperthyroid patients
    Variables: 60 scores for each patient on the Perception
    of Parent Behavior Scale
    Method: 1) Cluster analysis, and 2) inverse cluster
    analysis, for each parent
    Results: 1) Nine clusters, identical for father and mother
    data, were extracted but not labeled. 2) Two clusters
    (Consideration-respect for the rights of others; Sex)
    differentiated between groups.
    Barker, H. R., Fowler, R. D., & Peterson, L. P. Factor
    analytic structure of the short form MMPI items in a
    VA hospital population. Journal of Clinical Psychology,
    1971, 27, 228-223.
    Purpose: To identify dimensions underlying a short form
    of the MMPI for a VA hospital population
    Subjects: 1575 VA hospital inpatients
    Variables: Scores on MMPI items 1-365, plus 8 "new
    arrival" items
    Method: Principal axes condensation with varimax rotation
    Results: Nine factors emerged: I Psychotic triad (sug-
    gests conventional picture of schizophrenia); II
    Neurotic triad; III Predominantly L; IV Conformity to
    well-defined societal codes (F); V Mf; VI Gilberstadt
    and Duker profile; VII Paranoia (paranoid schizophrenia);
    VIII Cynicism; IX Encapsulation (anhedonia).

Blashfield, R. An evaluation of the DSM-II classification
of schizophrenia as a nomenclature. Journal of Abnormal
Psychology, 1973, 82, 382-389.
Purpose: To identify orthogonal second-order factors of
the Inpatient Multidimensional Psychiatric Scale (IMPS)
Subjects: Lorr, Klett, & McNair (1963) correlation matrix
(i.e., unspecified here)
Variables: 10 IMPS scales
Method: Unspecified condensation with varimax rotation
Results: Four second-order factors emerged: 1, 2, 3, 4.
Blunden, D., Spring, C., & Greenberg, L. M. Validation of
the Classroom Behavior Inventory. Journal of Consul-
ting and Clinical Psychology, 1974, 42, 84-88.
Purpose: To identify dimensions of behavior associated
with hyperactivity, as assessed by the Classroom
Behavior Inventory (CBI)
Subjects: 320 kindergarten boys, rated by their teachers
Variables: 40 CBI item scores
Method: Principal components condensation with varimax
rotation
Results: Four factors emerged: I Hyperactivity; II
Hostility; III Sociability; IV Uninterpretable.
Burton, D. A. A factor analysis of the Edwards Personal
Preference Schedule and 16PF in a psychiatric popula-
tion. Journal of Clinical Psychology, 1971, 27, 248-
251.
Purpose: To identify dimensions of the Edwards Personal
Preference Schedule (EPPS) and 16PF for a psychiatric
population
Subjects: 80 referrals for personality assessment
Variables: 15 EPPS percentile scores and 16 16PF sten
scores
Method: 1) Principal components condensation with varimax
rotation; 2) elementary linkage (cluster) analysis
Results: Twelve factors emerged from the first analysis:
I Cattell's second-order anxiety; II Not labeled (ex-
clusively EPPS); III Extraversion; IV Not labeled
(intelligence, with aggression negative); V 16PF M;
VI 16PF I; VII Conservatism and conformity; VIII Super-
ego functioning; IX EPPS Introception, Dominance, and
(negative) Abasement; X Willingness to admit need for
help; XI Extraversion subgroup; XII Heterosexuality.
Seven clusters emerged: I Q_3-Q_4; II F-A-Q_2; III Extraversion (peripheral system); IV C-O-L; V Aba-Dom-Agg;
VI Suc-N; VII Heterosexuality.
Chapman, L. J., Chapman, J. P., & Daut, R. L. Schizophrenic
response to affectivity in word definition. Journal of
Abnormal Psychology, 1974, 83, 616-622.
Purpose: To identify dimensions underlying affective and
neutral subtests of an expanded Stanford-Binet Vocabulary subtest (for nonpsychotic controls)
Subjects: 167 nonpsychotic Ss (firefighters, prison inmates, college students)

215

Variables: Scores on 29 items each of expanded S-B
Vocabularly 1) affective and 2) neutral subtests
Method: Principal components condensation with quartimax rotation, by subtest
Results: Eight unlabeled factors emerged for the affective subtest, and seven for the neutral subtest.
Clum, G. A. Relations between biographical data and patient
symptomatology. Journal of Abnormal Psychology, 1975,
84, 80-83.
Purpose: To identify dimensions underlying 1) a biographical inventory, and 2) Parts I and II of the Katz Adjust-
ment Scales (KAS) for psychiatric inpatients
Subjects: 117 university psychiatric service inpatients,
aged 18-65 years
Variables: Scores on 1) 73 biographical items, 2) unspecified number of KAS I, II scales
Method: Principal components condensation with orthogonal
rotation, by instrument
Results: Four biographical factors emerged: I Social and
familial alienation; II Age; III Social status; IV
Acting out. Six KAS factors emerged: I Hostility;
II Sadness-anxiety; III Depressive withdrawal-retarda-
tion; IV Retardation and confusion; V Verbal expansive-
ness; VI Loss of control.
Derogatis, L. R., Lipman, R. S., Covi, L., & Rickels, K.
Neurotic symptom dimensions as perceived by psychiatrists
and patients of various social classes. Archives of
General Psychiatry, 1971, 24, 454-464.
Purpose: To identify dimensions underlying Symptom Distress
Checklist (SCL) ratings of patients and psychiatrists
Subjects: 1-6) 1,066 anxious neurotic outpatients (3 social
classes); 7) 837 subsample
Variables: 58 SCL item scores 1-6) self-rated, 7) psychi-
atrist-rated
Method: Principal axes condensation with varimax rotation,
independently for females and total within social
classes (1-6), and for psychiatrists
Results: Five factors replicated across analyses: I
Somatization; II Obsessive-compulsive; III Irascibility;
IV Depression; V Anxiety.
Dielman, T. E., Cattell, R. B., & Rhoades, P. A. Cross-
validational evidence on the dimensions of problem
behavior in the early grades. Multivariate Behavioral
Research, 1972, 7, 33-40.
Purpose: To identify dimensions of problem behavior in
the early grades
Subjects: 147 first- through third-graders
Variables: Scores on 58 teacher-rated behavior problem
inventory items
Method: Principal axes condensation with promax rotation
Results: Five factors, of twelve extracted, were consid-
ered matches with those from a previous study: I Hyper-
activity; II Sluggishness; III Paranoic tendencies; IV
Social withdrawal; V Acting out.

Endicott, J., & Spitzer, R. L.   Current and Past Psychopath-
    ology Scales (CAPPS). Archives of General Psychiatry,
    1972, 27, 678-687.
    Purpose: To identify dimensions underlying the CAPPS
    Subjects: 800 individuals heterogeneous as to SES, etc.
    (43% inpatients)
    Variables: Scores on 1) 28 "current", 2) 78 "past" CAPPS
    psychopathology scales
    Method: Principal components condensation with varimax
    rotation, by section
    Results: Seven current factors emerged: I Reality testing-
    social disturbance; II Depression-anxiety; III Impulse
    control; IV Somatic concern-functioning; V Disorganiza-
    tion; VI Obsessive-guilt-phobic; VII Elation-grandiosity.
    Eighteen past factors emerged: I Depression-anxiety;
    II Impulse control; III Social-sexual problems; IV
    Reality testing; V Dependency; VI Somatic concern-
    functioning; VII Obsessive-compulsive; VIII Anger-
    excitability; IX Manic; X Sexual disturbance; XI Memory-
    orientation; XII Disorganized; XIII Organicity; XIV
    Neurotic childhood; XV Phobia; XVI Retardation-stubborn;
    XVII Hysterical symptoms; XVIII Intellectual performance.
Endicott, J., & Spitzer, R. L. What! Another rating scale?
    The Psychiatric Evaluation Form. Journal of Nervous
    and Mental Disease, 1972, 154, 88-104.
    Purpose: To identify dimensions of the Psychiatric Eval-
    uation Form (PEF)
    Subjects: 433 newly-admitted psychiatric inpatients
    Variables: 19 psychopathology scale scores from the PEF
    Method: Principal components condensation with varimax
    rotation
    Results: Six factors, or summary scales, emerged: I
    Disorganization; II Subjective distress; III Antisocial;
    IV Withdrawal; V Alcohol; VI Grandiosity-externalization.
Fitzgibbons, D. J., & Cutler, R. The factor structure of the
    Tennessee Self-Concept Scale among lower-class urban
    psychiatric patients. Journal of Clinical Psychology,
    1972, 28, 184-186.
    Purpose: To identify dimensions of the Tennessee Self-
    Concept Scale (TSCS) for a sample of lower-class urban
    psychiatric patients
    Subjects: 135 female and 117 male lower-class urban
    psychiatric patients
    Variables: 100 TSCS item scores
    Method: Principal components condensation with varimax
    rotation
    Results: Six factors emerged: I Ego strength; II
    Validity (L); III Active rejection of gross suggestions
    of inadequacy; IV Reports of self-depreciation with
    externally-directed hostility; V Negative self-evalua-
    tion associated with trivial moral transgressions; VI
    Reported feelings of alienation from one's family.

Fleiss, J. L., Gurland, B. J., & Goldberg, K. Independence
of depersonalization-derealization. Journal of Consulting and Clinical Psychology, 1975, 43, 110-111.
Purpose: To identify dimensions underlying a structured
mental state interview
Subjects: 866 hospitalized adult mental patients (from
New York, London), in samples of 500 and 366; analyses
on first sample reported by Fleiss, Gurland, & Cooper
(1971, non-scanned journal)
Variables: Scores on nearly 700 items from a structured
mental state interview
Method: Unspecified factor analysis, by sample
Results: In both cases one factor, of 25 extracted, was
labeled Depersonalization-derealization; it was relatively independent of both other factors and diagnosis.
Fontana, A. F., & Dowds, B. N. Assessing treatment and
outcome: I. Adjustment in the community. Journal of
Nervous and Mental Disease, 1971, 161, 221-230.
Purpose: To identify dimensions of a modified PARS-II
Subjects: Unspecified number of male psychiatric inpatients, rated by self (Time 1 N=171, Time 2 N=111,
Time 3 N=80) and significant other (Time 1 N=141,
Time 2 N=95, Time 3 N=64)
Variables: Scores on 41 modified PARS-II items plus 2
somatic and 1 alcohol item, administered at admission
and 1- and 6-months after discharge
Method: Principal components condensation with varimax
rotation
Results: Although it was possible to replicate seven factors found in other investigations, a five-factor
solution was considered optimal: I Symptomatology;
II Alcohol abuse; III Social involvement; IV Employment;
V Organizational participation.
Gurland, B. J., Yorkston, N. J., Goldberg, K., Fleiss, J. L.,
Sloane, R. B., & Cristol, A. H. The Structured and
Scaled Interview to Assess Maladjustment (SSIAM): II.
Factor analysis, reliability, and validity. Archives
of General Psychiatry, 1972, 27, 264-267.
Purpose: To identify dimensions underlying the SSIAM
Subjects: 164 adults judged acceptable for outpatient
psychotherapy
Variables: 33 SSIAM item scores
Method: Unspecified condensation with varimax rotation
Results: Six factors emerged: I Social isolation; II
Work inadequacy; III Friction with family; IV Dependence on family; V Sexual-dissatisfaction; VI Friction
outside family.
Guthrie, G. M., Verstraete, A., Deines, M. M., & Stern, R. M.
Symptoms of stress in four societies. Journal of Social
Psychology, 1975, 95, 165-172.
Purpose: To identify dimensions of stress symptoms in four
societies

Subjects: 94 male and 102 female French college students,
91 male and 108 female American college students, 119
male and 91 female Filipino college students, 112 male
and 106 female Haitian college students
Variables: Scores on 33 items concerning reactions to
stressful situations
Method: Unspecified condensation with unspecified rotation,
independently for total sample and each (8) subgroups
Results: Six factors were extracted for the total sample:
I Sympathetic; II Loss of control; III Cognitive disorganization; IV Affective; V Gastrointestinal; VI
Cardio-respiratory. Factor I appeared in all 8 subsamples; Factor IV did not appear in any. Factor II
did not appear for French males or American males;
Factor III did not appear for Haitian females. Factor
V did not appear for French males or females nor for
Haitian males; Factor VI did not appear for American
males. Factors not found in the total group emerged
for French males (A. Prolonged, diffuse muscle tension;
B. Depression; C. Oppression).
Hirschfeld, R., Spitzer, R. L., & Miller, R. G. Computer
diagnosis in psychiatry: A Bayes approach. Journal
of Nervous and Mental Disease, 1974, 158, 399-407.
Purpose: To identify dimensions of the Current and Past
Psychopathology Scales (CAPPS)
Subjects: 417 persons aged 15-44 (57% female), repre-
senting in-, out-, and non-patients
Variables: Scores on the 150-plus items of CAPPS
Method: Principal components condensation with varimax
rotation
Results: Eight current (I Reality testing-social distur-
bance; II Depression-anxiety; III Antisocial; IV
Somatic concern-functioning; V Disorganization; VI
Obsessive-guilt-phobic; VII Elation-grandiosity; VIII
Summary role) and eighteen past pathology factors (IX
Depression-anxiety; X Antisocial; XI Social-sexual
relations; XII Reality testing; XIII Dependency; XIV
Somatic concern-functioning; XV Obsessive-compulsive;
XVI Anger-excitability; XVII Manic; XVIII Sexual dis-
turbance; XIX Memory-orientation; XX Disorganized; XXI
Organicity; XXII Neurotic childhood; XXIII Phobia;
XXIV Retardation-stubborn; XXV Hysterical symptoms;
XXVI Intellectual performance) emerged.
Holmes, G. R., Rothstein, W., Stout, A. L., & Rosecrans,
C. J. Comparison of four factor analyses of the Fear
Survey Schedule. Journal of Clinical Psychology, 1975,
31, 56-61.
Purpose: To identify an oblique factor solution for the
Fear Survey Schedule (FSS)
Subjects: 50 male and 50 female psychiatric inpatients
Variables: 76 scores from the revised FSS
Method: Principal components condensation with varimax
rotation to oblique simple structure

Results: Sixteen factors were extracted: I Noninterpretable; II Fear of being socially unacceptable; III Fear of transportation resources; IV Fear of exposure to suffering; V Fear of nudity; VI Noninterpretable; VII Fear of animals; VIII Fear of personal harm; IX Fear of high places; X Fear of repulsive phenomena; XI Fear of personal attack; XII Noninterpretable; XIII Fear of pain; XIV Noninterpretable; XV Fear of grief; XVI Noninterpretable.

Conclusions: The present oblique solution was compared with previously reported solutions: orthogonal with psychiatric patients, oblique and orthogonal for a non-psychiatric population.

Hunter, S., Overall, J. E., & Butcher, J. N. Factor structure of the MMPI in a psychiatric population. <u>Multivariate Behavioral Research</u>, 1974, <u>9</u>, 283-301.

Purpose: To identify dimensions underlying the MMPI - 373 with a psychiatric population

Subjects: 339 psychiatric patients from a state or university hospital or university clinic

Variables: Scores on 1) 4 subsets of 124-129 MMPI items, 2) 24 factors identified in (1)

Method: Principal axes condensation with varimax rotation, for each set of scores

Results: Six replicating factors emerged: I Somatization; II Feminine interests; III Depression; IV Psychotic distortion; V Low morale; VI Acting out.

Kear-Colwell, J. J. The Bannister-Fransella Grid variables: Relationships to intelligence and personality in psychiatric patients. <u>Journal of Clinical Psychology</u>, 1972, <u>28</u>, 353-356.

Purpose: To identify relationships between Bannister-Fransella Grid (B-F Grid), intelligence and personality variables for psychiatric inpatients

Subjects: 106 psychiatric inpatients

Variables: Scores on Progressive Matrices (total), Mill Hill Vocabulary Scale (total), 2 B-F Grid subscales, 16 16PF scales, age, sex, social class

Method: Principal components condensation with varimax and promax rotation

Results: Eight factors emerged: I Anxiety vs. adjustment; II General or fluid intelligence; III Verbal ability; IV Aggression, social dominance, and impulsiveness vs. passivity, submissiveness, and shyness; V Tight construing of person-relevant constructs vs. loose- construing; VI Ability to make and maintain adequate interpersonal relationships; VII Self-absorption, self-direction of behavior, and oversensitivity vs. concern for practical reality, toughness, and group participation; VIII Controlled, socially sophisticated, and insightful behavior vs. expedient, uncontrolled, and socially naive behavior.

Conclusions: The grid factor (V) appeared independent of the other dimensions.

Kear-Colwell, J. J. The factor structure of the 16PF and the Edwards Personal Preference Schedule in acute psychiatric patients. *Journal of Clinical Psychology*, 1973, <u>29</u>, 225-228.
  Purpose: To identify relationships between the 16PF and the Edwards Personal Preference Schedule (EPPS) for acute psychiatric patients
  Subjects: 100 female and 74 male psychiatric inpatients
  Variables: Scores on 16 16PF and 15 EPPS scales
  Method: Principal components condensation with direct oblimin rotation, independently for 16PF, EPPS, and combined variables
  Results: Six 16PF factors emerged: I Anxiety; II Extraversion; III Super ego strength; IV Independence; V Cortertia; VI Radical intelligence. Six EPPS factors emerged: I Need for a carefully organized and socially diffident way of life; II Need for warm positive feelings toward other people and interpersonal relationships; III Need for a changing environment and independence from people; IV Need for extra-punitive aggression vs. need for intro-punitive aggression; V Need for a nondependent interest in peoples' feelings and motives; VI Need for achievement without the need to be aggressive. Ten factors emerged from the combined analysis: I Anxiety; II Extraversion; III EPPS Factor I with radicalism element; IV Interpersonal; V Need for a changing environment and independence from people; VI Extrapunitive aggression and social dominance; VII Super ego strength; VIII Pathemia vs. cortertia; IX Need for a nondependent interest in peoples' feelings and motives; X Need for achievement without being aggressive.
  Conclusions: These results indicate the two instruments provide complimentary information with a minimum of duplication.
Kear-Colwell, J. J. The structure of the Wechsler Memory Scale and its relationship to 'brain damage.' *British Journal of Social and Clinical Psychology*, 1973, <u>12</u>, 384-392.
  Purpose: To identify dimensions underlying the Wechsler Memory Scale (WMS)
  Subjects: 250 patients referred for evaluation of cognitive functioning
  Variables: 1) 7 WMS subscale scores; 2) WMS factor scores plus WAIS VIQ, WAIS PIQ, WAIS Full IQ, WAIS V-P discrepancy, age, and sex
  Method: Principal factors condensation with oblimin rotation
  Results: Three WMS factors emerged for a confirmed organic group (N=66), non-confirmed group (N=184), and combined group: I Learning and immediate recall of fairly complex novel information in both visual and auditory modalities; II Attention and concentration and ability

221

to process verbal non-semantic information; III Orientation in place and time and ability to recall simple long-established verbal information. Three factors emerged from the analysis of WMS factor scores with WAIS and other variables: I Intellectual ability; II V-P discrepancy; III Age.

Kerry, R. J., & Orme, J. E. A factorial study of the In-patient Multidimensional Psychiatric Scale. _Journal of Clinical Psychology_, 1973, _29_, 368-370.
Purpose: To identify dimensions underlying the In-patient Multidimensional Psychiatric Scale (IMPS)
Subjects: 100 psychiatric inpatients
Variables: 10 IMPS scale and 3 experimental syndrome scales, plus age
Method: Principal components condensation with varimax rotation
Results: Six factors emerged: I Depressive; II Paranoid schizophrenia; III Non-paranoid schizophrenia (retardation with motor disturbances involving repetitive and bizarre gestures); IV (Age negatively associated with hostile belligerence, paranoid projection, obsessive-compulsiority); V Manic states; VI Non-paranoid schizophrenia (disorientation with speech and thought disturbance).

Klerman, G. L. Clinical research in depression. _Archives of General Psychiatry_, 1971, _24_, 305-319.
Purpose: To identify dimensions of depression
Subjects: 220 patients from outpatient clinics, emergency services, day and hospital programs
Variables: Unspecified number of psychiatric interview ratings
Method: Unspecified factor analysis
Results: Three factors emerged: I Severity; II Neurotic vs. endogenous; III Anxiety vs. depressive mood.

Kolton, M. S., & Dwarshuis, L. A clinical factor analytic method for inferring construct meaning. _Educational and Psychological Measurement_, 1973, _33_, 653-661.
Purpose: To identify meaning dimensions of clinically inferred constructs
Subjects: 2 clinical psychologists
Variables: Scores (judgments) on 70 items from the Crowne-Marlowe, CPI, and MMPI Ego-Resilience Scale (scored for four constructs)
Method: Unspecified factor analysis
Results: Four factors emerged: I Openness-flexibility; II Acceptance of impulse expression; III Self-acceptance; IV Social inhibition.

Kupfer, D. J., Detre, T., & Koral, J. "Deviant" behavior patterns in school children, application of KDS[TM]-14. _Psychological Reports_, 1974, _35_, 183-191.
Purpose: To identify dimensions underlying a teacher's rating scale, KDS[TM]-14

Subjects: 20 children who might need psychological services (mean age=10.5 years, grade=5), rated by their teachers
Variables: 84 KDS^{TM}-14 item scores
Method: Principal components condensation with varimax rotation
Results: Three factors emerged: I Aggressivity; II Learning deficits; III Shy-withdrawn.

Lamont, J., & Tyler, C. Racial differences in rate of depression. Journal of Clinical Psychology, 1973, 29, 428-432.
Purpose: To identify dimensions of self-rated depression for 5 racial groups
Subjects: 44 Japanese-American, 43 Chinese-American, 81 Mexican-American, 243 white, and 47 black college students
Variables: 79 item scores from a depression self-report questionnaire
Method: Principal components condensation with varimax rotation, independently by racial group and for combined sample
Results: Two factors emerged in all analyses: I Depression; II Recall of parental behavior (except Japanese-Americans: II Social anxiety).

Leonard, C. V. Depression and suicidality. Journal of Consulting and Clinical Psychology, 1974, 42, 98-104.
Purpose: To identify the relationship between several measures of depression, suicidality and selected demographic variables
Subjects: 90 voluntary psychiatric hospital patients
Variables: Scores on 20 Self-Rating Depression Scale (SDS) items, SDS total, suicidality rating, drug use, alcohol use, age, sex, hospitalization length, education, 16 MMPI-derived scales
Method: Principal components condensation with orthogonal rotation
Results: Ten factors emerged: I Emptiness (SDS); II General psychopathology that will insure admission to a voluntary psychiatric hospital (MMPI); III Somatic and psychic disequilibrium (SDS); IV Naivete with a defensive and denying approach, overt hostility, and limited educational background; V MMPI Ma and Goldberg Index; VI Suicidality 1; VII Uninterpreted; VIII Suicidality 2; IX-X Uninterpreted.

Lessing, E. E., & Zagorin, S. W. Dimensions of psychopathology in middle childhood as evaluated by three symptom checklists. Educational and Psychological Measurement, 1971, 31, 175-198.
Purpose: To identify dimensions underlying the Peterson Problem Checklist (PPCL), Wichita Guidance Center Checklist (WGCCL), and IJR Symptom Checklist (IJRSCL)
Subjects: 102 children referred for child guidance services (aged 10-0 to 12-11)

223

Variables: Scores on 1) 58 PPCL, 2) 55 WGCCL (both by
mothers), 3) 36 IJRSCL (psychiatrist) item scores
Method: Principal axes condensation with binormamin
rotation, by instrument
Results: Four PPCL factors emerged: I Conduct problem;
II Personality problem-autism; III Personality problem;
IV Organic-somatic problem. Four WGCCL factors emerged:
I School failure; II Conduct problem; III Unhappiness-
unsociability; IV Conflict with parents. Three IJRCL
factors emerged: I Immature conduct problem; II Phobic-
suicidal; III Delinquent conduct.
Levenson, H. Multidimensional locus of control in psychiatric
patients. Journal of Consulting and Clinical Psychology,
1973, 41, 397-404.
Purpose: To identify dimensions of locus of control in
psychiatric patients
Subjects: 165 consecutive admissions to state mental
hospitals, tested within 5 days of admission
Variables: Scores on 24 internal, powerful others, and
chance scale items (Likert format)
Method: Principal components condensation with varimax
rotation
Results: Eight factors emerged: I Belief in fate or
chance happenings; II Power of other people over the
individual; III Competence of the individual in
planning and having his hard work rewarded; IV-VIII
One or two item-specific.
Mariotto, J. J., & Paul, G. L. A multimethod validation of
the Inpatient Multidimensional Psychiatric Scale with
chronically institutionalized patients. Journal of
Consulting and Clinical Psychology, 1974, 42, 497-508.
Purpose: To identify dimensions underlying the Inpatient
Multidimensional Psychiatric Scale (IMPS) for chronically
institutionalized patients
Subjects: 1 & 2) 80 chronically institutionalized mental
patients, 3 & 4) subset of 52
Variables: 1) 5 IMPS interview scores, 2) 7 NOSIE ward
scores, 3) 5 Time Sample Behavioral Checklist (TSBC)
scores, all taken at 2 times; 4) change scores
Method: Principal components condensation with varimax
rotation, independently for IMPS-NOSIE-TSBC (N=52)
and IMPS-NOSIE (N=80) at each time; multimethod factor
analysis, by same breakdown; principal components
condensation for residual change scores of N=80, N=52
Results: Four factors emerged for the first analysis
(both times): I General positive level of functioning;
II Excitement/irritability and observed cognitive distortion; III Interview-based anxiety and cognitive
distortion - irritability; IV Unstable. Three factors
emerged for the second (full sample) analysis: I Positive level of functioning; II Cognitive excitement/
irritability; III IMPS method. Six factors emerged for
the multimethod analyses: I IMPS schizophrenia disorganization; II IMPS anxious intropunitiveness 1; III

224

IMPS excited-retarded 1; IV IMPS cognitive distortion;
V IMPS anxious intropunitiveness 2; VI IMPS excited-
retarded 2. Four change factors emerged: I General
interview and ward-appropriate behavior change; II Ward
and observational change (high and low levels of appropriate behavior); III IMPS schizophrenic disorganization
-TSBC cognitive distortion and hostile belligerence;
IV IMPS method specific.

Martorano, R. D., & Nathan, P. E. Syndromes of psychosis
and non-psychosis: Factor analysis of a systems
analysis. Journal of Abnormal Psychology, 1972, 80,
1-10.
  Purpose: To identify dimensions underlying a behavior
  rating scale for psychotic and nonpsychotic inpatients
  Subjects: 924 psychotic and nonpsychotic inpatients
  Variables: 70 item scores from the Boston City Hospital
  Behavior Checklist
  Method: Unspecified condensation with varimax followed by
  oblimin rotation
  Results: Thirteen factors emerged: I Schizophrenia disorganization; II Depression; III Disordered memory and
  consciousness; IV Mania-hypomania; V Catatonia; VI
  Anxiety; VII Abnormal perceptual behavior; VIII "Reactive"
  depression; IX Autism with stupor; X "Endogenous" depression; XI Abnormal affective behavior; XII Paranoid
  process; XIII Abnormal cognitive behavior.

McNeil, T. F., & Wiegerink, R. Behavioral patterns and pregnancy and birth complication histories in psychologically
disturbed children. Journal of Nervous and Mental
Disease, 1971, 152, 315-323.
  Purpose: To identify dimensions of current behaviors in
  disturbed children
  Subjects: 61 children under treatment for behavioral or
  psychiatric disturbance
  Variables: Scores on a 17-item current behavior rating
  scale
  Method: Principal components condensation with varimax
  rotation
  Results: Three factors emerged: I Psychotic withdrawal;
  II Acting-out - aggression; III Organic signs.
  Conclusions: Factor scores were examined relative to
  maternal pregnancy and birth complications.

Mendels, J., Weinstein, N., & Cochrane, C. The relationship
between depression and anxiety. Archives of General
Psychiatry, 1972, 27, 649-653.
  Purpose: To identify dimensions underlying self-rating
  mood scales with a group of psychiatric inpatients
  Subjects: 100 female acute psychiatric service inpatients
  Variables: Scores from the Beck Depression Inventory,
  Zung Self-rating Depression Scale, 2 Costello-Comrey
  scales, 3 MAACL, 4 MMPI scales
  Method: Principal components condensation with varimax
  rotation

Results: Two factors emerged: I General psychiatric disturbance/distress; II Inhibition, over control, denial of hostility, repressive style of life.

Millimet, C. R., & Greenberg, R. P. Use of an analysis of variance technique for investigating the differential diagnosis of organic versus functional involvement of symptoms. Journal of Consulting and Clinical Psychology, 1973, 40, 188-195.
Purpose: To identify psychologist-judge dimensions of diagnoses of organic vs. functional involvement of symptoms
Subjects: 9 Ph.D. and 3 M.A. psychologists
Variables: 11-point rating of organic vs. functional involvement for 64 combinations of (6) clinical signs of organicity
Method: Inverse centroid condensation with varimax rotation
Results: Three factors emerged: I Importance of Symptom 5; II Importance of Symptoms 3 and 4; III Importance of Symptoms 2 and 6.

Newmark, C. S., Faschingbauer, T. R., Finch, A. J., & Kendall, P. C. Factor analysis of the MMPI-STAI. Journal of Clinical Psychology, 1975, 31, 449-452.
Purpose: To identify factors underlying MMPI and STAI scales
Subjects: 139 male and 186 psychiatric inpatients
Variables: 16 MMPI and 2 STAI scale scores
Method: Unspecified factor analysis
Results: Four factors emerged: I Adjustment; II Passivity; III Somatic concern; IV Anxiety proneness.
Conclusions: The results were compared with other MMPI factor-analytic studies.

Overall, J. E., Hunter, S., & Butcher, J. N. Factor structure of the MMPI-168 in a psychiatric population. Journal of Consulting and Clinical Psychology, 1973, 41, 284-286.
Purpose: To identify dimensions underlying the MMPI-168 in a psychiatric population
Subjects: 505 psychiatric patients (state hospital, university hospital & outpatient clinic)
Variables: Scores on the first 168 items of the MMPI, less 19 Mf items
Method: Approximate principal axes condensation with varimax rotation (and marker variable oblique factor analysis)
Results: Five factors emerged: I Somatization; II Low morale; III Depression; IV Psychotic distortion; V Acting out.
Conclusions: A third analysis using the additional 19 items verified the existence of Mf as a sixth factor.

Raskin, A., Boothe, H. H., Reatig, N. A., Schulterbrandt, J. G., & Odle, D. Factor analyses of normal and depressed patients' memories of parental behavior. Psychological Reports, 1971, 29, 871-879.

Purpose: To identify dimensions underlying an abbreviated
   Children's Reports of Parental Behavior Inventory (CRPBI)
   with three samples
Subjects: 1) 371, 2) 177 hospitalized depressives, 3) 254
   normal adults (66-71% female)
Variables: 90 CRPBI scores each for mother and for father
Method: Principal components condensation with varimax
   rotation, independently for mother and father within
   each sample
Results: Three factors emerged in all analyses: I Accepting and affectionate interest in the youngster and his
   activities; II Efforts to control children's behavior
   in negative and psychologically harmful ways; III Lax
   control.
Conclusions: Differences between samples, and with previous
   findings, are examined.
Rin, H., Schooler, C., & Caudill, W. A. Symptomatology and
   hospitalization: Culture, social structure and psychopathology in Taiwan and Japan. Journal of Nervous
   and Mental Disease, 1973, 157, 296-312.
Purpose: To identify dimensions of a symptom checklist
   with Taiwan Chinese and Japanese inpatients
Subjects: 890 Chinese Taiwan and 866 Japanese nonorganic
   inpatients
Variables: Scores on symptom checklist items (unspecified
   number less than total of 90)
Method: Principal components condensation with unspecified
   rotation, independently by sex, nationality, and combined sample
Results: Eight generally similar factors emerged: I
   Hostility; II Apathy; III Depression; IV Reality break;
   V Shinkeishitsu (absent in Taiwan sample); VI Hebephrenia;
   VII Hypochondria/headaches; VIII Somatization - gastro-
   intestinal/sleep disturbance.
Conclusions: Sex and cultural differences were examined
   through regression and MANOVA analyses.
Rothstein, W., Holmes, G. R., & Boblitt, W. E. A factor
   analysis of the Fear Survey Schedule with a psychiatric
   population. Journal of Clinical Psychology, 1972, 28,
   78-80.
Purpose: To identify dimensions of the Fear Survey Schedule
   (FSS) for a psychiatric population
Subjects: 50 male and 50 female psychiatric inpatients,
   aged 16-60 years
Variables: 76 FSS (revised, 1966) item scores
Method: Principal components condensation with varimax
   rotation
Results: Sixteen factors emerged, of which ten (I-X) were
   interpreted as highly specific object cathexes, and the
   remaining were seen as different fear constellations
   from those postulated by Wolpe and Lang: XI Fear of
   being socially unacceptable; XII Fear of exposure to
   suffering; XIII Fear of destructive agents; XIV Fear of
   repulsive phenomena; XV Fear of vast expanses; XVI Fear
   of authority.

Sacks, S., DeLeon, G., & Blackman, S. Psychological changes
   associated with conditioning functional enuresis.
   Journal of Clinical Psychology, 1974, 30, 271-276.
   Purpose: To identify dimensions underlying teachers'
   ratings of maladaptive behavior and academic achievement at school
   Subjects: 5½- to 14-year-old functional enuretics: 64 in
   conditioning group, 10 in psychotherapy-counseling, 9
   in control group, all rated by their teachers
   Variables: Scores on 17 4-point items related to maladaptive behavior in school and academic achievement estimates
   Method: Unspecified condensation with varimax rotation
   Results: Three factors emerged: I Psychological distur-
   bance; II Behavior problems; III Academic achievement
   Conclusions: Factor scores were used for between-group
   comparisons at follow-up.
Semrad, E. V., Grinspoon, L., & Fienberg, S. E. Development
   of an Ego Profile Scale. Archives of General Psychiatry,
   1973, 28, 70-77.
   Purpose: To identify dimensions of an ego profile scale
   Subjects: 63 hospital patients
   Variables: Scores on 45 ego-profile items
   Method: Unspecified condensation with unspecified rota-
   tion(s)
   Results: Eight factors emerged: I Hypochondriasis-somati-
   zation; II Projection; III Denial; IV Dissociation; V
   Neurasthenia; VI Compulsion; VII Not labeled; VII
   Anxiety alerts.
Shore, M. F., Clifton, A., Zelin, M., & Myerson, P. G. Pat-
   terns of masochism: An empirical study. British
   Journal of Medical Psychology, 1971, 44, 59-66.
   Purpose: To identify dimensions of masochism
   Subjects: 146 consecutive (unselected) psychiatric admis-
   sions
   Variables: 100 item scores from a masochism questionnaire
   Method: Principal-factors condensation with orthogonal
   varimax rotation
   Results: Ten factors emerged: I Suspicion; II Balancing;
   III-V Not interpreted; VI Compulsive; VII Negative fun;
   VIII-IX Not interpreted; X Stoicism, duty, forgiveness.
Sidman, J., & Moos, R. On the relation between psychiatric
   ward atmosphere and helping behavior. Journal of
   Clinical Psychology, 1973, 29, 74-78.
   Purpose: To identify dimensions underlying a Helping
   Scale (HS)
   Subjects: 226 psychiatric inpatients from 9 wards
   Variables: 7-point HS item scores: 70 patient and 70
   staff helping items
   Method: Minimum residual condensation with varimax rota-
   tion, independently for patient and staff items
   Results: Five factors emerged for patient helping
   behavior: I Friendship; II Directive teaching; III
   Enhancement of self-esteem; IV Interpretative psycho-
   analytic factor; V Condescension, avoidance, and

referral. Five factors emerged for staff helping be-
havior: I Friendship; II Directive teaching; III
Practical support; IV Avoidance of problems, referral,
and condescension; V Criticism and confrontation.
Conclusions: For both patient and staff helping behavior,
the first three factors were related to Ward Atmosphere
Scale subscales and discriminated between the 9 wards.
Starker, S., & Singer, J. L. Daydreaming and symptom pat-
terns of psychiatric patients: A factor-analytic study.
Journal of Abnormal Psychology, 1975, 84, 567-570.
Purpose: To identify relationships between some background
variables, symptoms, and daydreaming for psychiatric
patients
Subjects: 113 newly-admitted male VAH patients, age 20-
68 years
Variables: Scores on an unspecified number (20+?) measures
of background, symptoms, and daydreaming
Method: Principal components condensation with varimax
rotation
Results: Six factors emerged: I Variety of psychological
symptoms; II Anxious-distractible daydreaming; III
Alcoholism, indications of organiclike behavior; IV
Drug dependency; V Positive daydreaming; VI Schizoid
or schizophrenic symptoms.
Starker, S., & Singer, J. L. Daydream patterns and self-
awareness in psychiatric patients. Journal of Nervous
and Mental Disease, 1975, 161, 313-317.
Purpose: To identify relationships between measures of
daydreaming, symptomatology, and self-awareness
Subjects: 45 day hospital patients
Variables: Scores on 6 scales of an (abbreviated Imaginal
Processes Inventory) daydreaming measure, 5 scales of a
self-awareness measure, and 14 interview-derived ratings,
plus age
Method: Principal components condensation with orthogonal
varimax rotation
Results: Three factors emerged: I Psychiatric symptoms;
II Anxious-distractible daydreaming; III Unlabeled
daydream factor.
Steiner, J. A questionnaire study of risk-taking in psychi-
atric patients. British Journal of Medical Psychology,
1972, 45, 365-374.
Purpose: To identify dimensions underlying a risk-taking
questionnaire
Subjects: 212 psychiatric in- and 125 psychiatric out-
patients, 106 surgical or accident patients, 60 hospital
registrars
Variables: 20 item scores from a risk-taking questionnaire
Method: Principal components condensation, apparently
without rotation
Results: Five factors were reported: I General risk; II
Financial risk; III Risk related to driving and drink;
IV Social risk: sexual shyness; V Social risk: shy-
ness in public.

Vojtisek, J. E., & Magaro, P. A. The two factors present in
the Embedded Figures Test and a suggested short form for
hospitalized psychiatric patients. Journal of Consul-
ting and Clinical Psychology, 1974, 42, 554-558.
Purpose: To identify dimensions underlying the Embedded
Figures Test (EFT) with hospitalized psychiatric
patients
Subjects: 383 male hospitalized psychiatric patients
(aged 20-60 years)
Variables: 12 EFT item scores
Method: Principal axes condensation with varimax rotation
Results: Two factors emerged: I Pure embedded figures;
II Reversible perspective figures.
Wiggins, N. Individual differences in diagnostic judgments
of psychosis and neurosis from the MMPI. Educational
and Psychological Measurement, 1971, 31, 199-214.
Purpose: To identify dimensions in diagnostic MMPI
judgments
Subjects: 29 clinicians (Meehl, 1959)
Variables: Scores on 861 MMPI profiles
Method: Principal components condensation with hand rota-
tion to "idealized individuals"
Results: Three factors emerged: I Judge's model more
valid than his judgments; II Invalid judge; III Less
reliable judge.
Williams, J. D., & Dudley, H. K. Jr. Validity of the 16PF
and the MMPI in a mental hospital setting. Journal
of Abnormal Psychology, 1972, 80, 261-270.
Purpose: To identify dimensions underlying the 16PF and
MMPI in a mental hospital setting
Subjects: 201 new admissions to a state hospital
Variables: Scores on 16 16PF, and 3 validity and 9
clinical MMPI scales
Method: Principal axes condensation with (varimax and)
oblique rotation
Results: Six unlabeled factors emerged, of which three
appeared to be represented to some extent in both
instruments.
Zimmerman, R. L., Vestre, N. D., & Hunter, S. H. Validity
of family informants' ratings of psychiatric patients:
General validity. Psychological Reports, 1975, 37,
619-630.
Purpose: To identify dimensions underlying 1) Katz
Adjustment Scale (KAS) and 2) Interpersonal Checklist
(ICL)
Subjects: 149 university hospital (psychiatric ward)
patients (Vestre & Zimmerman, 1974; Zimmerman, 1970
unpublished dissertation)
Variables: (Unspecified here) numbers of 1) KAS and 2)
ICL scores
Method: Unspecified factor analysis
Results: Apparently 13 KAS factors emerged; the number of
ICL factors was not specified.

F. Behavior Disorder
  Capel, W. C., & Caffrey, B. "Faking good" as a problem in
    comparative studies of heroin addicts. Psychological
    Reports, 1974, 35, 859-864.
  Purpose: To identify dimensions underlying a socio-
    psychological attitude scale (Caffrey & Capel, 1970)
  Subjects: 43 individuals: personal bankrupts, poor
    credit risks, good credit risks, college faculty,
    black and white college students, random mail sample,
    orphaned auto club members, very rich persons, game-
    cock fighters, heroin addicts
  Variables: 44 item scores from a socio-psychological
    attitude scale
  Method: Principal components condensation
  Results: Five factors emerged: I Trust-optimism; II
    Authoritarianism; III Purposelessness; IV Health
    concern; V Control.
  Pishkin, V., & Thorne, F. C. A comparative study of the
    factorial composition of responses on the Life Style
    Analysis across clinical groups. Journal of Clinical
    Psychology, 1975, 31, 249-255.
  Purpose: To identify factors underlying the Life Style
    Analysis (LSA) for several clinical groups
  Subjects: 1) 190 incarcerated felons; 2) 74 incarcerated
    felons; 3) 89 alcoholic inpatients; 4) 155 students;
    5) 336 unwed mothers; 6) 159 college students; 7) 390
    chronic schizophrenics
  Variables: 200 LSA scores
  Method: Principal components condensation with orthogonal
    varimax rotation, independently for the seven groups
  Results: Five factors emerged for each group, and their
    composition compared across groups.
  Thorne, F. C., & Pishkin, V. A comparative study of the
    factorial composition of responses on the Existential
    Study across clinical groups. Journal of Clinical
    Psychology, 1973, 29, 403-410.
  Purpose: To identify the factorial structure of the
    Existential Study (ES) for different clinical groups
  Subjects: 193 incarcerated felons; 89 hospitalized
    alcoholics; 155 students of Ayn Rand; 336 unwed mothers;
    159 college students; 388 chronic undifferentiated
    schizophrenics
  Variables: 200 ES item scores
  Method: Principal components condensation with varimax
    rotation, independently by group
  Results: Five factors emerged in all groups; for felons
    and college students, all five factors were considered
    identical to those extracted for the combined nonpsy-
    chotic sample (Pishkin & Thorne, 1973): I Demoraliza-
    tion existential neurosis; II Religious dependency
    defenses; III Existential morale confidence; IV Self-
    actualization esteem; V Concern over the human condition.
    For alcoholics, Factors I-III and V were identical, but

Factor IV was labeled Poor fit and considerable ambivalence. For the Rand students, Factors I-III were
identical, but Factor IV was labeled Denial of religiosity
with probable atheistic philosophy, Factor V Joy in
living, pride in being a real person, and a good world
to live in, with rational liberalism. For unwed mothers,
Factors I-IV were identical, but Factor V emerged as
Social immaturity and reflection of immature self-
actualization. The five factors emerging for schizophrenics were: I Suicide and depression; II Religious
position; III Existential morale; IV Self-actualization;
V Self-status.

1. Drug Addiction

Berzins, J. I., & Ross, W. F. Locus of control among opiate
addicts. Journal of Consulting and Clinical Psychology,
1973, 40, 84-91.
Purpose: To identify dimensions underlying Rotter's
Internal-External Locus of Control (I-E) scale for
six samples
Subjects: 200 white male, 100 white female, 200 black
male, 100 black female opiate addicts; 400 male, 400
female college students (mostly white)
Variables: 23 I-E scale item scores
Method: Principal components condensation with varimax
rotation, by sample
Results: Two-factor (I Personal; II Sociopolitical) solutions appeared most interpretable in all analyses.

Berzins, J. I., Ross, W. F., English, G. E., & Haley, J. V.
Subgroups among opiate addicts: A typological investi-
gation. Journal of Abnormal Psychology, 1974, 83, 65-
73.
Purpose: To identify clusters of opiate addicts
Subjects: 1,500 opiate addicts in 10 subsamples (2 sexes
x 5 admission categories)
Variables: 13 MMPI scores
Method: Cluster analysis, by subsample
Results: Two clusters were replicated across analyses:
I Marked subjective distress, nonconformity, disturbed
thinking; II Single peak on Scale 4.

Naditch, M. P., Alker, P. C., & Joffe, P. Individual differences and setting as determinants of acute adverse
reactions to psychoactive drugs. Journal of Nervous
and Mental Disease, 1975, 161, 326-335.
Purpose: To identify dimensions of settings related to
psychoactive drug reactions
Subjects: 483 male drug users
Variables: 20 item scores from a questionnaire on drug
use
Method: Unspecified condensation with varimax rotation
Results: Five settings factors emerged: I Presence of
close knowledgeable friends; II Negative social sanctions; III Public place; IV Interpersonal difficulties
and pressure to use drugs; V Interpersonal difficulties
and adverse social facilitation.

2. Alcoholism
    Costello, R. M., Rice, D. P., & Schoenfeld, L. S. Attitudinal
        ambivalence with alcoholic respondents. Journal of
        Consulting and Clinical Psychology, 1974, 42, 303-304.
        Purpose: To identify dimensions underlying chronic
            alocholics' semantic differential (SD) ratings of a
            drinking scene and a pastoral scene
        Subjects: 30 chornic alcoholics in a rehabilitation
            center
        Variables: 7 4-point SD ratings each for a drinking and
            a pastoral scene
        Method: Principal components condensation with varimax
            rotation, independently by scene
        Results: Five alcohol factors emerged: I Harmful, danger-
            ous, safe; II Appetizing, pleasant, relaxing, tasty;
            III Good; IV Repulsive, harmful, unpleasant, alarming,
            distasteful; V Bad, unpleasant. Five non-alcohol fac-
            tors emerged: I Good, safe, pleasant, unpleasant,
            alarming, relaxing, distasteful; II Bad, harmless,
            harmful, dangerous, safe; III Repulsive, distasteful;
            IV Good, bad; V Appetizing, tasty.
    Hoffman, H., & Wefring, L. R. Sex and age differences in
        psychiatric symptoms of alcoholics. Psychological
        Reports, 1972, 30, 887-889.
        Purpose: To identify dimensions underlying the Brief
            Psychiatric Rating Scale (BPRS) for alcoholics (Hoffman,
            mimeo, no date cited)
        Subjects: Unspecified here
        Variables: 16 BPRS item scores (presumably)
        Method: Unspecified factor analysis
        Results: Three factors emerged: I Neuroticism; II Retar-
            dation; III Personality disorders.
    Hoffman, H., Wojtowicz, E. H., & Anderson, D. E. Analysis of
        demographic variables characterizing hospitalized male
        alcoholics. Psychological Reports, 1971, 29, 27-33.
        Purpose: To identify dimensions underlying demographic
            variables characterizing hospitalized male alcoholics
        Subjects: 214 male hospitalized alcoholics
        Variables: Scores on 37 demographic variables
        Method: Principal components condensation with varimax
            rotation
        Results: Twelve factors emerged: I Married; II Unmarried;
            III Economic status; IV Arrests; V Employment; VI Family
            size; VII Church attendance; VIII Drinking pattern; IX
            Residence; X Length of treatment; XI Widowed; XII Dis-
            charge.
    Hoffman, H., Wojtowicz, E. J., & Anderson, D. E. Analysis
        of drinking attitudes and drinking behavior in hospital-
        ized alcoholics. Psychological Reports, 1971, 28, 83-88.
        Purpose: To identify dimensions underlying drinking atti-
            tudes and behavior in hospitalized alcoholics
        Subjects: 211 male hospitalized alcoholics
        Variables: 41 item scores from an attitudes-toward-drinking-
            and-drinking-behavior inventory

Method: Principal components condensation with varimax
    rotation
Results: Twelve interpretable factors emerged: I Control;
    II Social situations; III Ego; IV Alcohol dependency;
    V Legal-employment; VI Trait evasion; VII Physical
    health; VIII Positive disposition; IX Family; X Conven-
    tional drinking; XI Conventional living; XII Problem
    avoidance.
Jansen, D. G., & Hoffman, H. Differences between the factor
    structure of alcoholics' moods prior to and after three
    weeks of treatment. Journal of Clinical Psychology,
    1972, 28, 593-595.
  Purpose: To identify dimensions of alcoholics' (self-
    descriptive) moods after three weeks of treatment and
    (retrospectively) just prior to admission
  Subjects: 120 consecutively admitted male alcoholics,
    mean age 48 years
  Variables: Scores on 52 5-step mood items, administered
    after three weeks of treatment under 1) "today" and
    2) "just prior to admission" instructions
  Method: Principal components condensation with varimax
    rotation, independently for current and retrospective
    data
  Results: Six factors emerged for "today" mood descriptions:
    I Depression; II Tension; III Guilt; IV Annoyance; V
    Hostility; VI Friendliness. Two interpretable factors
    emerged for "just prior to admission" descriptions: I
    General pathology; II Friendliness.
Kear-Colwell, J. J. Second structure personality factors
    found in psychiatric patients' responses to the 16PF.
    Journal of Clinical Psychology, 1972, 28, 362-365.
  Purpose: To identify second order deimensions of the
    16PF for psychiatric patients
  Subjects: 1) 230 general psychiatric inpatients, 2) 140
    male alcoholic inpatients
  Variables: 16 16PF scale scores
  Method: Principal components condensation with varimax
    and promax rotation, independently by group
  Results: Five "second-order" factors emerged in both
    analyses: I Anxiety vs. adjustment; II Impulsivity
    (related to Exvia-invia in normals); III Sociopathic
    deviance vs. obsessive-compulsivity; IV Related to
    exvia or extroversion; V Sensitive, self-absorption vs.
    tough practical realism.
Layden, T. A., & Smith, J. W. Nonmetric pattern analysis
    of behavioral and biological disease symptoms in
    alcoholism. Archives of General Psychiatry, 1973, 28,
    246-249.
  Purpose: To identify dimnnsions of behavioral and biolog-
    ical disease symptoms in alcoholism
  Subjects: 69 male (age 35-56) alcoholic inpatients
  Variables: Scores on 16 behavioral and 14 biological
    symptoms, plus age, days in withdrawal, and admission
    blood alcohol level

Method: Nonmetric factor analysis
Results: Ten factors emerged: I Emotional state; II Performance; III Enzyme; IV Color vision and liver function; V Steadiness; VI-X Not interpreted.

Nerviano, V. J. The second stratum factor structure of the 16PF for alcoholic males. Journal of Clinical Psychology, 1974, 30, 83-85.
Purpose: To identify second-order factors of the 16PF for alcoholic males
Subjects: 400 voluntary admissions to a VAH Alcoholism Treatment Program
Variables: 16 16PF sten scores
Method: Unspecified condensation with direct oblimin rotation
Results: Two second-stratum factors emerged: I Adjustment vs. anxiety; II Invia vs. exvia.
Conclusions: While Factor II appeared to be identical to that for the general population, loadings on Factor I indicated that for male alcoholics the level of source trait anxiety is related to that of Expediency vs. Conscientiousness.

Overall, J. E., & Patrick, J. H. Unitary alcoholism factor and its personality correlates. Journal of Abnormal Psychology, 1972, 79, 303-309.
Purpose: To identify dimensions underlying an alcohol behavior questionnaire
Subjects: 160 new admissions to a state hospital
Variables: Scores on 1) 135 alcohol behavior questionnaire items, 2) 42 items related most highly to the primary alcohol abuse factor identified in (1)
Method: 1) Orthogonal powered vector condensation with oblique rotation; 2) principal axes condensation, and principal components condensation
Results: One major alcohol abuse factor, plus 10 somewhat independent specific factors, emerged in the first analysis. Analyses of the 42-item set indicated the presence of a unitary alcoholism (severity) factor.

Skinner, H. A., Jackson, D. N., & Hoffman, H. Alcoholic personality types: Identification and correlates. Journal of Abnormal Psychology, 1974, 83, 658-666.
Purpose: To identify types of alcoholic psychiatric patients
Subjects: 94, 94, 94 alcoholic psychiatric patients
Variables: Unspecified number of Differentiated Personality Inventory (DPI) scores
Method: Unspecified Q-type factor analysis, by sample
Results: Eight clusters replicated across samples; they were discussed in terms of DPI and MMPI scores.

Whitelock, P. R., Overall, J. E., & Patrick, J. H. Personality patterns and alcohol abuse in a state hospital population. Journal of Abnormal Psychology, 1971, 78, 9-16.

Purpose: To identify 1) dimensions underlying an alcohol abuse questionnaire (AAQ), and 2) clusters of MMPI profiles

Subjects: 136 state hospital patients (in subsamples of 68 for (2))

Variables: Scores on 1) 80 AAQ items, 2) 12 MMPI scale scores

Method: 1) Principal axes and powered vector condensation; 2) oblique cluster-oriented factor analysis, by sub-samples of 68

Results: One AAQ factor emerged (in both analyses). Four profile patterns emerged in the second set of analyses (I-IV).

3. Suicide

Beck, A. T., Weissman, A., Lester, D., & Trexler, L. The measurement of pessimism: The Hopelessness Scale. Journal of Consulting and Clinical Psychology, 1974, 42, 861-865.

Purpose: To identify dimensions underlying clinical hope-lessness, as assessed by the Hopelessness Scale (HS)

Subjects: 294 hospitalized suicide attempters

Variables: 20 HS item scores

Method: Principal components condensation with varimax rotation

Results: Three factors emerged: I Affective; II Motivational; III Cognitive.

Dressler, D. M., Prusoff, B., Mark, H., & Shapiro, D. Clinician attitudes toward the suicide attempter. Journal of Nervous and Mental Disease, 1975, 160, 146-155.

Purpose: To identify dimensions of clinicians' attitudes toward the suicide attempter

Subjects: 20 second-year psychiatric residents, rating a total of 248 suicide attempters over a 6-month period

Variables: Scores on 23 mood adjective checklist items

Method: Unspecified condensation with varimax rotation

Results: Three affect factors emerged: I Warm and under-standing; II Anxious and confused; III Disinterest and annoyed.

Leonard, C. V. Self-ratings of alienation in suicidal patients. Journal of Clinical Psychology, 1973, 29, 423-428.

Purpose: To identify relationships between Alienation Self-rating Inventory (ASI) and MMPI scores for suicidal patients

Subjects: 34 male and 54 female voluntary psychiatric hospital inpatients (university medical school)

Variables: Scores on 15 ASI items and 16 MMPI scales (3 validity, 10 clinical, Ego Strength, Goldberg Index, Repression-Sensitization)

Method: Unspecified factor analysis

Results: Seven factors emerged: I General psychopathology; II Alienation; III Drug use; IV Open system; V Neurotic; VI Easy gratification; VII Social isolation.

4. Crime
  Blackburn, R. Dimensions of hostility and aggression in
    abnormal offenders. Journal of Consulting and Clinical
    Psychology, 1972, 38, 20-26.
    Purpose: To identify dimensions underlying hostility and
      personality measures for male (psychiatric) offenders
    Subjects: 165 male patients in a maximum security hospital
      for psychiatric offenders
    Variables: Scores on 17 hostility and personality scales,
      plus age
    Method: Principal components condensation with varimax
      (and promax) rotation
    Results: Four factors emerged: I Aggression; II Hostility;
      III Introversion-extraversion; IV Age.
  Carlson, K. A. Classes of adult offenders: A multivariate
    approach. Journal of Abnormal Psychology, 1972, 79,
    84-93.
    Purpose: To identify clusters, or classes, of adult
      reformatory inmates
    Subjects: 100 college males, 287 male correctional center
      inmates
    Variables: 15 Differential Personality Inventory scale
      scores
    Method: Cluster analysis, for half samples and total
    Results: Twelve clusters emerged and were described in
      terms of DPI profiles and additional demographic data.
  Holland, T. R., & Holt, N. Personality patterns among short-
    term prisoners undergoing presentence evaluations.
    Psychological Reports, 1975, 37, 827-836.
    Purpose: To identify dimensions of personality among
      short-term prisoners in presentence evaluation
    Subjects: 295 short-term prisoners in presentence evalua-
      tion (divided into two overlapping subsamples for repli-
      cation)
    Variables: K-corrected T scores (unspecified number) from
      MMPI
    Method: Cluster analysis (replicated with subsamples)
    Results: Four clusters emerged: I Mildly psychopathic
      mode of adjustment; II Attempted repressive control
      of conflictual emotions and consequent somatization
      tendencies; III Subjective distress coupled with
      notable degree of stimulus-seeking hyperactivity; IV
      Schizophrenic, borderline schizophrenic, or severe
      personality disturbance.
  Kahn, M. W. Murderers who plead insanity: A descriptive
    factor-analytic study of personality, social, and
    history variables. Genetic Psychology Monographs, 1971,
    84, 275-360.
    Purpose: To identify dimensions of personality and
      social/history variables for a sample of murders who
      plead insanity
    Subjects: 43 individuals who plead insanity to first-
      or second-degree murder

237

Variables: Scores on 4 past adjustment, 6 social class,
    3 demographic, 3 murder-related, 2 evaluation, 4 IQ,
    17 Rorschach primary process variables
Method: Principal components condensation with orthogonal
    varimax rotation
Results: Five strong factors (I Sanity-insanity--extent
    of primary process manifestations; II General intellec-
    tual level; III Social achievement--lower class and
    prior delinquency versus educational level and adjust-
    ment; IV Marital status and stability; V Hostile
    aggressive drive) and ten "weak" factors (VI Ethnic
    group versus educational level; VII Modulated libidinal
    drive and criminal history; VIII Weapon employed; IX
    Sibling status--aggression; X Multiple versus single
    killings; XI Male gender; XII Performance oriented
    cognitive ability versus age; XIII Violence of previous
    crimes; XIV Kinship to victim versus school adjustment
    and family size; XV Broken parental home) emerged.
Prisgrove, P. Group structure of prison officers' percep-
    tions of inmates. Journal of Consulting and Clinical
    Psychology, 1974, 42, 622.
Purpose: To identify dimensions underlying prison
    officers' perceptions of inmates
Subjects: 10 prison inmates, rated by 11 officers
Variables: Scores on 31 rating items related to person-
    ality, interpersonal relations, and institutional
    management and behavior
Method: Principal axes condensation with varimax rotation
Results: Three factors, of seven extracted, were inter-
    preted: I Assertively negative vs. more amenable
    inmate; II Inmates representing an institutional
    management problem vs. those who do not; III Inmates
    who attract an officer's friendship and respect vs.
    those who do not.
Silverstein, A. B., & Fisher, G. Cluster analysis of
    Personal Orientation Inventory items in a prison
    sample. Multivariate Behavioral Research, 1974, 9,
    325-330.
Purpose: To identify dimensions underlying the Personal
    Orientation Inventory (POI) with a prison sample
Subjects: 500 male prisoners
Variables: 150 POI item scores
Method: Elementary linkage analysis
Results: Five major clusters emerged: I Self acceptance;
    II Existentiality (A); III Nature of man; IV Existentiality
    (C); V Spontaneity.
Silverstein, A. B., & Fisher, G. Brief report: Second-order
    cluster analysis of Personal Orientation Inventory
    items in a prison sample. Multivariate Behavioral
    Research, 1975, 10, 503-506.
Purpose: To identify second-order clusters underlying the
    Personal Orientation Inventory (POI)
Subjects: 500 male prisoners

Variables: 31 first-order cluster (Silverstein & Fisher, 1974) scores
Method: Hierarchical cluster analysis
Results: Six second-order clusters emerged: I Self-regard; II Existentiality; III Nature of man; IV Acceptance of aggression; V Feeling reactivity; VI Self acceptance.

5. Juvenile Delinquency

Baer, D. J., Jacobs, P. J., & Carr, F. E. Instructors' ratings of delinquents after Outward Bound survival training and their subsequent recidivism. Psychological Reports, 1975, 36, 547-553.

Purpose: To identify dimensions underlying instructors' ratings of delinquents after Outward Bound training (OB)
Subjects: 60 delinquents participating in OB, rated by 25 instructors
Variables: 40 Instructor Rating Scale item scores
Method: Principal components condensation with varimax rotation
Results: Six factors were identified: I Maturity; II Extraversion; III Leadership; IV Confidence; V Effort; VI Physical ability.

Gibson, H. B. The factorial structure of juvenile delinquency: A study of self-reported acts. British Journal of Social and Clinical Psychology, 1971, 10, 1-9.

Purpose: To identify dimensions of self-reported juvenile delinquency
Subjects: 402 boys, aged 14-15 years
Variables: Binary scores for 37 delinquent acts, plus a social handicap score computed 4 years previously, and a criminal conviction score
Method: Principal components condensation with oblique promax rotation
Results: Twelve first-order factors emerged: I General delinquency; II Active theft v. cheating and pilfering incidental to normal activities, and disorderly acts which are not dishonest; III Maturity v. immaturity; IV Violence, possibly admissions of bravado; V-XII Uninterpreted specific factors. Three second-order factors emerged: I Social handicap; II Criminal conviction; III Disorderly but not necessarily dishonest.

Woodbury, R., & Shurling, J. Factorial dimensions of the Jesness Inventory with black delinquents. Educational and Psychological Measurement, 1975, 35, 979-981.

Purpose: To identify personality dimensions underlying the Jesness Inventory (JI) with black male delinquents
Subjects: 250 black male delinquents (adjudicated), mean age of 14.6 years
Variables: 155 JI item scores
Method: Principal components condensation with varimax rotation
Results: Three factors emerged: I Self-estrangement; II Social isolation; III Immaturity.

6. Homosexuality and Sexual Deviation
G. Mental Disorder
  1. Neurosis and Emotional Disorder
    Andrews, G., Kiloh, L. G., & Neilson, M.   Patterns of depres-
        sive illness: The compatability of disparate points of
        view. Archives of General Psychiatry, 1973, 29, 670-
        673.
      Purpose: To identify dimensions of depression
      Subjects: 145 neurotic or endogeneous depressives (in
        Australia)
      Variables: Scores on 1) 47, 2) 44 Feelings and Concerns
        Checklist items
      Method: Principal components condensation 1) without rota-
        tion, 2) with varimax rotation
      Results: Four factors emerged for 47 items: I Hopelessness
        and loss of self-esteem and a self-image of badness; II
        Not labeled; III Guilt; IV Anxiety. Four factors emerged
        for 44 items: I Anxiety-depression; II Self-pity; III
        Guilt; IV Dispairing-hopeful.
    Crisp, A. H., & Fransella, F.   Conceptual changes during
        recovery from anorexia nervosa. British Journal of
        Medical Psychology, 1972, 45, 395-405.
      Purpose: To identify dimensions of patients' repertory
        grids across treatment
      Subjects: 2 female anorexia nervosa patients
      Variables: Scores from 7 testings on each of two repertory
        grids (1 with 15 constructs, 1 with 18)
      Method: Principal components condensation, independently
        for each grid type at each testing for each subject
      Results: Two unlabeled factors were interpreted to ac-
        count for most of the variance in all grids of each
        type for both subjects. Consistencies (and changes)
        within grid type were examined across testing times.
    Derogatis, L. R., Lipman, R. S., Covi, L., & Rickels, K.
        Factorial invariance of symptom dimensions in anxious
        and depressive neuroses. Archives of General Psychiatry,
        1972, 27, 659-665.
      Purpose: To identify dimensions underlying pre-treatment
        self-ratings on the Symptom Distress Checklist (SDCL)
        for two patient groups
      Subjects: 641 anxious patients, 251 depressed neurotics
      Variables: 58 SDCL item scores
      Method: Principal components condensation with normal
        varimax rotation, by sample
      Results: Five factors emerged in both analyses: I Soma-
        tization; II Obsessive-compulsive; III Interpersonal
        sensitivity; IV Anxiety (anxious neurotics), or
        hostility (depressed neurotics); V Depression.
    Derogatis, L. R., Serio, J. C., & Cleary, P. A.   An empirical
        comparison of three indices of factorial similarity.
        Psychological Reports, 1972, 30, 791-804.
      Purpose: To identify dimensions underlying the Symptom
        Distress Checklist (SCL) for various subsamples

Subjects: 1) 210, 215 lower SES neurotic patients; 2) 219
    upper-middle, 422 working social status neurotic patients
Variables: Scores on 58 SCL scales
Method: Principal components condensation with varimax
    rotation, for subsamples within samples
Results: Five factors emerged for both analyses in the
    first sample: I G.N.F.; II Somatization; III Ob-Comp;
    IV Anxiety; V Anxiety II. Five factors emerged for
    both subsamples in the second set: I-IV Same as above;
    V Int-Sen.
Feldman, M. M. The body image and object relations: Explor-
    ation of a method utilizing repertory grid techniques.
    British Journal of Medical Psychology, 1975, 48, 317-
    332.
Purpose: To identify dimensions of a body image repertory
    grid
Subjects: 2 normal females, and 2 female patients diagnosed
    as anorexia nervosas
Variables: 15 ratings each for a 50-element (10 body parts,
    5 persons) repertory grid
Method: Principal components condensation without rotation,
    for each subject
Results: The first two factors, unlabeled in all cases,
    were used to examine the grid data; body parts tended
    to be less clustered together between than within per-
    sons.
Forbes, A. R. Some differences between neurotic and psychotic
    depressives. British Journal of Social and Clinical
    Psychology, 1972, 11, 270-275.
Purpose: To identify dimensions underlying the Runwell
    Symptom-Sign Inventory (SSI); data collected by Forbes,
    1970 (unpublished dissertation)
Subjects: 172 (non-chronic) depressive inpatients
Variables: Unspecified number of SSI item scores
Method: Unspecified condensation with varimax rotation
Results: Thirteen factors emerged: I Reduced behavioral
    efficiency; II Paranoid ideation; III Hypomania; IV
    Somatic complaints; V Obsessional orderliness; VI
    Depersonalization; VII Tension/restlessness; VIII Not
    interpreted; IX Hopelessness; X Obsessional symptoms;
    XI-XII Not interpreted; XIII Self-devaluation.
Conclusions: Scores on the ten interpretable factors were
    used in the present study.
Goldberg, H. L., Finnerty, R. J., Nathan, L., & Cole, J. O.
    Doxepin in a single bedtime dose in psychoneurotic
    outpatients. Archives of General Psychiatry, 1974, 31,
    513-517.
Purpose: To identify dimensions underlying the Hamilton
    Anxiety Scale (HAS)
Subjects: 93 psychoneurotic outpatients
Variables: 13 HAS item scores
Method: Unspecified condensation with orthogonal rotation

Results: Three factors emerged: I Nonphysical symptoma-
tology; II Physical complaints; III Patient's behavior
at interview.
Kerry, R. J., & Orme, J. E. Varieties of depression. Journal
of Clinical Psychology, 1975, 31, 607-609.
Purpose: To identify dimensions of depression as measured
by selected Inpatient Multidimensional Psychiatric
Scale (IMPS) items with depressives
Subjects: 51 depressed female inpatients
Variables: 22 IMPS item scores, age, diagnosis (neurotic
or endogenous depression)
Method: Principal components condensation with varimax
rotation
Results: Five factors emerged: I Loss of motivation and
energy; II Feelings of depression and hopelessness;
III Feelings of guilt and sin; IV Retarded behavior;
V Anxiety.
Conclusions: Four separate depressive factors (I-IV) were
obtained; neither age nor diagnostic variables appeared
to be important.
Meikle, S., & Mitchell, M. C. Factor analysis of the Fear
Survey Schedule with phobics. Journal of Clinical
Psychology, 1974, 30, 44-46.
Purpose: To identify dimensions of the Fear Survey
Schedule III (FSS-III) with phobics
Subjects: 95 male and 20 female phobics living in the
community
Variables: 79 FSS-III item scores (2 items were expanded
to 2 and 3 items)
Method: Unspecified condensation with varimax rotation
Results: Twenty-one factors emerged: I Fear of social
disapproval; II Fear of witnessing physical assaults;
III Fear of natural events and animals; IV Fear of
surface travel; V Fear of strangers; VI Fear of repul-
sive phenomena; VII Fear of heights; VIII Fear of
assault; IX Fear of being enclosed; X Fear of nude
bodies; XI Fear of noises; XII Fear of car accidents;
XIII Fear of sick people; XIV Fear of flying animals;
XV Fear of open spaces; XVI Fear of doctors; XVII
(Related to fear of noise or to painful experiences?
sirens, doctors); XVIII Fear of strange experiences;
XIX Fear of dogs (singlet); XX Fear of being alone;
XXI Fear of losing control (singlet).
Paykel, E. S., & Prusoff, B. A. Relationships between
personality dimensions: Neuroticism and extraversion
against obsessive, hysterical and oral personality.
British Journal of Social and Clinical Psychology, 1973,
12, 309-318.
Purpose: To identify dimensions common to neuroticism-
extraversion and obsessive-hysterical-oral personality
Subjects: 131 depressed patients
Variables: Scores on three dimensions (obsessive, hyster-
ical, oral) derived from the LaZare-Klerman Trait Scales

and from a relative interview, and on two dimensions
(neuroticism, extraversion) derived from the MPI
    Method: Principal components condensation without, and
    with unspecified rotation
    Results: Two unrotated factors emerged: I Neurotic,
    introverted, oral personality without obsessionality;
    II Hysteric extrovert without obsessional traits.
    Rotation indicated obsessives to be non-neurotic with
    small element of introversion, hysterics to be neurotic
    and extraverted, and oral personalities to be neurotic
    and introverted.
Paykel, E. S., Prusoff, B. A., Klerman, G. L., & DiMascio,
    A. Self-report and clinical interview ratings in
    depression. Journal of Nervous and Mental Disease,
    1973, 156, 166-182.
    Purpose: To identify dimensions of depression
    Subjects: Varied sample of 207 depressed patients, rated
    by self or one psychiatrist
    Variables: 1) 28 symptom ratings from a clinical interview;
    2) 18 self-report items
    Method: Principal components condensation, apparently
    without rotation
    Results: Only the first factor was reported from the
    clinical interview analysis (I Overall severity of
    depression) and from the self-report analysis (I
    Overall symptom severity).
    Conclusions: Factor scores were related to various other
    variables for this and an additional sample.
Paykel, E. S., Weissman, M., Prusoff, B. A., & Tonks, C. M.
    Dimensions of social adjustment in depressed women.
    Journal of Nervous and Mental Disease, 1971, 152, 158-
    172.
    Purpose: To identify dimensions of social adjustment in
    depressed women
    Subjects: 40 depressed and 40 normal females, mean age
    42 years
    Variables: 37 item scores from a semistructured interview
    related to work, social and leisure, extended family,
    marital as spouse, parental, and marital family unit
    areas
    Method: Principal components condensation with varimax
    rotation
    Results: Six factors emerged: I Work performance; II
    Interpersonal friction; III Communication; IV Submissive
    dependency; V Degree of attachment or orientation to
    family; VI Satisfactions and feelings.
Raskin, A., Crook, T. H., & Herman, K. D. Psychiatric his-
    tory and symptom differences in black and white depressed
    inpatients. Journal of Consulting and Clinical Psychology,
    1975, 43, 73-80.
    Purpose: To identify dimensions of psychopathology in de-
    pressed inpatients; reported 1971 by Raskin & McKeon
    (journal not scanned for this project)

Subjects: 159 black, 555 white hospitalized depressives
Variables: Scores on 69 first-order factors derived from
10 instruments (symptom rating, mood, ward behavior
rating scales from psychologist, patient, or nurse),
plus unspecified (here) number of other scores
Method: Unspecified factor analysis
Results: Fourteen "superfactors" emerged: I Emotional
withdrawal; II Depression; III Guilt-worthlessness;
IV Hostility (psychiatrist); V Hallucinations; VI Hostility (patient); VII Hypochondriasis; VIII Social
participation; IX Sick set (patient); X Anxiety-tension;
XI Sleep disturbances; XII Irritability; XIII Cognitive
disturbances; XIV Paranoid and grandiose delusions.

Rump, E. E. Cluster analysis of personal questionnaires
compared with principal components analysis. British
Journal of Social and Clinical Psychology, 1974, 13,
283-292.
Purpose: To identify a cluster analytic solution for
Shapiro's (1969, Case A) pre-post personal questionnaire
data
Subjects: One patient diagnosed as neurotic depressive
with morbid jealousy
Variables: Score on 17 symptoms on a total of 24 occasions
(pre, post each of 12 treatment sessions)
Method: Modified linkage cluster analysis
Results: Four clusters emerged: I Tension and depression;
II Anger, and jealousy toward wife; III Feeling bad in
evening and being hurt easily (doublet); IV Worry about
future and not feeling like ever working again (doublet).
Conclusions: Changes in cluster scores over treatment
were computed, and compared with Slater's (1970) principal components solution.

Schless, A. P., Mendels, J., Kipperman, A., & Cochrane, C.
Depression and hostility. Journal of Nervous and
Mental Disease, 1974, 159, 91-100.
Purpose: To identify dimensions of expression of hostility
in depression
Subjects: 37 depressed inpatients
Variables: 7 Buss-Durkee Inventory, 18 MMPI, 3 Symptom
Check List, Beck Depression Inventory, and 6 semantic
differential scores, plus age, sex
Method: Principal components condensation with varimax
rotation
Results: Five factors, of the 19 extracted, were labeled:
I Anxious, depressed, resentful patient with evidence
of obsessive-compulsive defenses, who feels s/he must
control hostility, turn anger inward, and who feels
guilty; II Hostility-out and release of inhibition;
III Behavior conformity (probably men more likely to
show this inhibition); IV Anxious, suspicious, resentful,
control of hostile feelings; V Assaultive, verbally
abusive outward expression of hostility.

Conclusions: The out vs. in patterns of hostility (I &
IV, II & V) were examined relative to severity of
depression and other variables.
Schroeder, H., & Craine, L. Relationships among measures of
fear and anxiety for snake phobics. Journal of Consul-
ting and Clinical Psychology, 1971, 36, 443.
Purpose: To identify relationships between fear and anxiety
measures for snake phobics
Subjects: 107 female college student snake phobics
Variables: Scores on 9 fear and anxiety measures (MAS,
SRIA, FSS, SNAQ, FT, BAT, Em, SRIA-s, Welsch A)
Method: Principal components condensation with varimax
rotation
Results: Two factors emerged: I Generalized, pervasive
anxiety; II Specific fears.
Schulterbrandt, J. G., Raskin, A., & Reatig, N. Further
replication of factors of psychopathology in the interview, ward behavior and self-reported ratings of hospitalized depressed patients. Psychological Reports,
1974, 34, 23-32.
Purpose: To identify dimensions of psychopathology in
interview (I), ward behavior (W), and self-report
ratings (SR) for a new sample
Subjects: 325 depressed patients
Variables: 43 item scores from Inventory of Psychic and
Somatic Complaints (IPSC) for psychiatrist; IPSC scores
for patient; 52 Mood Scale items each for nurse and for
patient; 141 W items (nurse)
Method: Principal components condensation with varimax
rotation, independently for each set of scores
Results: Ten PISC psychiatrist and nine IPSC patient
factors emerged; seven nurse and eight patient mood
factors, and 15 W factors emerged. These results were
compared with those for two previous studies. Ten
major areas of psychopathology appeared stable: I
Depressed mood; II Hostility; III Guilt-ashamed; IV
Anxiety; V Sleep disturbances; VI Interest and involvement in activities; VII Paranoid projections; VIII Cognitive loss; IX Apathy - motor retardation; X Hypochon-
driasis.
Vath, R., Miranda, M., Becker, J., & Gibson, S. Attempted
validation of a "pragmatic classification of depression."
Psychological Reports, 1972, 30, 287-290.
Purpose: To identify dimensions of observer ratings of
depression
Subjects: 75 hospitalized depressives
Variables: Observer ratings on 57 (Blinder) traits
Method: Principal components condensation with orthogonal
rotation
Results: Three factors, of 27 extracted, were interpreted:
I Schizo-affective depression; II Anxiety; III Physiologic
retarded depression.

Weckowicz, T. E., Cropley, A. J., & Muir, W. An attempt to replicate the results of a factor analytic study in depressed patients. Journal of Clinical Psychology, 1971, 27, 30-31.

Purpose: To identify dimensions of depression as measured by Hamilton's rating scale (H)

Subjects: 52 depressed male psychiatric inpatients

Variables: 17 H item scores

Method: Principal axes condensation with varimax rotation

Results: Four factors emerged; they were unlabeled but very dissimilar to the four factors found by Hamilton (1960).

Weckowicz, T. E., Yonge, K. A., Cropley, A. J., & Muir, W. Objective therapy predictors in depression: A multivariate approach. Journal of Clinical Psychology, 1971, 27, 3-29.

Purpose: To identify dimensions underlying a battery of therapy predictors (and thereby isolate dimensions of depression)

Subjects: 170 psychiatric inpatients (scoring above 17 on the Beck inventory on admission)

Variables: Scores on 21 Beck inventory subscales, 2 Maudsley Personality Inventory scales, TMAS (Q data); resident depression rating, 22 Hamilton rating scales (L); 7 measures of psychomotor functions, 4 measures of physiological functions (T); age, sex, and previous admissions (binary)

Method: Principal axes condensation with orthogonal varimax rotation (and promax for second-order solution) for all variables; independent alpha condensation with varimax rotation for L-, Q-, and T-data.

Results: Twenty first-order factors emerged: I Insomnia; II Psychomotor retardation; III Self depreciation and suicidal tendencies; IV Gastric disturbance; V Neurotic anxiety; VI Guilty depression; VII Agitation; VIII Autonomic (sympathetic) reactivity; IX Fatigue; X Diurnal variation; XI Somatization or hypochondriasis; XII Disturbance of sexual function; XIII (Poorly defined; sex); XIV Tearful depression; XV Chronicity; XVI (Poorly defined doublet); XVII (Poorly defined); XVIII (Poorly defined); XIX Withdrawal; XX (Poorly defined). Six second-order factors emerged: I Somatic factor of retarded psychotic depression; II Atypical depression, or involutional dimension of depression with schizoid features; III Typical or guilty depression; IV Fatigue, exhaustion, or neurasthenia; V Neurotic secondary somatization; VI Bodily functions and neurotic symptoms. For L-data, five of eight extracted factors were interpreted: I Insomnia; II Agitation and anxiety; III Somatization; IV Depersonalization and paranoid symptoms; V Guilt and suicidal tendencies. For Q-data, eight of nine extracted factors were interpretable: I Neurotic anxiety; II Guilty depression; III Retardation; IV

246

Self-depreciation; V Schizoid; VI Distortion of body
image; VII Involutional setting; VIII Not reported.
For T-data, three of five factors were interpretable:
I Sympathetic nervous system reactivity; II-III Psychomotor speed 1 and 2.

2. Psychosis

Fleiss, J. L., Lawlor, W., Platman, S. R., & Fieve, R. R.
On the use of inverted factor analysis for generating
typologies. Journal of Abnormal Psychology, 1971, 77,
127-132.

Purpose: To identify clusters, or types, of manic-
depressive inpatients (MDs)

Subjects: 52 MDs

Variables: Scores on 7 measures of quantity/quality of
sleep, 8 MMPI, 2 MAACL, 10 of interview behavior

Method: 1) Inverse, 2) principal components condensation
with varimax rotation

Results: Three types emerged: I Mild severity, little
sleep disturbance; II Most severely ill, moderate sleep
disturbance; III Little evidence of obvious psychotic
features, greatest sleep disturbance. Three factors
emerged in the second analysis: I Manic-depressive; II
Overall psychoticism; III Sleep disturbance.

Lorr, M., & Hamlin, R. M. A multimethod factor analysis of
behavioral and objective measures of psychopathology.
Journal of Consulting and Clinical Psychology, 1971,
36, 136-141.

Purpose: To identify dimensions underlying 1) interview
behavior (IB) and Ward behavior syndrome (WBS) variables,
2) IB, WBS, rated verbalizations (RV), cognitive and
psychomotor (C&P) tests

Subjects: 125 functional psychotics from 5 hospitals

Variables: Scores on 1) 10 IB and 12 WBS variables, 2)10
IB, 12 WBS, 26 RV and C&P tests

Method: Principal components condensation with varimax
rotation, independently for the two data sets

Results: Ten factors emerged in the first analysis: I
Excitement; II Hostile belligerence; III Paranoid
projection; IV Grandiosity; V Perceptual distortion;
VI Anxious depression; VII Retardation; VIII Disorien-
tation; IX Motor disturbances; X Conceptual disorgani-
zation. Fourteen factors emerged in the second analysis:
I-IV, VII, VIII, IX, X Same as above; V Perceptual
distortion - interview; VI Anxious depression - interview;
XI Perceptual distortion - ward; XII Seclusiveness - ward;
XIII-XIV Uninterpreted singlets.

Mullens, B. N. A cognitive and perceptual taxonomy of
character disorders. Journal of Clinical Psychology,
1972, 28, 9-13.

Purpose: To identify clusters of character-disordered
males on the basis of cognitive and personality tests

Subjects: 76 males in the Armed Forces diagnosed as
character disorders

Variables: Congruency scores on 7 personality and 7 cognitive measures
Method: Modified B coefficient cluster analysis
Results: Five clusters, including 59 subjects, emerged:
I Poor intellectual functioning (mentally retarded, or
constitutionally unfit); II Withdrawn, paranoid, and
similar to conventional schizoid-schizophrenic; III
Classic neurotic; IV Social desirability as a projective
mechanism or cognitive style; V Psychopath.
Murphy, D. L., & Beigel, A. Depression, elation, and lithium
carbonate responses in manic patient subgroups. *Archives
of General Psychiatry*, 1974, *31*, 643-648.
Purpose: To identify the relationship between depression,
elation, and manic behavior
Subjects: 30 manic patients
Variables: Scores on 26 Manic-State Rating Scale items
Method: Unspecified factor analysis
Results: Four factors emerged: I Core manic behavior;
II P-D vs. E-G; III Sexually related behavior; IV
Uninterpreted (depression, impulsive behavior).
3. Schizophrenia
Alumbaugh, R. V., & Sweney, A. B. Factor analytic contribu-
tions to the study of information processing in schizo-
phrenia. *Journal of Clinical Psychology*, 1973, *29*, 437-
439.
Purpose: To identify dimensions of irrelevancy (overinclu-
sion) in schizophrenics' information processing
Subjects: 30 chronic (and hospitalized continuously more
than 2 years) schizophrenics, 30 tubercular patients,
and 30 community citizens
Variables: Scores for 16 combinations of relevant/irrele-
vant visual stimuli, plus age, WAIS Digit Span scores,
and Raven Progressive Matrices score
Method: Unspecified condensation with varimax, and promax,
rotation
Results: Four factors emerged for both varimax and pro-
max solutions: I Figural irrelevancy; II Nonirrelevancy;
III Qualitative irrelevancy; IV Perceptual intelligence.
Steer, R. A. Relationship between the Brief Psychiatric
Rating Scale and the Multiple Affect Adjective Check
List for schizophrenic women. *Psychological Reports*,
1974, *35*, 79-82.
Purpose: To identify relationships between the Brief
Psychiatric Rating Scale (BPRS) and selected Multiple
Affect Adjective Check List (MAACL) scales
Subjects: 75 schizophrenic women
Variables: 18 BPRS symptom scores, MAACL Anxiety, Depres-
sion, and Hostility scores
Method: Principal components condensation with varimax
rotation
Results: Seven factors emerged; the three MAACL scores
all loaded on the second factor, the only other definer
of which was hallucinatory behavior.

248

4. Schizophrenia Treatment

Astrachan, B. M., Brauer, L., Harrow, M., & Schwartz, C.
Symptomatic outcome in schizophrenia. Archives of
General Psychology, 1974, 31, 155-160.
Purpose: To identify dimensions of mental status (symptoms)
for schizophrenics at 2-3 years post discharge
Subjects: 132 schizophrenic patients 2-3 years post dis-
charge
Variables: Scores on 20 Gurin Mental Status Index, 21 New
Haven Schizophrenia Index, 21 Psychiatric Evaluation
Form items
Method: Unspecified condensation with unspecified rotation
Results: Five factors emerged: I Neurotic; II Schizo-
phrenia; III Motor retardation; IV Hallucinations and
delusions; V Turbulent.

Daly, D. L., Bath, K. E., & Nesselroade, J. R. On the confounding of inter- and intraindividual variability in
examining change patterns. Journal of Clinical
Psychology, 1974, 30, 33-36.
Purpose: To identify dimensions of change (in self-ratings
of patients) through P- and chain-P technique factor
analysis
Subjects: 3 female schizophrenics hospitalized in a token
economy ward ("matched" on demographics)
Variables: 6-point self-ratings on 24 of Cattell's (1946)
adjectives, on each of 60 consecutive days
Method: Principal axes condensation with Meredith's (1964)
rotation, independently for each subject and for pooled
data (chain-P)
Results: Eight factors emerged in all four analyses but
were not named; the individual P analyses indicated at
least three common factors, but only one of them was
also common to the chain-P results.

Fitzgibbons, D. J., & Shearn, C. R. Concepts of schizophrenia
among mental health professionals: A factor-analytic
study. Journal of Consulting and Clinical Psychology,
1972, 38, 288-295.
Purpose: To identify dimensions underlying mental health
professionals' concepts of schizophrenia
Subjects: 183 clinical psychologists, psychiatrists, and
psychiatric social workers
Variables: Scores on 94 opinion items related to beliefs
about cause, prognosis, and descriptive characteristics
of schizophrenia
Method: Principal components condensation with unspecified
rotation
Results: Eight factors were interpreted: I Interpersonal
etiology; II Bleulerian phenomenology; III Disease concept of schizophrenia; IV Poor prognosis; V Poor understanding of schizophrenia; VI Schizophrenia as thinking
disorder; VII Adaptive symptomatology; VIII Irreversi-
bility.

Lorei, T. W., & Gurel, L. Use of a biographical inventory to
predict schizophrenics' posthospital employment and re-
admission. Journal of Consulting and Clinical Psychology,
1972, 38, 238-243.
Purpose: To identify dimensions underlying the Palo Alto
Social Background Inventory (PASBI) with about-to-be-
discharged schizophrenics
Subjects: 720 VAH about-to-be-discharged schizophrenics
Variables: 91 PASBI item scores
Method: Principal components condensation with varimax
(and promax) rotation
Results: Ten factors emerged: I Perception of disability;
II Marital affiliation; III Alienation; IV Chronicity;
V Education; VI Uncritical optimism; VII Social isola-
tion; VIII Irregular employment; IX Drinking problem;
X Socioeconomic status.
Schwartz, C. C., Myers, J. K., & Astrachan, B. M. Psychiatric
labeling and the rehabilitation of the mental patient.
Archives of General Psychiatry, 1974, 31, 329-334.
Purpose: To identify dimensions of mental status (symptoms)
for schizophrenics at 2 years after discharge
Subjects: 132 schizophrenic patients 2-3 years post dis-
charge
Variables: Scores on 20 Gurin Mental Status Index, 21 New
Haven Schizophrenia Index, 21 Psychiatric Evaluation
Form items
Method: Unspecified condensation with unspecified rotation
Results: Five factors emerged: I Neurotic; II Schizo-
phrenia; III Motor retardation; IV Hallucinations and
delusions; V Turbulent.
Schwartz, C. C., Myers, J. K., & Astrachan, B. M. Concordance
of multiple assessments of the outcome of schizophrenia.
Archives of General Psychiatry, 1975, 32, 1221-1227.
Purpose: To identify factors common to three instruments
assessing mental status
Subjects: 132 diagnosed and treated (discharged) schizo-
phrenics
Variables: 62 item scores from the Psychiatric Evaluation
Form, New Haven Schizophrenia Index, and Gurin Mental
Status Index
Method: Unspecified condensation with unspecified rotation
Results: Five factors emerged: I Neurotic; II Schizophrenia;
III Motor retardation; IV Hallucinations and delusional;
V Turbulent.
H. Psychosomatic Disorder
Coursey, R. D., Buchsbaum, M., & Frankel, B. L. Personality
measures and evoked responses in chronic insomniacs.
Journal of Abnormal Psychology, 1975, 84, 239-249.
Purpose: To identify dimensions underlying a battery of
personality measures
Subjects: 13 male; 5 female chronic insomniacs, matched
with 18 non-insomniacs on age, sex
Variables: Scores on 24 scales from a personality battery

Method: Principal components condensation with orthogonal
rotation
Results: Five factors emerged: I Anxious worriers; II
Extraversion; III Concern with internal vs. external
stimuli; IV Test honesty; V Other MMPI scales.
Gering, R. C., & Mahrer, A. R. Difficulty falling asleep.
Psychological Reports, 1972, 30, 523-528.
Purpose: To identify dimensions of a psychological prob-
lems inventory for insomniac and non-insomniac patients
Subjects: 180 insomniac, 131 non-insomniac male hospitalized
psychiatric patients
Variables: Scores on 60 Psychological Problem Inventory
items discriminating between groups
Method: Centroid condensation with varimax rotation, by
group
Results: Four factors emerged for insomniacs: I Somati-
zation; II Depression; III Dependent rejection; IV
Decreased control. Four factors emerged for non-
insomniacs: I Verbal interaction problems; II Somati-
zation; III Manic depression; IV Depression.
I. Case History
J. Mental Health and Rehabilitation
Eddy, G. L., & Sinnett, E. R. Behavior setting utilization
by emotionally disturbed college students. Journal of
Consulting and Clinical Psychology, 1973, 40, 210-216.
Purpose: To identify relationships among behavior setting
utilization (for activities), clinicians' action-
oriented ratings, and MMPI scale scores for emotionally
disturbed rehabilitation students
Subjects: 46 client members of a rehabilitation living
unit (college campus)
Variables: Scores on (utilization of) 25 behavior
settings, 4 clinicians' action-oriented ratings, 14
MMPI scales, diagnosis, sex
Method: Principal components condensation with varimax
rotation
Results: Four factors were identified: I Action-oriented;
II Ego strength vs. neuroticism; III Action-oriented
out-of-living-area, indicating unconventional action
or behavior; IV Sex differences.
Stone, L. A., Coles, G. J., Sinnett, E. R., & Sherman, G. L.
Multidimensional scaling used to evaluate students
residing in a rehabilitation unit. Psychological
Reports, 1971, 28, 879-886.
Purpose: To identify 1) judge-observer and 2) similarity
dimensions for similarity judgments of rehabilitation
students
Subjects: 4 mental health professionals and 2 volunteers
serving as judges
Variables: Similarity judgments of 9 clients and 6 volunteers residing in a rehabilitation unit
Method: Principal components condensation (and inverse)
with varimax rotation

Results: Only one judge factor emerged. Four judgment
    factors emerged: I Action-oriented vs. overcontrolled
    (reversed); II Sex difference; III Adjustment vs.
    severity of disturbance; IV Uninterpreted.
Triandis, H. C., Feldman, J. M., Weldon, D. E., & Harvey,
    W. M. Ecosystem distrust and the hard-to-employ.
    Journal of Applied Psychology, 1975, 60, 44-56.
Purpose: To identify concept and person dimensions of
    ecosystem distrust among hard-core blacks
Subjects: 83 white female college students; 43 white
    high school, 60 black high school, and 41 black hard-
    core males
Variables: 573 attribute/concept ratings on 6 dimensions
Method: Three-mode factor analysis
Results: Five stimulus person (stereotype) factors
    emerged: I Outgroup violent men; II Thrivers on status
    quo; III Ingroup or safe dependable people; IV People
    with status and authority; V Socially aggressive people.
    Five attribute factors emerged: I Go-getting people;
    II Brutes; III Opportunistic; IV Unimportant; V Trust-
    worthy and hardworking. Six unlabeled subject factors
    emerged.
Wright, L. Components of positive mental health. Journal
    of Consulting and Clinical Psychology, 1971, 36, 277-280.
Purpose: To identify dimensions underlying "positive
    mental health"
Subjects: 256, 393 college students (each in various
    subsamples by sex and type of residence)
Variables: 30 "positive mental health" item scores
Method: Principal components condensation with varimax
    rotation, by sample (and subsample)
Results: Four factors emerged in all analyses: I Task
    and perceptual effectiveness; II Autonomy and self-
    actualization; III Commitment; IV Openness.
Conclusions: These factors were discussed relative to
    those hypothesized by Jahoda (1958).
K. Counseling and Guidance
   1. Marriage and Family
Elton, C. F., & Rose, H. A. The counseling center: A mirror
    of institutional size. Journal of Counseling Psychology,
    1973, 20, 176-180.
Purpose: To identify dimensions underlying counseling
    center census-type variables
Subjects: 157 counseling centers
Variables: Scores on 19 census-type variables
Method: Principal components condensation with varimax
    rotation
Results: Four factors emerged: I Traditional model; II
    Psychotherapy model; III Vocational guidance model;
    IV Training model.
Farnsworth, K. E., Lewis, E. C., & Walsh, J. A. Counseling
    outcome criteria and the question of dimensionality.
    Journal of Clinical Psychology, 1971, 27, 143-145.

Purpose: To identify dimensions of client behaviors indicative of successful counseling outcome
Subjects: 107 counseling center staff members and advanced graduate students from college centers across the country
Variables: 51 (rationally derived from 180 items) content areas of client behaviors, rated for relevance as indicator of successful counseling outcome
Method: Principal components condensation with varimax rotation
Results: Six factors emerged: I Objective vs. subjective criteria; II Maturity; III Inner- vs. other-directed behavior; IV Conventionality; V Energetic and socially-approved behavior; VI Ability to deal with reality.

Jansen, D. G., Robb, G. P., & Bonk, E. C. Peer ratings and self-ratings on twelve bipolar items of practicum counselors ranked high and low in competence by their peers. Journal of Counseling Psychology, 1973, 20, 419-424.
Purpose: To identify dimensions underlying peer and self-ratings of practicum counselors
Subjects: 173 graduate students who had completed a counseling practicum, rated by self (S) and peers (P)
Variables: Scores on 12 bipolar adjective scales
Method: Principal axes condensation with varimax rotation, independently for S and P
Results: Two unlabeled factors emerged in each analysis, but were different for S and P.

Miller, A., & Thompson, A. Factor structure of a goal checklist for clients. Psychological Reports, 1973, 32, 497-498.
Purpose: To identify dimensions underlying the Goal Checklist (GC)
Subjects: 548, 446 clients of a university counseling center
Variables: 39 GC item scores
Method: Principal components condensation with varimax rotation, by group
Results: Four factors replicated across samples (a fifth emerged in the first analysis): I Social-personal security; II Vocational-educational; III Academic skill; IV Marital-close relationships.

Rose, H. A., & Elton, C. F. Identification of potential personal-problem clients. Journal of Counseling Psychology, 1972, 19, 8-10.
Purpose: To identify dimensions underlying Omnibus Personality Inventory (OPI) scales for University Counseling and Testing Center nonclients
Subjects: 50 male, 50 female nonclients of a university counseling center
Variables: 10 OPI scores
Method: Principal components condensation, by sex
Results: Two unlabeled factors emerged and were used in further analyses with personal-problem clients.

Sinnett, E. R., Stone, L. A., & Matter, D. E. Clinical
    judgment and the Strong Vocational Interest Blank.
    Journal of Counseling Psychology, 1972, 19, 498-504.
Purpose: To identify dimensions of clinical judgments
    of agreement between Strong Vocational Interest Blank
    (SVIB) scores and expressed vocational choice.
Subjects: 40 Ph.D., Ed.D., or M.A. counselors
Variables: 25 judgments of agreement between SVIB scores
    and expressed vocational choice
Method: Inverse principal components condensation; uni-
    dimensional cluster analysis
Results: The first judge factor accounted for 71% of
    interjudge variance. Four clusters emerged: I Strong
    support; II Moderate support; III Low support; IV Very
    low (or no) support.
2. Social Casework
Ryle, A., & Breen, D. Change in the course of social-work
    training: A repertory grid study. British Journal of
    Medical Psychology, 1974, 47, 139-147.
Purpose: To identify dimensions of social work students'
    repertory grids related to their relationships with
    two clients
Subjects: 12 social work students
Variables: 16-element, 20-construct repertory grids re-
    lated to student-client relationships
Method: Principal components condensation, independently
    for each student and combined group
Results: An unspecified number of factors were extracted
    in each case; the first two factors were used for
    examining changes across a two-year training period.
L. Physical Handicap
1. Blindness and Visual Disorder
2. Deafness and Hearing Disorder
Bolton, B. A factor analytic study of communication skills
    and non-verbal abilities of deaf rehabilitation clients.
    Multivariate Behavioral Research, 1971, 6, 485-501.
Purpose: To identify dimensions underlying communication
    skills and nonverbal abilities of deaf clients
Subjects: 1) 159 profoundly deaf rehabilitation clients,
    2) subsample of 87 aged 16-25
Variables: Scores on 1) 10 communication, 2) plus 14 non-
    verbal ability variables
Method: Principal components condensation with varimax
    rotation, for each S-variable set
Results: Two factors emerged for the total sample: I
    Manual communication; II Oral communication. Four
    factors emerged for the subsample: I Nonverbal
    ability; II Manual communication; III Oral communication;
    IV Psychomotor skill.
Bolton, B. Psychometric validation of a clinically derived
    typology of deaf rehabilitation clients. Journal of
    Clinical Psychology, 1972, 28, 22-25.

254

Purpose: To psychometrically validate a clinically-derived
typology of deaf rehabilitation clients
Subjects: 42 prelingually deafened rehabilitation clients
Variables: A 42-item Q-sort for global personality
description (at least 2 per client)
Method: Principal components condensation with varimax
rotation
Results: Four factors emerged: I Creative behavioral
style; II Rigid-inhibited; III Undisciplined; IV
Acceptance-anxious.
Bolton, B. Quantification of two projective tests for deaf
clients. Journal of Clinical Psychology, 1972, 28,
554-556.
Purpose: To identify dimensions underlying the Bender-
Gestalt (BG) and Draw-A-Person (DAP) tests for deaf
rehabilitation clients
Subjects: 57 young adult deaf rehabilitation clients
Variables: Scores on 8 BG and 8 DAP rating scales
Method: Principal components condensation with promax
rotation
Results: Five factors emerged: I General adjustment; II
(Latent) aggression; III Extraversion; IV Impulsiveness; V Anxiety-drive.
Conclusions: Relationships between these factors and
several nonpersonality variables, including outcome
measures, were also investigated.
Bolton, B. An alternative solution for the factor analysis
of communication skills and nonverbal abilities of
deaf clients. Educational and Psychological Measure-
ment, 1973, 33, 459-463.
Purpose: To identify dimensions of communication skills
and nonverbal abilities of deaf clients with an en-
larged sample
Subjects: 192 profoundly deaf young adults
Variables: 10 rated communication abilities, 5 Ravens
Progressive Matrices, 6 Revised Beta, and 4 Purdue
Pegboard subtest scores, plus total score from
Minnesota Paper Form Board
Method: 1) Principal components condensation with varimax rotation (Little Jiffy (LJ)), 2) Little Jiffy III
(LJIII)
Results: The first three factors for both solutions were
identified as I Nonverbal reasoning; II Psychomotor
skill; III Manual communication. The LJ solution produced two other factors (IV Oral-verbal communication;
V Speed of response (Beta-specific)), which were different from the LJIII factors (IV Oral-verbal communication; V Residual hearing).
Bolton, B., Donoghue, R., & Langbauer, W. Quantification
of two projective tests for deaf clients: A large
sample validation study. Journal of Clinical Psychology,
1973, 29, 249-250.

Purpose: To identify dimensions underlying Bender-Gestalt
(B-G) and Draw-A-Person (DAP) rating scales for deaf
clients
Subjects: 229 deaf rehabilitation clients
Variables: Scores on 16 B-G and DAP rating scales
Method: Principal components condensation with varimax
rotation
Results: Six factors emerged: I Overall level of func-
tioning efficiency; II-VI Uninterpreted.
Conclusions: Six composite (factor) scores were examined
relative to communication and ability variables, global
personality ratings, and outcome for six subgroups of
clients.
M. Speech Disorder
N. Neurological Disorder
Bachrach, H., & Mintz, J. The Wechsler Memory Scale as a
tool for the detection of mild cerebral dysfunction.
Journal of Clinical Psychology, 1974, 30, 58-60.
Purpose: To identify dimensions underlying Wechsler
Memory Scale (WMS) subtests
Subjects: 42 individuals with mild to moderate cerebral
dysfunction, and 42 individuals judged free of neuro-
logical impairment (with no significant differences
between groups on age, IQ, patient status, range of
ego weakness)
Variables: Scores on 7 WMS subtests
Method: Principal components condensation with varimax
rotation
Results: Three factors emerged: I Designs, Story Recall,
Paired Associates; II Information and Orientation; III
Mental Control and Digit Span (loading above .80).
Conclusions: Scores on the highest loading subtest for
each factor were used in a regression analysis.
Barnes, G. W., & Lucas, G. J. Cerebral dysfunction vs.
psychogenesis in Halstead-Reitan tests. Journal of
Nervous and Mental Disease, 1974, 158, 50-60.
Purpose: To identify dimensions of the Halstead-Reitan
battery (H-R) for a psychogenic and an organic group
of inpatients
Subjects: 39 psychogenic and 77 organic VA inpatients
Variables: 14 scores from the H-R, age, Damage Index
Method: Principal axes condensation with orthogonal varimax (and graphical, general factor retained) rotation,
independently for the two groups
Results: Six "varimax factors" emerged for each group:
I Biological intelligence, or basic adaptive ability
dependent on cortical integrity; II TPT; III-V Test-
specific (Trails, Time Sense Visual, unclear); VI
Language (psychogenic only). The general-factor
solutions yielded five similar factors for organics
and six, of which one was clearly different (VI Pure
motor speed), for psychogenics.
Conclusions: Discrepancies between these results and
Halstead's are discussed.

Crinella, F. M., & Dreger, R. M.   Tentative identification
    of brain dysfunction syndromes in children through
    profile analysis. Journal of Consulting and Clinical
    Psychology, 1972, 38, 251-260.
  Purpose: To identify brain dysfunction syndromes in
    children, and to identify clusters of children with
    brain dysfunction
  Subjects: 39 children aged 8-6 to 12-6 (12 with known
    brain dysfunction, 12 with symptomatology but no hard
    evidence, 15 controls)
  Variables: 1) 35 scores from a neuropsychological test
    battery, 2) scores on 8 factors from (1)
  Method: 1) Principal components condensation with vari-
    max and rotoscope rotation for all Ss; 2) multiple-
    group cluster analysis with oblique rotation for 24 Ss
  Results: Eight unlabeled factors emerged in the first
    analysis. Three clusters of subjects were identified:
    I High visual-motor integrity and fast manual reaction
    speed, but relatively less skill in serial synthetic
    and tactual motor abilities; II Severely deficient
    "across the board," III Phasicality in profile con-
    figuration.
Goldstein, G., & Shelly, C. H. Statistical and normative
    studies of the Halstead Neuropsychological Test Battery
    relevant to a neuropsychiatric hospital setting.
    Perceptual and Motor Skills, 1972, 34, 603-620.
  Purpose: To identify dimensions underlying the Halstead
    Neuropsychological Test Battery (Halstead) for a
    neuropsychiatric population
  Subjects: 619 (VAH) neuropsychiatric inpatients
  Variables: 8 Halstead, 11 WAIS, and 6 alternative test
    scores
  Method: Alpha factor condensation with varimax rotation
  Results: Four factors emerged: I Language skills; II
    Finger agnosia and finger-tip number-writing; III
    Complex, primarily non-verbal problem-solving ability;
    IV Motor speed.
Page, J. G., Janicki, R. S., Bernstein, J. E., Curran, C. F.,
    & Michelli, F. A. Pemoline (Cylert) in the treatment
    of childhood hyperkinesis. Journal of Learning Dis-
    abilities, 1974, 7, 498-503.
  Purpose: To identify factors of a parent and a teacher
    questionnaire administered in conjunction with a drug
    study
  Subjects: 238 hyperkinetic MBD children participating in
    a drug study, rated by 1) a parent and 2) a teacher
  Variables: 1) 48 and 2) 28 item scores from questionnaires
    concerned with behavioral characteristics associated
    with MBD
  Method: Unspecified factor analysis
  Results: Eight parent factors emerged: I Aggressive/
    antisocial; II Emotionalism; III Restless/inattentive;

257

IV Somaticism/aches and pains; V Nervous habits; VI
Fearfulness; VII Sociopathic; VIII Somaticism/G.I.
problems. Five teacher factors emerged: I Aggressive/
antisocial; II Restless/hyperactive; III Emotionalism;
IV Distractibility; V Immaturity.
Conclusions: Change scores were computed on these factors.
1. Brain Damage
   Crinella, F. M. Identification of brain dysfunction syndromes
   in children through profile analysis: Patterns associated with so-called "minimal brain dysfunction."
   Journal of Abnormal Psychology, 1973, 82, 33-45.
   Purpose: To identify 1) dimensions underlying a neuropsychological battery for brain lesion (BD) or suspected
   "minimal brain dysfunction" (MBD) children, and 2)
   clusters of children
   Subjects: 53 BD or MBD children, 37 controls ((1) only)
   Variables: Scores on 89 measures from a neuropsychological battery, plus CA
   Method: 1) Principal factors condensation with oblique rotation; 2) multiple-group condensation (inverse)
   Results: Four factors, of 16 extracted, were interpreted
   in the first analysis: I Cognitive maturation; II
   Spatial orientation; III Cortertia vs. cerebroasthenia;
   IV Kinetic mobility vs. pathological inertia. Eight
   clusters of children were identified in the second
   analysis.
   Goebel, R. A., & Satz, P. Profile analysis and the abbre-
   viated Wechsler Adult Intelligence Scale: A multi-
   variate approach. Journal of Consulting and Clinical
   Psychology, 1975, 43, 780-785.
   Purpose: To identify clusters of WAIS profiles
   Subjects: 55 VAH psychiatric inpatients, 118 brain-
   damaged Ss
   Variables: WAIS profiles (long form)
   Method: Modified hierarchical grouping analysis, by group
   Results: Fourteen profile types emerged for psychiatric
   patients, 18 for the brain-damaged individuals; none
   were labeled.
   Lansdell, H. A general intellectual factor affected by tem-
   poral lobe dysfunction. Journal of Clinical Psychology,
   1971, 27, 182-184.
   Purpose: To identify dimensions underlying Differential
   Aptitude Tests (DAT) and Wechsler-Bellevue Intelligence
   Scale (WB) subscales for a sample of surgical (neurolo-
   gical) patients
   Subjects: 42 neurological patients who underwent unilateral
   temporal lobe surgery
   Variables: Scores on 9 DAT and 11 WB subtests
   Method: Principal axes condensation with unspecified
   rotation
   Results: One general (I General) and three common fac-
   tors (II Perceptual; III Verbal; IV Speed) emerged.

Conclusions: Separate analyses by sex apparently produced
  differing factor structures; the results were examined
  as a function of extent and hemisphere of lesion.
Lansdell, H., & Smith, F. J. Asymmetrical cerebral function
  for two WAIS factors and their recovery after brain
  injury. Journal of Consulting and Clinical Psychology,
  1975, 43, 923.
Purpose: To identify dimensions underlying WAIS subtest
  scores for brain injured Ss
Subjects: 119 men with clearly lateralized head injury
  and localized dysfunction, 74 men with damage on the
  right, and 76 with damage on the left
Variables: Scores on WAIS subtests
Method: Unspecified condensation with varimax rotation
Results: Two factors emerged and were described in terms
  of subtest having highest loading: I Vocabulary; II
  Object Assembly.
Russell, E. W. WAIS factor analysis with brain-damaged
  subjects using criterion measures. Journal of Con-
  sulting and Clinical Psychology, 1972, 39, 133-139.
Purpose: To identify dimensions underlying the WAIS
  with brain-damaged subjects
Subjects: 21 lateralized left-, 16 lateralized right-
  hemisphere, 40 diffuse brain-damaged patients, and
  26 non-brain damaged controls
Variables: 11 WAIS subtest scores
Method: Principal components condensation with varimax
  rotation, independently for total group and for
  lateralized Ss only
Results: In both analyses, three factors emerged: I
  Verbal; II Performance; III (Digit Span).
Conclusions: Brain damage was found to affect WAIS performance but not factor structure.
Russell, E. W. Reanalysis of Halstead's biological intelligence factor matrix. Perceptual and Motor Skills,
  1973, 37, 699-705.
Purpose: To reanalyze Halstead's biological intelligence
  factor matrix
Subjects: Patients recovering from closed head trauma
  (number unspecified here); Halstead, 1947
Variables: Scores on 1) 13, 2) 8 neuropsychological
  indicators
Method: 1) Principal components condensation with varimax and direct oblique oblimin, biquartimin rotation,
  2) principal axes condensation with varimax rotation
  (for 8 variables)
Results: The analyses with 13 variables produced four
  factors: I C; II A; III P; IV D. The eight-variable
  analysis produced three factors: I C; II A; III D.
Russell, E. W. The effect of acute lateralized brain
  damage on Halstead's biological intelligence factors.
  Journal of General Psychology, 1974, 90, 101-107.

259

Purpose: To identify dimensions of Halstead's (1947)
    battery for acute lateralized brain lesion patients
Subjects: Patients with acutely lateralized brain
    lesions, 35 each hemisphere
Variables: 12 test scores from Halstead
Method: Principal components condensation with orthogonal
    varimax rotation
Results: Four factors emerged: I Abstraction performance;
    II Verbal; III Motor - left hemisphere; IV Uninterpre-
    table ability related to right-hemisphere impairment.
Conclusions: These results were compared to those for
    chronic diffuse brain damaged subjects, for whom only
    one of Halstead's biological intelligence factors ap-
    peared.
Swiercinsky, D. P., & Hallenbeck, C. E. A factorial approach
    to neuropsychological assessment. Journal of Clinical
    Psychology, 1975, 31, 610-618.
Purpose: To identify dimensions of a test battery related
    to brain damage
Subjects: 1) 292 adult brain-damaged patients, 2) random
    subgroups of 140 and 152
Variables: 8 WAIS scores, chronological age, years of
    education, and scores for Finger Tapping Dominant
    Hand, Finger Tapping Nondominant Hand, Finger Tip
    Number Writing (right only), Seashore Rhythm Test,
    Wisconsin Card Sort, Finger Agnosia Right, Finger Agnosia
    Left, Speech Perception, Tactual Performance Test, and
    Form Discrimination tests
Method: Principal components condensation with normal
    orthogonal varimax rotation, independently for random
    and total groups
Results: Five factors emerged in all three analyses: I
    Motor; II Perceptual; III Verbal-cognitive; IV Spatial-
    orientation; V Concept-reasoning.
  2. Epilepsy
    Lin, Y., & Rennick, P. M. WAIS correlates of the Minnesota
        Percepto-Diagnostic Test in a sample of epileptic
        patients. Differential patterns for men and women.
        Perceptual and Motor Skills, 1973, 37, 643-646.
    Purpose: To identify dimensions common to WAIS subtests
        and the Minnesota Percepto-Diagnostic Test (MPDT)
    Subjects: 117 male, 60 female epileptics
    Variables: Scores on 11 WAIS subtests and MPDT
    Method: Principal components condensation with orthoblique
        rotation, by sex
    Results: Two unlabeled factors emerged for males and
        three for females.
O. Mental Retardation
    Guarnaccia, V. J., & Weiss, R. L. Factor structure of fears
        in the mentally retarded. Journal of Clinical Psychology,
        1974, 30, 540-544.
    Purpose: To identify dimensions of the mentally retarded's
        (MR) fears

260

Subjects: 102 trainable MRs, aged 6-21 years, IQ 15-65
(mean=43)
Variables: Parents' responses on 56 items of the Louis-
ville Fear Survey for Children, plus age, sex, IQ
Method: Principal components condensation with varimax
rotation
Results: Four factors emerged: I Separation; II Natural
events; III Injury; IV Animals.
Conclusions: These factors were discussed relative to
those for subjects of average intelligence.
Harrison, R. H., & Budoff, M. A factor analysis of the
Laurelton Self-concept Scale. American Journal of
Mental Deficiency, 1972, 76, 446-459.
Purpose: To identify self-report personality dimensions
for mentally retarded persons
Subjects: 172 special class and institutionalized EMR
children
Variables: 23 Bialer IE Scale item scores, 137 modified
Laurelton Self-Concept Scale item scores
Method: Principal components condensation with normal
varimax rotation of 11, 29, and 38 factors
Results: Factors from the first solution were labeled as
follows: I Positive concept of the physical self; II
Maladjustment; III Combativeness; IV Scholastic inadequacy; V Narcissism; VI Obedience; VII Depression; VIII
Positive motivation; IX Identification with authority;
X Social adjustment at school; XI Social inferiority.
Factors from the second (29-factor) solution were
labeled as follows: I Positive concept of the physical
self; II Neuroticism; III Social desirability lie scale;
IV Altruism; V Getting ahead; VI Excitable aggressiveness; VII Social inferiority; VIII Responsibility for
interpersonal behavior; IX Easygoing friendliness-
adults; X Social competence with elders; XI Disdainful
narcissim - looks; XII Willfulness; XIII Immature
weakness; XIV Future control; XV Sheltered optimism;
XVI Bravado, fears of rejection; XVII Insensitivity to
disapproval; XVIII Fun-loving popularity; XIX Athletic
and social incompetence; XX Depressed self criticism;
XXI Dependent inadequacy; XXII Narcissim-intellect;
XXIII Narcissism-looks at home; XXIV Positive school
motivation; XXV Modest self respect; XXVI Inner locus
of control; XXVII Friendlessness; XXVIII Depression;
XXIX Blame taking.
Factors from the third (38-factor) solution were labeled
as follows: I Positive concept of the physical self;
II Neuroticism; III Social desirability lie scale; IV
Altruism - positive motivation; V Self-abasement; VI
Being approved of at home; VII Social inferiority; VIII
Social competence with peers; IX Temper control; X
Social competence with elders; XI Body narcissism; XII
Manipulation through guilt; XIII Excitable aggressiveness;
XIV Future control; XV Sheltered optimism; XVI Bravado;

XVII Positive school motivation; XVIII School social
adjustment - peers; XIX Physical health - others' opinion;
XX Social competence - others' opinion; XXI Dependent
inadequacy; XXII Narcissism - intellect; XXIII Self-
satisfaction; XXIV Friendlessness; XXV Responsibility
for interpersonal behavior; XXVI Inner locus of control;
XXVII Passive, fantasied control; XXVIII Bitterness;
XXIX Intropunitiveness; XXX Tearfulness; XXXI Depressed
alienation; XXXII Unrelatedness; XXXIII Defensive self-
respect in school; XXXIV Fantasy of self-destruction;
XXXV Easygoing friendliness; XXXVI Mischievousness;
XXXVII Feeling rejected at home; XXXVIII Fantasy of
control while weak.

McCormick, M., Balla, D., & Zigler, E. Resident-care practices
in institutions for retarded persons: A cross-institu-
tional, cross-cultural study. American Journal of
Mental Deficiency, 1975, 80, 1-17.
Purpose: To identify dimensions underlying a Child Management Inventory used in a cross-cultural study of institutions for the retarded
Subjects: Unspecified total number of living-unit senior
care attendants in 1) 19 institutions for retarded
persons in the U.S., 2) 11 institutions for the retarded
in Scandinavia
Variables: 30 item scores from the Child Management Inventory
Method: Principal components condensation with varimax
rotation, independently for the two countries
Results: Four U.S. factors emerged: I Toileting practices;
II Bathing practices; III Feeding practices; IV Regimen-
tation of either visiting or residents. Six Scandinavian
factors emerged: I Toileting and bedtime practices; II
Bathing practices; III Handling of clothing; IV Uninterpreted; V Feeding practices; VI Permissibility of
personal items.
Conclusions: Factor structure was compared between institutions as well as countries.

Ross, R. T. Factor groupings of problem behaviors. American
Journal of Mental Deficiency, 1971, 76, 136-137.
Purpose: To identify dimensions underlying problem
behaviors as assessed by the Fairview Problem Behavior
Record (FPBR)
Subjects: 617 state hospital for the mentally retarded
inpatients
Variables: Scores on 26 triadic FPBR items
Method: Principal components condensation with orthogonal
rotation
Results: Five factors emerged: I Hostile, destructive,
and outward-oriented; II Hyperactivity; III Sexually
oriented behaviors; IV Covert behaviors; V Major
annoyances (displaced aggression?).
Conclusions: Analyses by sex yielded the same results.

1. Learning and Motor Ability
   Neeman, R. L. Perceptual-motor skills of mental retardates:
      A factor-analytic study. Perceptual and Motor Skills,
      1971, 33, 927-934.
      Purpose: To identify dimensions underlying the Purdue
      Perceptual-motor Survey)PPMS) with mental retardates
      (MRs)
      Subjects: 1) 99 child and young adult MRs; 2) subsample
      of 66 non-mongoloid Ss
      Variables: 19 PPMS scores, IQ, CA, sex, and in (1)
      mongolism
      Method: Unspecified condensation with orthogonal rota-
      tion, by sample
      Results: Seven factors, similar across analyses (except
      for V), emerged: I Postural dimensionality; II
      Shoulder-arm movement; III Laterality; IV Ocular con-
      trol; V Intelligence; VI Developmental; VII (possibly)
      Cultural sex bias.
   Silverstein, A. B., Brownlee, L., Hubbell, M., & McLain, R.
      E. Comparison of two sets of Piagetian scales with
      severely and profoundly retarded children. American
      Journal of Mental Deficiency, 1975, 80, 292-297.
      Purpose: To identify dimensions underlying selected
      Corman-Escalona (C-E) and Uzgiris-Hunt (U-H) scales
      Subjects: 64 severely and profoundly retarded persons
      (mean CA=14)
      Variables: Scale scores for the C-E Object Permanence
      and Spatial Relationships scales, and the U-H Object
      Permanence and Spatial Relationships scales
      Method: 1) Principal components condensation, 2) multi-
      method factor analysis
      Results: 1) One general principal component, accounting
      for 85% of the total variance, emerged. 2) Jackson's
      method generated two factors: I Object permanence;
      II Spatial relationships.
2. Training and Vocational Rehabilitation
   Domino, G., & McGarty, M. Personal and work adjustment of
      young retarded women. American Journal of Mental
      Deficiency, 1972, 77, 314-321.
      Purpose: To identify dimensions of personal and work
      adjustment of young retarded women
      Subjects: 35 MR young adults working in a sheltered
      workshop
      Variables: Scores on 5 Sonoma Checklist scales, 8 Work
      Adjustment Rating Form scales, on 1 interview rating
      Method: Principal components condensation with varimax
      rotation
      Results: Three factors emerged: I General adjustment
      (WARF); II Personal adjustment; III Motivation.
   Hanson, D. L., & Stone, L. A. Multidimensional perceptions
      regarding actual and potential job-placement success
      of mentally retarded adults. Perceptual and Motor
      Skills, 1974, 38, 247-254.

Purpose: To identify dimensions underlying counselors'
perceptions related to job placement for mentally
retarded adults (MRs)
Subjects: 1) 4, and 2) 4 professional counselors as
judges
Variables: Similarity judgments for 1) potential job-
placement success of 10 institutionalized male MRs,
2) actual job-placement success of 15 male MRs considered successful
Method: Principal components condensation with varimax
rotation, by S-variable set
Results: Three factors emerged in the first analysis:
I Socialization; II Independent functioning; III
(Singlet). Three factors emerged in the second
analysis: I Employment stability; II Inappropriate
independence; III (Outcome opposite of prediction).
P. Physical Illness
Brand, C. R. Relations between emotional and social be-
haviour: A questionnaire study of individual differences. British Journal of Social and Clinical
Psychology, 1972, 11, 10-19.
Purpose: To identify dimensions underlying scales of a
"personality" questionnaire
Subjects: 99 adult males aged 19-45 found in casualty
clinic waiting rooms
Variables: Age, plus fearfulness, aggressiveness, dependence, social extraversion, behavioral extraversion,
neuroticism, and lie scale scores from a 69-item
questionnaire
Method: Principal components condensation with orthogonal
graphic rotations (2 solutions)
Results: The marginally preferable (on the basis of
simple structure) solution was defined by three factors:
I Aggression related to social approach; II Inadequacy;
III Defensiveness, or defendedness. The second solution was also defined by three factors: I Neuroticism,
with aggressiveness; II Inadequacy; III Extraversion,
with aggressiveness.
Quinlan, D. M., Kimball, C. P., & Osborne, F. The experience of open heart surgery: IV. Assessment of disorientation and dysphoria following cardiac surgery.
Archives of General Psychiatry, 1974, 31, 241-244.
Purpose: To identify dimensions of 1) preoperative interview and 2) a behavior checklist for postoperative
(major cardiac surgery) patients
Subjects: 1) 76 preoperative, 2) 58 postoperative major
cardiac surgery patients
Variables: Scores on 1) 23 interview ratings, 2) 11
behavior checklist items
Method: 1) Unspecified factor analysis; 2) unspecified
condensation with varimax rotation
Results: Two pre-op interview factors emerged: I Emotional stability; II Level of anxiety. Two post-op
factors emerged: I Orientation; II Dysphoria.

Stamper, D. A., Sterner, R. T., & Linsman, R. A. Symptoma-
tology subscales for the measurement of Acute Mountain
Sickness. Perceptual and Motor Skills, 1971, 33, 735-
742.
   Purpose: To identify dimensions underlying General High
   Altitude Questionnaire (GHAQ) responses
   Subjects: 30 U.S. Army enlisted men
   Variables: 22 GHAQ item scores (means for final 48 hrs at
   low and initial 48 hrs at high altitude)
   Method: Key cluster analysis (orthogonal and oblique solu-
   tions)
   Results: Four clusters emerged: I Arousal level; II
   Somatic discomfort; III Tired; IV Mood.
Vitale, J. H., Pulos, S. M., Wollitzer, A. O., & Steinhelber,
   J. C. Relationships of psychological dimensions to im-
   pairment in a population with cerebrovascular insuffi-
   ciency. Journal of Nervous and Mental Disease, 1974,
   158, 456-467.
   Purpose: To identify dimensions underlying a psychological
   battery for cerebrovascular insufficient patients (CVIs),
   and types of CVIs
   Subjects: 138 CVIs
   Variables: 1) 213 scores from a 26-test psychological
   battery; 2) scores on V-Clusters I-IV
   Method: 1) V cluster analysis; 2) O-type cluster analysis
   Results: Six V-type clusters emerged: I Verbal ability;
   II Numerical ability; III Visual-motor ability; IV
   Somatic concerns; V Conformity; VI Defensiveness. Fif-
   teen unlabeled subject types emerged.
Q. Community Services
   Benfari, R. C., Beiser, M., Leighton, A. H., & Mertens, C.
   Some dimensions of psychoneurotic behavior in an urban
   sample. Journal of Nervous and Mental Disease, 1972,
   155, 77-90.
   Purpose: To identify dimensions of psychoneurotic behavior
   in an urban sample
   Subjects: 531 adults from selected NYC census tracts
   Variables: 1) scores on 93 items of a physical-complaint
   and psychological status questionnaire; 2) 53 psychological status item scores from the questionnaire
   Method: 1) Unspecified factor analysis; 2) principal
   components condensation with varimax rotation
   Results: The initial factor analysis split the 93-item
   inventory into two parts: I Physical complaints; II
   Psychological status. Five factors emerged from the
   analysis of the psychological items: I Physiological
   anxiety; II Topically oriented depression; III Physiological process disturbance; IV Mixed anxiety/depression;
   V Low esteem/low self-confidence versus self-confidence/
   high esteem.
   Benfari, R. C., Beiser, M., Leighton, A. H., Murphy, J., &
   Mertens, C. The manifestation of types of psychological
   states in an urban sample. Journal of Clinical Psychology, 1974, 30, 471-483.

Purpose: To identify person types, by psychological states, in a natural (community) setting
Subjects: 530 urban adults drawn in a geographically (and hence SES) stratified urban sample (45% male, age 20-70+ years)
Variables: T-scores on (interview categories, apparently factor-analytically derived) physiological anxiety (10 items), topically oriented depression (12), physiological process disturbance (8), noncognitive depression (8), alienation vs. involvement (5).
Method: Cluster analysis (BC Try)
Results: Six types, and a residual group, emerged: I Low symptomatology - high involvement; II Low symptomatology - high alienation; III Average symptomatology - high involvement; IV Moderately high somatic depression and moderately high alienation; V Topical depressives, plus physiological concomitants, plus involvement; VI Anxious type, plus all manifestations of depression alienation.

Cowen, E. L., Lorion, R. P., & Caldwell, R. A. Nonprofessionals' judgments about clinical interaction problems. Journal of Consulting and Clinical Psychology, 1975, 43, 619-625.
Purpose: To identify dimensions of discomfort ratings of nonprofessional child aides' interaction with young maladapting school children
Subjects: 49 nonprofessional child aides
Variables: 56 item scores from the Aide-Experience Questionnaire
Method: Principal components condensation with varimax rotation
Results: Four factors emerged: I Aide discomfort about child aggression and family conflict; II Discomfort about child's desire to be close to aide; III Discomforts about child's need for exclusive ownership of aide; IV Discomfort about child's limit-testing behaviors.

Graham, J. R., Lilly, R. S., Paolino, A. F., Friedman, I., & Konick, D. S. Measuring the adjustment of ex-patients in the community: A comparison of the factor structures of self-ratings and ratings of others. Journal of Clinical Psychology, 1972, 28, 380-384.
Purpose: To identify dimensions of ex-patient adjustment in the community, as assessed by self-rating on a Katz Adjustment Scale modified for use with ex-patients (KAS-Pl)
Subjects: 469 ex-patients from an urban short-term psychiatric hospital
Variables: 127 KAS-Pl item scores
Method: Principal components condensation with varimax rotation
Results: Eight factors emerged: I Agitation-depression; II Social conformity; III Belligerence; IV Paranoid alienation; V Loss of control; VI Immaturity; VII Disorientation; VIII Psychomotor retardation.

Conclusions: These results were compared with those for relatives' ratings (same sample rated).

Lubin, B., Bangert, C. J., & Hornstra, R. K. Factor structure of psychological assessment t a community mental health center. Psychological Reports, 1974, 35, 455-460.

Purpose: To identify dimensions of psychological assessment at a community mental health center (CMHC)

Subjects: 387 adult applicants to a CMHC (and their relatives)

Variables: Scores on 40 assessment variables

Method: Principal components condensation with varimax rotation

Results: Two factors emerged: I Relatives' report of the patients' overall emotional-behavioral disturbance; II Patients self-report of depression.

Lubin, B., Hornstra, R. J., & Love, A. Initial treatment assignment as a function of patient self-assessment and/or family assessment of patient. Psychological Reports, 1974, 35, 495-498.

Purpose: To identify dimensions underlying a series of intake assessment measures for community mental health center (CMHC) applicants

Subjects: Cohort of 611 CMHC applicants and 433 of their family members

Variables: Scores on unspecified number of standardized, interview-derived intake assessment measures

Method: Principal components condensation with varimax rotation

Results: Two factors emerged: I Relatives' report of patients' illness; II Patients' report on own illness.

Lyon, K. E., & Zucker, R. A. Environmental supports and post-hospital adjustment. Journal of Clinical Psychology, 1974, 30, 460-465.

Purpose: To identify clusters of environmental supports as measured by the Environmental Support Questionnaire (ESQ)

Subjects: 38 ex-inpatients (max stay=30 days) of a community mental health center, interviewed at one, two, and three months after discharge

Variables: Scores on 13 ESQ items, plus 3 3-month outcome measures

Method: Cluster analysis (PACKAGE: ordered similarity matrix, multiple groups analysis)

Results: Seven clusters emerged: I Stable home life; II Home responsibilities; III Professional and medication involvement; IV Employment; V Visitors; VI Low social class; VII Outcome.

Spiro, H. R., Siassi, I., & Crocetti, G. M. What gets surveyed in a psychiatric survey? A case study of the MacMillan Index. Journal of Nervous and Mental Disease, 1972, 154, 105-114.

267

Purpose: To identify dimensions of an abbreviated form of the MacMillan Index (MI)
Subjects: 1) 888 UAW workers or their wives (undiagnosed); 2) 30 UAW members of spouses who were active patients in an outpatient clinic
Variables: Scores on 13 MI items
Method: Unspecified condensation with unspecified rotation, independently for undiagnosed and "ill" groups
Results: Two factors emerged for the undiagnosed group: I Impaired work and social function accompanied by respiratory, cardiac, and diffuse somatic complaints; II Anxious with some fear of "breakdown." Three factors emerged for the "ill" group: I Cold sweats, recreational impairment, and nervousness; II Frequency of stomach upset, loss of appetite, and shortness of breath; III Damp and clammy hands.

Tischler, G. L., Henisz, J. E., Myers, J. K., & Boswell, P. C. Utilization of mental health services: II. Mediators of service allocation. Archives of General Psychiatry, 1975, 32, 416-418.
Purpose: To identify 1) factors underlying a mental status index and selected demographic variables, and 2) clusters of person attributes useful in identifying characteristics mediating allocation of services
Subjects: 938 adults from a defined geographic area and 808 residents of the same area who were admitted to a community mental health center
Variables: 1) mental health center use, age, marital status, employment status, social class, education, religion, race, symptom level, income, household composition, distance from mental health center, welfare status, marital disruption, student status, sex, and 7 other variables scores; 2) all of the above variables except mental health center use
Method: 1) Unspecified condensation with normal varimax rotation; 2) unspecified condensation with unspecified rotation
Results: Seven factors emerged in the first analysis; one was labeled (I Patienthood) and considered demonstrative of the existence of a patienthood factor. Seven factors emerged in the second (22-variable) analysis: I Youth; II Social status; III White Roman Catholic; IV Female-headed household; V Singles; VI Two-person family; VII Student.

R. Geriatrics

X. EDUCATIONAL PSYCHOLOGY

Khan, S. B., & Roberts, D. M. Structure of academic attitudes
and study habits. Educational and Psychological Measure-
ment, 1975, 35, 835-842.
Purpose: To identify dimensions underlying the Survey of
Study Habits and Attitudes (SSHA)
Subjects: 243 high school seniors and 603 college freshmen
Variables: 100 SSHA item scores
Method: Principal components condensation with orthogonal
rotation
Results: Five varimax factors emerged: I Delay avoidance;
II Work methods; III Teacher approval; IV Education ac-
ceptance; V Not discussed. One second-order factor
emerged: I Study orientation.
Langhorne, J. E. Jr., Stone, L. A., & Coles, G. J. Multi-
dimensional scale analysis of elementary teachers'
impressions of selected social reinforcers. Perceptual
and Motor Skills, 1973, 36, 291-297.
Purpose: To identify dimensions underlying teachers' perceptions of classroom social reinforcers
Subjects: 15 elementary school teachers
Variables: Similarity judgments for 12 classroom social
reinforcers
Method: Inverse principal components condensation, and
principal components condensation with varimax rotation
Results: The inverse analysis yielded three factors, but
the size of the first indicated a common point-of-view
across judges. Three factors emerged in the second
analysis: I Teacher-preferred reinforcement; II Student-
preferred reinforcement; II Be first.
Morrison, T. L. The Classroom Boundary Questionnaire: An
instrument to measure one aspect of teacher leadership
in the classroom. Educational and Psychological Measure-
ment, 1975, 35, 119-134.
Study One
Purpose: To identify dimensions of the Classroom Boundary
Questionnaire (CBQ)
Subjects: 62 graduate education students
Variables: 30 CBQ item scores
Method: Unspecified condensation with varimax rotation
Results: One unlabeled factor emerged.
Study Two
Purpose: To identify relationships between selected teacher
behavior and classroom movement variables
Subjects: 32 fourth-, fifth-, and sixth-grade classrooms
Variables: Observer ratings on 5 teacher behavior and 4
classroom movement variables
Method: Principal components condensation with varimax
rotation
Results: Two factors emerged: I Amount of child activity;
II Unlabeled (teacher behavior).

269

Steele, J. M., House, E. R., & Kerins, T.   An instrument for
     assessing instructional climate through low-inference
     student judgments.   American Educational Research Journal,
     1971, 8, 447-466.
     Purpose: To identify dimensions of the Class Activities
          Questionnaire (CAQ)
     Subjects: 2071 sixth- through twelfth-graders (some in
          gifted programs)
     Variables: 25 CAQ item scores
     Method: Principal components condensation with varimax
          rotation
     Results: Ten factors emerged: I Application and synthesis;
          II Memory and test/grade stress; III Interpretation; IV
          Discussion and enthusiasm; V Analysis and translation;
          VI Humor (singlet); VII Translation; VIII Lecture; IX
          Evaluation 1; X Evaluation 2.
A. Attitude and Adjustment
     Bean, A. G., & Covert, R. W.   Prediction of college persis-
          tence, withdrawal, and academic dismissal: A discriminant
          analysis.   Educational and Psychological Measurement, 1973,
          33, 407-411.
     Purpose: To identify dimensions underlying the Runner Studies
          of Attitude Patterns - College Form (RSAP); reported by
          Bean (unpublished dissertation, 1971)
     Subjects: 1125 freshmen college students
     Variables: Scores on 10 RSAP scales
     Method: Principal components condensation without rotation
     Results: Three factors emerged: I Independence; II Ac-
          quiescence; III Nonassertiveness.
     Biggs, J. B., & Das, J. P.   Extreme response set, internality-
          externality and performance.   British Journal of Social
          and Clinical Psychology, 1973, 12, 199-210.
     Study One
     Purpose: To identify dimensions of a study behavior ques-
          tionnaire (SBQ)
     Subjects: 90 upper-division summer session students in
          educational psychology
     Variables: Unspecified number of SBQ scores
     Method: Unspecified condensation with procrustes rotation
     Results: Four factors, of 13 extracted, were considered
          relevant to the subsequent analysis: I Academic interest;
          II Meaning assimilation; III Fact-rote; IV Openness.
     Study Two
     Purpose: To identify dimensions of three extreme response
          set (ERS) measures
     Subjects: The same 90 above
     Variables: Scores from the Personal Friends Questionnaire,
          Dogmatism Scale, and Study Behavior Questionnaire (ERS-
          SBQ)
     Method: Principal components condensation with varimax
          rotation
     Results: Two factors emerged: I Intensity of academic/
          ideological attitudes/beliefs; II Intensity of attitudes
          toward friends.

Birnbaum, R. Student attitudes toward 2- and 4-year colleges.
Journal of Educational Research, 1972, 65, 369-374.
Purpose: To identify dimensions of high school students'
attitudes toward collegiate institutions
Subjects: 390 high school juniors
Variables: Mean ratings (2-year, 4-year, ideal college) on
40 School Rating Scale items
Method: Principal components condensation without rotation
Results: Three factors emerged: I Social activities; II
Supportive interpersonal environment; III Intellectual
climate.
Bowen, D. D., & Kilmann, R. H. Developing a comparative mea-
sure of the learning climate in professional schools.
Journal of Applied Psychology, 1975, 60, 71-79.
Purpose: To identify dimensions underlying the Learning
Climate Questionnaire (LCQ)
Subjects: 1) 125 full-time MBA students; 2) 453 professional
school students
Variables: 26 LCQ item scores
Method: Principal factors condensation with varimax rotation
(and measet analysis), by sample
Results: Five factors emerged in both analyses: I Grading
process; II Physical environment; III Task relationships
with faculty; IV Social relationships with faculty; V
Course material presentation.
Butzow, J. W., & Williams, C. M. The content and construct
validation of the Academic-Vocational Involvement Scale.
Educational and Psychological Measurement, 1973, 33,
495-498.
Purpose: To identify dimensions underlying the Academic-
Vocational Involvement Scale (AVIS)
Subjects: 196 college freshmen
Variables: 40 semantic differential ratings for 6 AVIS items
Method: Unspecified condensation with varimax rotation
Results: Three factors emerged: I Valuing; II Striving;
III Enjoying.
Centra, J. A. Comparison of three methods of assessing college
environments. Journal of Educational Psychology, 1972,
63, 56-62.
Purpose: To identify dimensions descriptive of college
environments
Subjects: 103 colleges
Variables: 53 college variables assessed via student self-
report (Questionnaire on Student and College Character-
istics), published objective data, and perceptual data
Method: 1) Multimethod factor analysis; 2) principal axes
condensation with equamax rotation
Results: Ten multimethod factors emerged: I Female cultural
vs. male, athletic; II Faculty contact with students; III
Academic stimulation; IV Campus political-social activity;
V High regulated (and religious) campus sending few students to graduate school; VI Not interpreted; VII Frater-
nity-sorority; VIII Not interpreted; IX Religiously

271

affiliated colleges; X Involvement in science/lab facilities. Six factors emerged in the traditional analysis;
they were not labeled, but three were reported as highly
similar to Factors I-III from the multimethod analysis.
Coles, G. J., & Stone, L. A. Instructors' multidimensional
perceptions of their students. Perceptual and Motor
Skills, 1973, 36, 13-14.
 Purpose: To identify dimensions underlying teachers' per-
 ceptions of their students
 Subjects: 11 college students, rated by their lecture and
 laboratory instructors
 Variables: Similarity judgments between students
 Method: Principal components condensation with varimax
 rotation
 Results: Three factors emerged: I Academic achievement, or
 grade; II Sex; III Liberal-conservative attitudes-behavior.
Creager, J. A. Academic achievement and institutional environ-
ments: Two research strategies. Journal of Experimental
Education, 1971, 40, 9-23.
 Purpose: To identify dimensions of Astin's (1968) data on
 academic achievement and measures of student input characteristics and college environment
 Subjects: 669 college students
 Variables: 45 scores from Astin's 179 variables (3 GRE
 scores, 103 student input characteristics scores, 70 measures of college environment, and 3 NMSQT-derived measures)
 Method: Principal components condensation with normalized
 varimax rotation
 Results: Seventeen factors were interpreted: I Achievement
 pretest input; II College environment - pragmatic EAT
 orientations; III Environment-large, impersonal, realistic EAT; IV Masculinity-femininity, both input and environment; V Environment - severity of administrative
 policy; VI-VII Input - Consistency of career choice and
 related major field; VIII Environment - extraversion of
 instructor; IX Rural, non-art major input; X Orientation
 toward clinical career input; XI-XII Social sciences vs.
 math/accountancy input; XIII Uninterpreted input; XIV
 Social science vs. business input; XV Biological orientation input; XVI Abstract dialectic interest input;
 XVII Clerical input.
Cruickshank, D. R., Kennedy, J. J., & Myers, B. Perceived
problems of secondary school teachers. Journal of
Educational Research, 1974, 68, 154-159.
 Purpose: To identify dimensions of bothersomeness and fre-
 quency of problems perceived by secondary school teachers
 Subjects: 310 secondary school teachers
 Variables: 105 Teacher Problems Check List items, scored
 for 1) bothersomeness and 2) frequency (both binary scores)
 Method: Principal axes condensation with oblique promax
 rotation, independently for the bothersomeness and fre-
 quency data
 Results: Seven bothersomeness factors emerged: I Efficiency;
 II Support; III Invigoration; IV Control; V Inclusion;

272

VI Nurturance; VII Influencing. Seven frequency factors
emerged: I Security; II Remediation; III Invigoration;
IV Control; V Satisfaction; VI Support; VII Time.
Cunningham, W. G. The impact of student-teacher pairings on
teacher effectiveness. American Educational Research
Journal, 1975, 12, 169-189.
Purpose: To identify types of teachers and types of students
Subjects: 110 5-year-old kindergarteners; 108 kindergarten
teachers
Variables: Students: age, sex, race, HIS, PSI, DAM, TOBE
(L & M), CBI1-6; teachers: age, sex, race, years of
teaching experience, TSB1-5
Method: Unspecified (Q-mode) condensation with orthogonal
and oblique rotation, independently for students and
teachers
Results: Four student factors emerged: I Young/advantaged;
II Extroverted/black/female; III Introverted/disadvantaged/
white/female; IV Slow/alienated/male. Four teacher fac-
tors emerged: I White/subject integration; II Black/
experienced; III Inexperienced/student-centered/empathetic;
IV White/experienced/subject-centered.
Conclusions: The effects of student-teacher pairings was
investigated.
Dziuban, C. D., & Harris, C. W. On the extraction of components and the applicability of the factor model.
American Educational Research Journal, 1973, 10, 93-99.
Purpose: To identify, via several procedures, dimensions
of Shaycroft's (unpublished AERA paper, 1970) data
Subjects: 3689 12th-grade boys in Project Talent
Variables: Scores on 10 interest (TALENT) tests plus 4
random deviates
Method: 1) Principal components (by Shaycroft), 2) image-
analysis, 3) uniqueness rescaling factor, 4) alpha-factor
condensation, and 5) Joreskog's UMLFA, all with normal
varimax rotation
Results: The five solutions yielded three, seven, seven,
four, and five factors, respectively. The factors were
not labeled but were examined in terms of random-deviate
loadings.
Conclusions: Discussion centered about the potential for
interpreting random relationships as meaningful.
Feather, N. T. Coeducation, values, and satisfaction with
school. Journal of Educational Psychology, 1974, 66,
9-15.
Purpose: To identify dimensions of Australian students'
own and perceived school values
Subjects: 1307 Australian high school seniors from 8 schools
Variables: Average value systems (median rank) by school
and for "own" and "school" (16x16 matrix) for 1) 18
terminal and 2) 18 instrumental values
Method: Principal components condensation with varimax
rotation, independently for terminal and instrumental
values

Results: Two terminal factors emerged: I School values;
II Own values. Three factors emerged for instrumental
values: I School values; II Female own values; III
Male own values.
Feild, H. S., Holley, W. H., & Armenakis, A. A. Graduate
students' satisfaction with graduate education: Intrinsic
versus extrinsic factors. Journal of Experimental Educa-
tion, 1974, 43, 8-15.
Purpose: To identify dimensions of overall satisfaction
criteria of the Graduate Education Questionnaire (GEQ)
designed to measure both intrinsic and extrinsic factors
Subjects: 62 graduate business students
Variables: 5 overall satisfaction criteria scores from the
GEQ
Method: Principal axes condensation without rotation
Results: One factor emerged: I Overall satisfaction.
Conclusions: The intrinsic and extrinsic items from the
remainder of the GEQ were examined relative to overall
satisfaction.
Feild, H. S., Lissitz, R. W., & Schoenfeldt, L. F. The
utility of homogeneous subgroups and individual informa-
tion in prediction. Multivariate Behavioral Research,
1975, 10, 449-461.
Study One
Purpose: To identify dimensions of the Biographical Infor-
mation Blank (BIB)
Subjects: 1037 male and 897 female college freshmen (Feild,
unpublished dissertation, 1969)
Variables: 375 BIB item scores
Method: Principal axes condensation with varimax rotation
by six
Results: Thirteen (unlabeled here) factors emerged for
males and 15 for females.
Study Two
Purpose: To identify dimensions of the College Experience
Inventory (CEI)
Subjects: 743 college seniors (Feild, unpublished disserta-
tion, 1969)
Variables: 88 CEI item scores
Method: Principal axes condensation with varimax rotation
Results: Twelve factors emerged: I Participation in
religious activities; II Satisfaction with college; III
Participation in athletic activities; IV Participation
in organized extracurricular activities; V Married vs.
single; VI Physical health; VII Participation in
fraternity/sorority activities; VIII Reading activities;
IX Academic conscientiousness; X Interpersonal effectiveness; XI Personal financing of education; XII Participation in cultural activities.
Conclusions: Factor scores were used in the present investigation.

Feild, H. S., & Schoenfeldt, L. F. Development and application of a measure of students' college experiences. Journal of Applied Psychology, 1975, 60, 491-497.
Purpose: To identify dimensions underlying the College Experience Inventory (CEI)
Subjects: 743 college seniors (in residence for 3 previous years)
Variables: 87 CEI item scores
Method: Principal components condensation with varimax rotation
Results: Twelve factors emerged: I Participation in religious activities; II Satisfaction with college; III Participation in athletic activities; IV Participation in organized extracurricular activities; V Married versus single; VI Physical health; VII Participation in fraternity/sorority activities; VIII Reading activities; IX Academic conscientiousness; X Interpersonal effectiveness; XI Personal financing of education; XII Participation in cultural activities.
Feshback, S., Adelman, H., & Fuller, W. W. Early identification of children with high risk of reading failure. Journal of Learning Disabilities, 1974, 7, 639-644.
Purpose: To identify dimensions of the Kindergarten Student Rating Scale (KSRS)
Subjects: 888 kindergarteners (participating in a larger study)
Variables: 41 KSRS item scores, rated by 1 of 32 teachers
Method: Unspecified condensation with varimax rotation
Results: Five factors emerged: I Impulse control or classroom management; II Verbal ability and language development; III Perceptual discrimination; IV Recall of necessary classroom information; V Perceptual-motor skills.
Conclusions: Factor scores were related to other measures from the first study year.
Gluskinos, U. M., & Wainer, H. A multidimensional analysis of school satisfaction. American Educational Research Journal, 1971, 8, 423-434.
Purpose: To identify dimensions of school satisfaction
Subjects: 120 college students, half of whom ranked the variables in terms of satisfaction, and half of whom ranked them in terms of dissatisfaction
Variables: Rank ordering of 12 variables (grades, student-teacher relations, registration, participation in university extracurricular activities, friends at the university, recognition for performance, school spirit, transportation, achievement of non-grade goals, school policy, school facilities, school-related finances)
Method: 1) Principal components condensation without rotation, 2) hierarchical cluster analysis, independently for subgroups and total
Results: Two factors emerged from both analyses for all groups: I Personal-impersonal; II Motivator-hygiene.

275

Goldstein, K. M., & Tilker, H. A. Attitudes toward A-B-C-D-F
and Honors-Pass-Fail grading systems. Journal of Educational Research, 1971, 65, 99-100.
Purpose: To identify dimensions underlying an attitudes-
toward-grading instrument designed to tap merits of 3-
and 5-point grading systems
Subjects: 242 college juniors, 201 graduate education
students, 24 faculty
Variables: 64 item-scores from an attitudes-toward-grading
questionnaire
Method: Unspecified condensation with varimax rotation
Results: Three factors accounting for 31% of the total
variance emerged but were not interpretable in terms
of the eight intended dimensions; the first two factors
probably represented system-specific acquiescent response
sets.
Conclusions: Results of an alternative method for checking
on scale appropriateness of the items was also reported.
Greenhaus, J. H., & Ribaudo, M. The structure of college
goals. Psychological Reports, 1972, 31, 902.
Purpose: To identify dimensions of college students' goals
Subjects: 172 college students
Variables: Importance ratings (5-point) of 20 college goals
Method: Principal axes condensation with varimax rotation
Results: Three factors emerged: I Intellectual-social
college goals; II Career preparation involving external
referents; III Goal-directed career preparation.
Jaeger, R. M., & Freijo, T. D. Race and sex as concomitants
of composite halo in teachers' evaluative rating of
pupils. Journal of Educational Psychology, 1975, 67,
226-237.
Purpose: To identify dimensions underlying teachers'
ratings of pupil behavior changes
Subjects: Over 8,000 second-, fourth-, and sixth-grade
teachers, in 8 subgroups
Variables: Scores on 21 items of pupil behavior change
Method: Principal components condensation, by subgroup
Results: An unspecified number of factors emerged; the
various sex x race subgroups were compared on the first
factor.
Kane, R. B. Minimizing order effects in the semantic differential. Educational and Psychological Measurement, 1971,
31, 137-144.
Purpose: To identify the factorial structure of semantic
differential (SD) ratings of curriculum-related concepts
across several ordering conditions
Subjects: 50 college students in each of three experiments
(ordering conditions)
Variables: 14 SD ratings for 9 concepts, in 54 matrices
(2 treatments x 3 experiments x 9 concepts)
Method: Unspecified condensation with varimax rotation,
independently for each matrix
Results: The proportion of variance accounted for by the
first two factors ranged from .45-.82, and the third

factor contributed more than 10% of the total variance in only 1 of 54 cases.

Conclusions: Factor scores for the first two factors were compared across treatments, experiments, and concepts.

Kohn, M., & Cohen, J. Emotional impairment and achievement deficit in disadvantaged children--fact or myth? Genetic Psychology Monographs, 1975, 92, 57-78.

Purpose: To identify dimensions of classroom behavior, as assessed via the Classroom Behavior Inventory (CBI) by Schaefer (with Aaronson, unpublished form, 1966; in 1971 Minnesota Symposium series)

Subjects: Unspecified number of elementary school children

Variables: Unspecified number of CBI item scores

Method: Unspecified factor analysis

Results: Three factors emerged: I Extroversion versus introversion; II Love versus hostility; III High versus low task orientation.

Conclusions: Scores on Factors I and II, seen as equivalent to Kohn's preschool factors, were used in the present study.

Lambert, N. M. Intellectual and nonintellectual predictors of high school status. Journal of Special Education, 1972, 6, 247-259.

Purpose: To identify dimensions of high school behavior

Subjects: Unspecified number (apparently more than 150) high school pupils

Variables: 19 counselor evaluation ratings, 5 measures of contacts with school services, and 4 measures of ability and achievement in high school

Method: Key cluster analysis

Results: Four clusters emerged: I High school adjustment status; II High school scholarship; III Independent-aggressive high school behavior; IV High school guidance problems.

Conclusions: Subjects' cluster scores were used to examine additional data.

Langhorne, J. E. Jr., Stone, L. A., & Coles, G. J. Multidimensional scaling of elementary teachers' impressions regarding a set of behavior problems. Perceptual and Motor Skills, 1974, 38, 267-274.

Purpose: To identify dimensions underlying teachers' perceptions of behavior problems

Subjects: 15 elementary school teachers

Variables: Similarity judgments for 12 student behavior problems

Method: Inverse principal components condensation, and principal components condensation with varimax rotation

Results: The first Q-factor accounted for 52% of the interjudge variance, the second 7%. Three factors emerged in the second analysis: I Aggression problems; II Individualized emotional problems vs. more social nonemotional problems; III Verbal vs. nonverbal problems.

Magnusson, D. An analysis of situational dimensions. Perceptual and Motor Skills, 1971, 32, 851-867.
Purpose: To identify dimensions underlying similarity judgments of a set of (academic training) situations
Subjects: 3 students, one of whom provided 2 sets of judgments
Variables: Similarity judgments for 36 academic training situations
Method: Principal components condensation with varimax rotation, independently for judges as a group, each judge individually, and for Judge A's retest matrix
Results: Five factors emerged for the group: I Positive and rewarding; II Negative; III Passiveness; IV Social interaction; V Activity. Five dimensions also emerged for each judge, and for Judge A's retest, and could be labeled like the group factors.
Marshall, J. C., Watson, E. P., & Sokol, A. P. Dimensions of teacher expectations and student behavior in high school social studies classes. Journal of Educational Research, 1972, 66, 61-67.
Purpose: To identify dimensions of teacher expectations and student behavior from teachers' and students' perspectives
Subjects: 29 high school social studies teachers, and 90 (randomly sampled from 1,733) of their students
Variables: 71 Watson Analysis Schedule (WAS) scores for each of three instructional sets (students are 1) told to, 2) actually expected to, 3) actually do)
Method: Principal axes condensation (Q technique) with varimax rotation, independently for each set of WAS scores for students and teachers
Results: Seven factors emerged for teachers Part I: I Toward teacher initiated and directed problem-solving, with general "panacean outcomes"; II Toward teacher as exclusive decision-maker with non-critical acceptance by students; III Away from recall, toward critical student participation confined to school without consideration for affective (personal) characteristics; IV Away from critical-creative participation by students, toward uniformity; V Toward student as decision-maker with individualized evaluation; VI Toward student as decision-maker, with relevant activity; VII Toward student-directed problem-solving, with significant decision-making. Seven factors emerged for teachers Part II: I Toward teacher initiated and directed problem-solving, allowing for student participation and decision-making-- general "panacean outcomes" and consideration for affective (personal) characteristics; II Away from recall of teacher-determined content, toward student-centered problems; III Away from current problems, toward active discussion of past issues; IV Away from non-critical, teacher-centered recall, toward self-directed learning; V Toward student as decision-maker in data gathering; VI Away from critical participation by students, toward

278

uniformity; VII Uninterpreted. Nine factors emerged
for teachers Part III: I Toward teacher identified
problems, with student-centered problem-solving and
decision-making; II Toward teacher as exclusive decision-
maker with non-critical acceptance by students; III
Toward recall of teacher determined content, not relevant
out of school; IV Away from teacher-centered recall and
means for continued learning, toward significant activity
and teacher empathy; V Away from problem identification
by students, toward decision-making in data gathering
and evaluation; VI Toward critical thinking, with significant decision-making; VII-IX Uninterpreted.

In all three student analyses, 22 factors were extracted;
in each case, however, only seven factors were found to
account for more than 2% of the variance each. Six factors were named from Part I for students: I Toward
teacher directed problem-solving, allowing for student
participation and decision-making--not limited to school
applications; II Away from student critical participation
in problem-solving, toward past oriented recall and
uniformity; III Away from student participation in
problem-solving, and relevant learnings--toward non-
critical acceptance; IV Toward student as decision-maker,
within an open and accepting climate; V Toward teacher
directed learning, away from critical participation and
understanding of society; VI Toward student critical
participation in problem-solving decisions. Seven factors were named from Part II for students: I Toward
teacher as decision maker, with teacher directed problem-
solving, allowing for student participation; IV Toward
acquisition of factual content, limited to the past, for
future utility; II Toward student critical participation
in problem-solving and decision-making, with "self"-
social learning; III Toward student-teacher interaction,
insignificant activity; V Away from problems based on
students' interests and experiences; VI Away from self-
evaluation; VII Open climate for significant decision-
making by students, excepting topic for study. Six
factors were named from Part III for students: I Away
from recall and uncritical acceptance of teachers'
statements, while accepting authority of text and
lectures--toward societal understanding and decision-
making with enjoyment of activities; II Toward student
critical participation in problem-solving, with an open
climate--skill not limited to school application; (III
Uninterpreted;) IV Away from significant activities,
outcomes, and interests of students; V Away from problem-
solving and meaning, toward better citizenship; VI Away
from problem-solving, toward study limited to school
applications and uniformity; VII Toward student participation in problem development, with self evaluation.

McCallon, E. L., & Brown, J. D. A semantic differential in-
strument for measuring attitude toward mathematics.
Journal of Experimental Education, 1971, 39, 69-72.
Purpose: To identify dimensions of an attitudes toward
mathematics measure
Subjects: 68 doctoral (non-math major) students
Variables: 15 semantic differential scale scores for the
concept "mathematics"
Method: Unspecified condensation with varimax rotation
Results: Two factors emerged: I Evaluative; II Potency.
Conclusions: This instrument was compared with a Likert-
type measure developed by Aiken and Dreger (1961).
Mehra, N. Standardized versus unstandardized factor analysis
in a study of "organizational climate." Journal of
Experimental Education, 1973, 42, 60-67.
Purpose: To identify dimensions of the Organizational
Climate Description Questionnaire Form IV (OCDQ), as an
illustration of standardized vs. unstandardized R-analysis
Subjects: 668 randomly sampled secondary schools in State
of Delhi, India
Variables: 1) Scores on all OCDQ items (unspecified number);
2) scores on subtests identified in (1): normative and
ipsative
Method: 1) Cluster and factor analyses used by Halpin &
Croft (1963); 2) principal components condensation with
varimax rotation
Results: Eight factors, subsequently used as subtests,
emerged from the first analysis, and can be categorized
as teacher behavior (I Disengagement; II Aviscidity;
III Esprit; IV Intimacy) or principal behavior (V Con- ·
trols; VI Hindrance; VII Thrust; VII Task-orientation).
Three uninterpretable factors emerged for the normative
data. Three ipsative factors emerged: I Organizational
esprit; II Social needs; III Social control.
Conclusions: These data support the view that standardized
and unstandardized R-techniques yield different factors.
Moos, R. H., VanDort, B., Smail, P., & DeYoung, A. J. A
typology of university student living groups. Journal
of Educational Psychology, 1975, 67, 359-367.
Purpose: To identify clusters of university student living
groups
Subjects: 100 university living groups
Variables: Scores on 10 University Residence Environment
Scale subscales
Method: Carlson (1972) cluster analysis
Results: Five clusters emerged: I Competition oriented;
II Supportive achievement oriented; III Independence
oriented; IV Relationship oriented; V Traditionally
socially oriented.
Pascarella, E. T. A factor analytic comparison of faculty
and students' perceptions of students. Journal of
Experimental Education, 1975, 44, 26-32.

Purpose: To identify dimensions of faculty and students'
perceptions of students
Subjects: 1) 410 college seniors, 2) 306 faculty members
and teaching assistants
Variables: 26 semantic differential scale scores
Method: Principal components condensation with orthogonal
varimax rotation, independently for student and faculty
samples
Results: Six student factors and five faculty factors
emerged but were not labeled.
Conclusions: Factor structures for the two samples were
found moderately similar.
Paton, S. M., Walberg, H. J., & Yeh, E. G. Ethnicity, environment control, and academic self-concept in Chicago.
American Educational Research Journal, 1973, 10, 85-99.
Purpose: To identify dimensions underlying a questionnaire
related to students' academic self-concept, environment
control, and ethnicity
Subjects: 429 metropolitan ninth- and tenth-graders
Variables: Scores on a questionnaire consisting of 3
academic self-concept, 14 family background, and 1
ethnicity item
Method: Principal components condensation (with unsuccess-
ful varimax rotation), independently for males and
females
Results: Three uninterpreted factors emerged in each
analysis.
Pfeifer, C. M. Jr., & Schneider, B. University climate
perceptions by black and white students. Journal of
Applied Psychology, 1974, 59, 660-662.
Purpose: To identify dimensions underlying black and
white students' perceptions of the university climate
Subjects: 138 black and 550 white college students
Variables: Scores on 115 university climate items
Method: Principal factors condensation with varimax
rotation, by race
Results: Five factors emerged for whites: I Impersonal
academic atmosphere; II Administrative neglect; III
Social interaction; IV Racism; V Racial Separatism.
Six factors emerged for blacks: I Institutional
racism; II Nonacademic atmosphere; III Social isolation;
IV Personal racism; V Nonclassroom related activities;
VI Attempts at communication.
Price, E., & Rosemier, R. Some cognitive and affective
outcomes of same-sex versus coeducational grouping in
first grade. Journal of Experimental Education, 1972,
40, 70-77.
Purpose: To identify dimensions underlying the Classroom
Behavior Inventory (CBI)
Subjects: 21 children (10 girls, 11 boys) assigned to a
coeducational first grade and 37 (20 girls, 17 boys)
assigned to class by sex

Variables: 60 CBI item scores
Method: Principal components condensation with varimax
rotation
Results: Three factors emerged: I Verbal expressiveness,
social withdrawal-gregariousness, self-consciousness;
II Kindness, irritability, considerateness, resentful-
ness; III Hyperactivity, perserverance, distractibility,
concentration.
Conclusions: Factors scores were used in analyses of covariance.

Purkey, W. W., Cage, B. N., & Graves, W. The Florida Key:
A scale to infer learner self-concept. Educational and
Psychological Measurement, 1973, 33, 979-984.
Purpose: To identify dimensions underlying the Florida
Key
Subjects: 357 fifth- and sixth-graders
Variables: 18 Florida Key item scores
Method: Principal axes condensation with varimax rotation
Results: Four factors emerged: I Relating; II Asserting;
III Investing; IV Coping.

Romine, S. Student and faculty perceptions of an effective
university instructional climate. Journal of Educational
Research, 1974, 68, 139-143.
Purpose: To identify dimensions of an effective university
instructional environment
Subjects: 1) 1,237 college students, 2) 268 faculty
Variables: 71 item scores from a questionnaire consisting
of attributes of instructional climate
Method: Unspecified condensation with normal varimax
rotation, independently for students and faculty
Results: Seven factors emerged and were apparently highly
similar between groups: I Instructor's personality;
II Instructional preparation and organization; III Instructional outcomes; IV Classroom presentation; V
Evaluation, feedback, and reinforcement; VI Supplemental
student assistance; VII Student learning obligations.

Rubenstein, G., & Fisher, L. A measure of teachers' obser-
vations of student behavior. Journal of Consulting and
Clinical Psychology, 1974, 42, 310.
Purpose: To identify dimensions underlying the Rochester
School Competency Schedule (RSCS)
Subjects: 360 male elementary school students, rated by
their teachers
Variables: 63 RSCS item scores
Method: Principal components condensation with varimax
rotation
Results: Four factors emerged: I Cognitive competence;
II Social compliance or rule-following behavior; III
Motivational orientation; IV Social competence.

Rutkowski, K., & Domino, G. Interrelationship of study
skills and personality variables in college students.
Journal of Educational Psychology, 1975, 67, 784-789.

282

Purpose: To identify relationships between study skills and habits and personality
Subjects: 201 college freshmen
Variables: 4 Survey of Study Skills and Habits scale, 18 CPI, and 2 SAT scale scores
Method: Principal components condensation with normalized varimax rotation
Results: Five interpretable factors emerged: I General adjustment; II Social poise or social confidence; III Intellectual resourcefulness; IV Traditional study orientation; V Unconventionality.

Sadker, D. Dimensions of the elementary school environment: A factor analytic study. Journal of Educational Research, 1973, 66, 441-442, 465.
Purpose: To identify dimensions of the elementary school environment
Subjects: 5,412 elementary school students
Variables: 25 scores on one of two (Form A, Form B) 40-item instruments related to elementary school environments
Method: Unspecified condensation with oblique rotation, independently for the two forms
Results: Nine factors for Form A and six for Form B were extracted; judges reached a consensus on nine factors and found that three factors were common to the two forms. Six factors were labeled: I Alienation; II Humanism; III Autonomy; IV Morale; V Opportunism; VI Resources.
Conclusions: On the basis of these results, the instrument, now called the Elementary School Environment Survey, was revised.

Spivack, G., & Swift, M. The classroom behavior of children: A critical review of teacher-administered rating scales. Journal of Special Education, 1973, 7, 55-89.
Purpose: To review reports on teacher-administered rating scales related to children's classroom behavior. (The five factor-analytic reports abstracted here are those which appeared in journals or technical reports not included in the Taxonomy Project.)
Study One (Pimm, Quay, & Werry, 1967)
Subjects: 1,445 Ottawa first-graders (data collected by Pimm & McClure, 1966)
Variables: 36 item scores from the School Behavior Survey
Method: Unspecified factor analysis
Results: Four factors emerged: I Behavioral immaturity; II Verbal overacting; III Conduct problem; IV Personality problem.
Study Two (Walker, 1967)
Subjects: 534 fourth- through sixth-graders
Variables: 50 item scores from the Walker Problem Behavior Identification Checklist
Method: Unspecified factor analysis
Results: Five factors emerged: I Acting-out syndrome; II Withdrawal syndrome; III Distractability; IV Disturbed

283

peer relations; V Immaturity.

Study Three (Davidson & Greenberg, 1967)
Subjects: Unspecified number of achieving and nonachieving
   disadvantaged fifth-graders
Variables: 26 item scores from the School Behavior Rating
   Scale
Method: Unspecified factor analysis
Results: Three factors emerged: I Academic effort; II
   Conformity to authority demands; III Personal qualities.
   These three factors loaded on the same second-order factor
   in another factor analysis with unspecified variables.

Study Four (Rubin, Simson, & Betwee, 1966)
Subjects: 83 children referred for early identification of
   maladjustment at elementary school level
Variables: 39 item scores from a behavior checklist
Method: Unspecified factor analysis
Results: Seven factors emerged: I Disorientation and mal-
   adaptation to environment; II Antisocial behavior; III
   Unassertive-overconforming; IV-V Infantile, immature
   behaviors; VI Evidence of neglect; VII Irresponsible
   behavior.

Study Five (Dayton, 1967)
Subjects: 3,138 second- through sixth-graders, rated by
   109 teachers
Variables: 24 Pupil Classroom Behavior Scale item scores
Method: Unspecified factor analysis
Results: Three factors emerged and were replicated in
   separate analyses 1) by grade level and 2) 343 high
   school students: I Achievement orientation; II Socio-
   academically creative; III Socio-cooperative.

Spivack, G., Swift, M., & Prewitt, J.   Syndromes of disturbed
   classroom behavior: A behavioral diagnostic system for
   elementary schools.   Journal of Special Education, 1971,
   5, 269-292.
Purpose: To identify patterns of classroom behavior that
   typify children
Subjects: 101 kindergarten, 121 first-grade, 118 second-
   grade, 107 third-grade, 132 fourth-grade, 108 fifth-
   grade, 122 sixth-grade children
Variables: 11 factor scores from the Devereaux Elementary
   School Behavior Rating Scale
Method: Cluster analysis, independently by grade
Results: Six clusters emerged for kindergarten, first-
   grade, third-grade, fourth-grade, and fifth-grade children;
   five clusters emerged for second- and sixth-graders.   Two
   types account for almost half of the children between the
   first and sixth grades.

Sweet, P. R., & Nuttall, R. L.   The effects of a tracking
   system on student satisfaction and achievement.   American
   Educational Research Journal, 1971, 8, 511-520.
Purpose: To identify dimensions of the Pupil Attitude about
   the Tracking System questionnaire (PATS)

Subjects: 386 ninth-graders
Variables: 36 PATS item scores
Method: Principal components condensation with varimax
rotation
Results: One factor, of the eight extracted, was designated
as the PATS factor.
Terranova, C. A method to determine the congruity of dimen-
sions across instruments. Educational and Psychological
Measurement, 1974, 34, 247-251.
Purpose: To identify dimensions underlying the high school
and the college experience
Subjects: 168 high school students who had been accepted
for college, and 110 subjects after they had completed
college orientation
Variables: 40 semantic differential ratings (summed across
several concepts related to either the high school or the
college experience)
Method: Principal components condensation with varimax
rotation, independently for each sample
Results: Four high school components emerged: I Evaluation;
I Intellectual arrogance; III-IV Unlabeled. Three factors
emerged for college: I Evaluation; II Informality; III
Intellectual arrogance.
Thistlethwaite, D. L. Accentuation of differences in values
and exposures to major fields of study. Journal of
Educational Psychology, 1973, 65, 279-293.
Purpose: To identify dimensions underlying college press
and attitude scales
Subjects: 1) 50, 2) 1,036, and 3) 822 college males (2
freshmen, 3 juniors)
Variables: Scores on 33 college press and attitude scales
Method: Principal axes condensation with oblique rotation,
by subject group
Results: Five similar factors emerged across analyses:
I Scientism; II Faculty rapport; III Liberalism; IV
Political participation; V Estheticism. At least for
the third sample, three other factors emerged as well:
VI Gregariousness; VII Conventionalism; VIII Benevolence.
Wentworth, D. R., & Lewis, D. R. An evaluation of a simula-
tion game for teaching introductory economics in junior
colleges. Journal of Experimental Education, 1973, 42,
87-96.
Purpose: To identify dimensions of student attitudes related
to an economics course
Subjects: 149 college students enrolled in economics courses
Variables: 13 semantic differential scores
Method: Unspecified condensation with normal varimax rotation
Results: Four factors emerged: I Capitalism syndrome; II
Instructional process; III Student learning behaviors;
IV Command economy.
Conclusions: This analysis was conducted in relation to a
study of the effects of an economics game on classroom
learning.

Williams, F., Whitehead, J. L., & Miller, L. Relations between language attitudes and teacher expectancy. American Educational Research Journal, 1972, 9, 263-277.
  Purpose: To identify dimensions of teachers' semantic differential (SD) ratings of pupils' behavior
  Subjects: 175 elementary school teachers, stratified by experience
  Variables: 10 SD ratings (of videotapes) of children's behavior
  Method: Unspecified condensation with varimax rotation
  Results: Two factors emerged: I Confidence-eagerness; II Ethnicity-non-standardness.
  Conclusions: Factor scores were computed and used as dependent variables in the major portion of the study.
B. Testing
  Abeles, H. F. A factet-factorial approach to the construction of rating scales to measure complex behaviors. Journal of Educational Research, 1973, 10, 145-151.
  Purpose: To identify dimensions of a scale to assess clarinet music performance
  Subjects: 32 instrumental music teachers enrolled in graduate music courses, each judged by 14 raters
  Variables: Scores on the 1) 94-item version of the Clarinet Performance Rating Scale (CPRS) and 2) revised 30-item CPRS
  Method: Unspecified condensation with unspecified rotation, independently for the two forms of the CPRS
  Results: Six factors emerged in both analyses: I Interpretation; II Intonation; II Rhythm-continuity; IV Tempo; V Articulation; VI Tone.
  Conclusions: On the basis of the first analysis, the 30-item CPRS was developed.
  Biggs, J. B., & Braun, P. H. Models of evaluation and their relation to student characteristics. Journal of Educational Measurement, 1972, 9, 303-309.
  Purpose: To identify dimensions underlying five evaluations of student performance
  Subjects: 60 upper-division college students
  Variables: Scores on 5 different measures of student performance (midterm, term paper, object final, essay final, short-answer test)
  Method: Unrestricted maximum likelihood condensation with quasi-procrustean oblique rotation
  Results: Two factors emerged: I Objective; II Essay. Dimensions of study behavior and habits were also examined in relation to evaluation performance.
  Diamond, J. J., & Evans, W. J. An investigation of the cognitive correlates of test-wiseness. Journal of Educational Measurement, 1972, 9, 145-150.
  Purpose: To identify dimensions underlying test-wiseness (TW)

286

Subjects: 95 sixth-graders
Variables: 5 subscale and 1 total score from a TW instrument, 3 Lorge-Thordike Intelligence Test scores, 7 Iowa Test of Basic Skills scores
Method: Unspecified factor analysis
Results: The number of factors extracted was not specified, but it was reported that the TW scales did not load together and hence that TW is multidimensional.
Hunt, D., & Randhawa, B. S. Relationship between and among cognitive variables and achievement in computational science. Educational and Psychological Measurement, 1973, 33, 921-928.
Purpose: To identify dimensions underlying a battery of tests thought to tap abilities necessary for success in a computational science class
Subjects: 119 college students in an introductory computers class
Variables: Scores on 7 KIT tests for cognitive factors, five critical thinking tests, and final achievement mark in the course
Method: Principal factors condensation with varimax rotation, Schmid Leiman hierarchical condensation with varimax rotation
Results: Five first-level (I Critical thinking; II Achievement; III Ideational fluency; IV Inductive-deductive reasoning; V Object-number specific) and two second-level factors (VI General cognitive; VII Specific rote memory) emerged.
Powell, J. C., & Isbister, A. G. A comparison between right and wrong answers on a multiple choice test. Educational and Psychological Measurement, 1974, 34, 499-509.
Purpose: To identify dimensions underlying right and wrong responses on a multiple-choice test
Subjects: 307 college freshmen
Variables: 3 right and five wrong answer subscores from a multiple-choice test
Method: Principal components condensation with varimax rotation
Results: Five factors emerged but were not labeled.
Reilly, R. R., & Jackson, R. Effects of empirical option weighting on reliability and validity of an academic aptitude test. Journal of Educational Measurement, 1973, 10, 185-193.
Purpose: To identify factors of an academic aptitude test for two different scoring methods
Subjects: 2500 students taking the GRE (randomized block group from spaced sample of one administration)
Variables: Items (unspecified number) from both subtests of the GRE, scored by two different weighting methods
Method: Principal components condensation with varimax rotation, independently for the two keying methods
Results: An unspecified number of uninterpreted factors emerged.

287

Conclusion: The first few factors (especially the first) revealed sharp increases in variance accounted for with empirical (vs. formula) weighting, and item intercorrelations also underwent changes.

Tamir, P. The relationship among cognitive preference, school environment, teachers' curricular bias, curriculum, and subject matter. American Educational Research Journal, 1975, 12, 235-264.

Purpose: To identify dimensions underlying the Biology Cognitive Preference Test (BCPT) for Israeli twelfth-graders

Subjects: 989 Israeli twelfth-graders, stratified by type of high school

Variables: 4 BCPT subscale scores

Method: Unspecified condensation with varimax rotation

Results: Two factors emerged: I Recall principles (critical questioning); II Application and critical questioning (recall).

Conclusions: The results are compared with those from two previous studies in chemistry cognitive preferences.

C. Special Education

Durojaiye, M. O. A., & Such, M. Predicting educational suitability of children in an assessment unit. Journal of Experimental Education, 1971, 40, 27-36.

Purpose: To identify dimensions of the assessment (for special education placement) of children

Subjects: 36 children, aged 6 to 7 years, being assessed for special educational placement

Variables: Ratings from 3 intelligence tests, 4 school adjustment assessments, 7 school skills, 3 general activities, 4 domestic activities, 2 health measures, school placement recommendation

Method: Principal components condensation with normal varimax and oblique promax rotation

Results: Seven factors emerged: I School skills; II Intellectual ability and placement; III Social adequacy; IV Physical appearance; V Sensory adequacy; VI Self-help, toilet skills; VII Self-help, dietary skills.

1. Gifted

Payne, D. A., & Halpin, W. G. Use of a factored biographical inventory to identify differentially gifted adolescents. Psychological Reports, 1974, 35, 1195-1204.

Purpose: To identify dimensions of a biographical questionnaire (Owens, unpublished inventory, 1969)

Subjects: Unspecified number of boys and girls

Variables: 389 Biographical Data Inventory item scores

Method: Principal components condensation with varimax rotation by sex

Results: Thirteen factors emerged for boys: I Parental warmth; II Intellectualism; III Academic achievement; IV Social introversion; V Scientific interest; VI Socioeconomic status; VII Independence; VIII Parental control; IX Positive academic attitude; X Sibling friction; XI Religious activity; XII Athletic interest; XIII Social desirability.

Fifteen factors emerged for girls: I Maternal warmth;
II Social leadership; III Academic achievement; IV Par-
ental control; V Cultural-literary interests; VI
Scientific-artistic interests; VII Socioeconomic status;
VIII Expression of negative emotions; IX Athletic par-
ticipation; X Conformity to female role; XI Maladjustment;
XII Popularity with opposite sex; XIII Positive academic
attitude; XIV Daddy's girl; XV Social maturity.
Conclusions: A 118-item version of the instrument was used
in the present study.

2. Remedial Education

Cicirelli, V. G. A note on the factor analysis of disadvan-
taged children's Illinois Test of Psycholinguistic
Abilities and achievement test scores. Journal of
Experimental Education, 1972, 41, 5-8.
Purpose: To identify dimensions of the Illinois Test of
Psycholinguistic Abilities (ITPA) and achievement test
scores for disadvantaged children
Subjects: 1) 537 disadvantaged first-, 527 second-, 380
third-graders; 2) subgroup of 274 whites; 3) subgroup
of 186 blacks
Variables: 10 ITPA subtest and 6 either Metropolitan
Readiness Test or Stanford Achievement Tests scores
Method: Principal components condensation with orthogonal
rotation, independently for total sample and whites and
blacks
Results: Two factors emerged in all analyses: I Achievement; II Language.
Conclusions: Differences in factor loadings for the three
analyses are discussed.

McKerracher, D. W., & Orritt, C. P. Prediction of vocational
and social skill acquisition in a developmentally handicapped population: A pilot study. American Journal
of Mental Deficiency, 1972, 76, 574-580.
Purpose: To identify dimensions underlying WAIS and vocational program performance for developmentally handicapped
trainees
Subjects: 75 developmentally handicapped vocational program
trainees, aged 16-40 years
Variables: Scores on 12 WAIS subtests, WAIS VIQ, WAIS Full
IQ, sex, age, duration in program, and credits earned
(performance) in program
Method: Principal components condensation (with varimax
and promax rotation)
Results: Five factors emerged and were labeled without
rotation: I Intelligence; II Age effects on intelligence;
III Low social and vocational achievement 1; IV Duration
of training; V Low social and vocational achievement 2.

Sewell, T. E., & Severson, R. A. Learning ability and intelligence as cognitive predictors of achievement in first-
grade black children. Journal of Educational Psychology,
1974, 66, 948-955.

289

Purpose: To identify dimensions underlying a battery of
ability and achievement measures
Subjects: 52 black first-graders
Variables: Sex, age, WISC VIQ, PIQ, Full Scale IQ, Raven's
CPM pre and posttest, and 9 achievement variables
Method: Principal components condensation with varimax
rotation
Results: Six factors emerged: I IQ, diagnostic teaching,
achievement; II CPM gain, with achievement and verbal
intelligence; III Not discussed; IV Paired associate
tasks (specific); V Nonverbal IQ; VI Developmental.
Stoll, P. D. A study of the construct and criterion-related
validity of the Stanford Diagnostic Reading Test.
Journal of Educational Research, 1972, 66, 184-189.
Purpose: To identify dimensions of the construct "reading,"
as measured by the Stanford Diagnostic Reading Test
(SDRT)
Subjects: 143 fifth-graders in remedial reading programs
Variables: 7 SDRT Level II and 2 SDRT Level I scores
Method: Unspecified condensation with varimax rotation
Results: Four factors emerged: I Word attack; II Auditory;
the ability to distinguish likenesses and differences in
sounds; III Verbal meaning, or auditory vocabulary; IV
Speed or rate of silent reading.
Conclusions: These, and other results reported here, indicated that the internal organization of the SDRT, with
the exception of the rate subtest, are consistent with
the authors' analysis.
3. Sensory and Physical Handicaps
Becker, J. T., & Sabatino, D. A. Frostig revisited. Journal
of Learning Disabilities, 1973, 6, 180-184.
Purpose: To identify factors common to the Frostig Developmental Test of Visual Perception (FDTVP), Bender Visual-
Motor Gestalt Test (BVMGT), Visual Discrimination Test
of Words (VDTW), and American School Intelligence Test
(ASIT)
Subjects: 154 kindergarteners
Variables: Total scores from the ASIT, VDTW, and BVMGT,
plus 5 FDTVP scores
Method: Principal components condensation with orthogonal
varimax rotation
Results: Three factors emerged: I Visual-motor; II Figure-
ground; III Visual discrimination skills.
Chissom, B. S., & Thomas, J. R. Comparison of factor structures for the Frostig Developmental Test of Visual
Perception. Perceptual and Motor Skills, 1971, 33,
1015-1019.
Purpose: To identify factors underlying the Frostig, as
collected by 12 investigators: 1) Corah & Powell (1963);
2) Hepburn & Donnelly (1968); 3) Ayres (1965); 4) Ayres
(1965); 5) Hueftle (1967); 6) Ohnmacht & Olson (1968);
7) Sprague (1963); 8) Trussell (1969); 9) Shepard (1969);
10) Olson (1966); 11) Fretz (1970); 12) Allen (1968)

Subjects: 1) 40 50-76 month-old normals; 2) 112 5-6.5
    year-old normals; 3) 100 70-96 month-olds with selected
    difficulties; 4) 50 68-95 month-old normals; 5) 52 83-
    month-old normals; 6) 232 normal semi-rural first-graders;
    7) 111 normal first-graders; 8) 75 normal first- and
    second-graders; 9) 47 bedridden second-graders; 10) 71
    normal second-graders; 11) 68 101-month-old males with
    motor difficulties; 12) 36 10-16 year-old EMRs.
Variables: 5 Frostig subtest scores
Method: Principal components condensation with orthogonal
    rotation, for each study
Results: A single factor (unlabeled) emerged for 10 of the
    12 analyses; the Shepard and Hueftle data were exceptions
    in that two factors were extracted.
McKinney, J. D. Factor analytic study of the Developmental
    Test of Visual Perception and the Metropolitan Readiness
    Test. Perceptual and Motor Skills, 1971, 33, 1331-1334.
Purpose: To identify the relationship between the DTVP and
    MRT
Subjects: 45 boys and 30 girls, mean age=70.69 months
Variables: Scores on 5 DTVP and 6 MRT subtests, plus CA
Method: Principal components condensation with varimax
    rotation
Results: Six factors emerged: I Psychomotor; II Evaluation;
    III Verbal-linguistic; IV CA; V Perceptual organization;
    VI Cognition of figural transformations (tentative).
Pate, J. E., Webb, W. W., Sell, S. H., & Gaskins, F. M. The
    social adjustment of post-meningitic children. Journal
    of Learning Disabilities, 1974, 7, 21-25.
Purpose: To identify dimensions of that part of a test
    battery found to discriminate between children who had
    survived (without sequelae) acute bacterial meningitis
    prior to age 4 (PMs) and a control group (Cs)
Subjects: 25 PMs and 25 Cs, matched on age, sex, SES,
    classroom membership
Variables: Scores on battery subtests found to discriminate
    PVs from Cs (apparently 16 scores)
Method: Unspecified factor analysis
Results: Five factors emerged: I Instructional receptivity;
    II Student image; III Response to extra-ocular movement;
    IV Gross motor coordination; V Visual orientation.
Smith, P. A., & Marx, R. W. Some cautions on the use of the
    Frostig test: A factor analytic study. Journal of
    Learning Disabilities, 1972, 5, 357-362.
Purpose: To identify factors common to the Frostig Develop-
    mental Test of Visual Perception (FDTVP), WISC, and a
    reading achievement measure
Subjects: 43 elementary school children referred for educa-
    tional assessment
Variables: 1) 5 FDTVP subtest scores; 2) 5 FDTVP scores,
    age, IQ, WRAT scores; 3) 5 FDTVP, IQ, WRAT scores
Method: Principal components condensation, with varimax
    rotation for 2) and 3)

Results: A large general factor emerged in the Frostig analysis; the possibility of a small second factor was also raised. Two factors emerged in the analysis of all variables, and was replicated without age: I Frostig; II IQ and reading achievement.

Thomas, J. R., & Chissom, B. S. Note on factor structure of the Frostig Developmental Test of Visual Perception. Perceptual and Motor Skills, 1973, 36, 510.

Purpose: To identify dimensions underlying the DTVP (for a new sample)

Subjects: 38 kindergarteners

Variables: 5 DTVP subtest scores

Method: Principal components condensation with orthogonal rotation

Results: One factor had an eigenvalue exceeding unity.

4. Mental Retardation and Learning Disorder

Ayres, A. J. Improving academic scores through sensory integration. Journal of Learning Disabilities, 1972, 5, 338-343.

Purpose: To identify dimensions of sensory integrative dysfunction

Subjects: 148 learning-disabled children

Variables: Scores on a battery of neuromuscular tests (number unspecified), portions of the Southern California Sensory Integration Tests, ITPA, all WRAT subtests, and Slosson Oral Reading Test

Method: Unspecified factor analysis

Results: Five factors emerged: I Disorders in postural, ocular, and bilateral integration; II Praxis; III Functions of the left side of the body; IV Form and space perception; V Auditory-Language function.

Conclusions: These results, congruent with previous factor-analytic studies, were the theoretical basis for a remediation program.

Beatty, J. R. The analysis of an instrument for screening learning disabilities. Journal of Learning Disabilities, 1975, 8, 180-186.

Purpose: To 1) reduce an 80-item screening measure, and 2) identify dimensions underlying the shortened instrument

Subjects: 400 learning-disabled first-, second-, and third-graders

Variables: 1) Scores on an 80-item instrument for screening learning disabilities, and 2) scores on 48 items found to load on factors isolated in the first analysis

Method: Principal axes condensation, with orthogonal varimax rotation for the 48-item analysis

Results: Ten uninterpreted factors emerged for the 80-item analysis. Ten factors emerged for the 48-item analysis: I Speech/auditory disability; II Hyperactive-aggressive; III Reading disability; IV Neurological disability; V Drawing/writing disability; VI Anxiety; VII Visual Disability; VIII Conceptualization incompetencies; IX Inactivity/lack of concentration; X Laterality disability.

Conclusions: The Classroom Screening Instrument (CSI) may
be useful for screening learning disabilities.
Belford, B., & Blumberg, H. M. Factor analytic study of the
Revised Illinois Test of Psycholinguistic Abilities
(ITPA). Perceptual and Motor Skills, 1975, 40, 153-154.
Purpose: To identify dimensions underlying the Revised
ITPA
Subjects: 102 learning disabled children with average intel-
ligence, aged 5 to 10-3.
Variables: 10 ITPA subtest scores
Method: Principal axes condensation with varimax rotation
Results: Four factors emerged but were not labeled.
Bell, D. B., Lewis, F. D., & Anderson, R. F. Some personality
and motivational factors in reading retardation. Journal
of Educational Research, 1972, 65, 229-233.
Purpose: To identify relationships between motivational,
personality, and selected variables and reading achieve-
ment retardation
Subjects: 50 white and 50 black junior high school boys
Variables: Scores on 1 reading criterion, 14 HSPQ, 11 SMAT,
6 WISC, 2 parental achievement, 1 (number of) siblings,
1 (number of) persons in household, 1 McGuire-White, 1
CA, 1 Lincoln-Oseretsky, 1 Bean Symbol Substitution Test,
1 Durrell Visual Memory Subtest, 1 Myklebust Total Words
Count, and 1 racial variables
Method: Unspecified condensation with varimax rotation
Results: Five factors, of the fifteen extracted, were related
to reading difficulties: I Verbal deficit; III Aggres-
siveness; VI The caucasion reader; V Low socioeconomic
status; XI Passivity.
Blackman, L. S., & Burger, A. L. Psychological factors re-
lated to early reading behavior of EMR and nonretarded
children. American Journal of Mental Deficiency, 1972,
77, 212-229.
Purpose: To identify dimensions of reading readiness in
EMR and nonretarded children
Subjects: 94 EMR and 78 nonretarded first-grade children
Variables: Scores on a reading criterion (MAT Word
Knowledge), intellectual evaluation (PPVT MA), and 17
reading readiness measures
Method: Principal components condensation with varimax
rotation, independently by group
Results: Six factors emerged for the EMR group: I Visual
perception; II Auditory-visual integrative ability; III
Auditory perception and memory; IV Conceptual; V Criterion, visual orientation; VI Learning and visual perception. Seven factors emerged for the nonretarded
group: I Visual perception; II Reading criterion,
auditory perception and memory; III Visual-tactile
cross-modality; IV Auditory-visual integration; V Conceptual; VI Language ability; VII Auditory processes.

Bryan, T. S., & McGrady, H. J. Use of a teacher rating scale. Journal of Learning Disabilities, 1972, 5, 199-206.
  Purpose: To identify the factorial structure of the Pupil Behavior Rating Scale (PBRS)
  Subjects: 183 potential disabled learners and 176 control children (third- through sixth-graders)
  Variables: 24 teacher ratings from the PBRS
  Method: Principal components condensation with unspecified rotation, independently for combined group and the two subgroups
  Results: Four factors emerged in all analyses: I Auditory comprehension and listening; II Motor; III Spoken language, orientation, cooperation and attention behavior; IV Remaining behavior items.
  Conclusions: Differences in factor structure between disabled learners and controls were discussed.
Gottlieb, J., Gampel, D. H., & Budoff, M. Classroom behavior of retarded children before and after integration into regular classes. Journal of Special Education, 1975, 9, 307-315.
  Purpose: To identify dimensions of a behavioral observation scheme
  Subjects: 110 EMR children
  Variables: Ratings on 12 behavior categories, averaged for 10 times and 2 raters
  Method: Principal components condensation with varimax rotation
  Results: Three factors emerged: I Prosocial behavior; II Verbally hostile behavior; III Physically hostile behavior.
Kerr, B. A., McKerracher, D. W., & Neufeld, M. Motor assessment of the developmentally handicapped. Perceptual and Motor Skills, 1973, 36, 139-146.
  Purpose: To identify dimensions underlying a motor assessment battery
  Subjects: 33 EMRs, mean age 21.9 years, and IQ 71.2
  Variables: 10 test scores from a motor assessment battery, and 7-day retest scores
  Method: Principal components condensation with varimax rotation, independently for test and retest data
  Results: Three factors emerged for both analyses: I General (Stronger Ss unable to run fast); II Strength; III General (unlabeled).
Rugel, R. The factor structure of the WISC in two populations of disabled readers. Journal of Learning Disabilities, 1974, 7, 581-585.
  Purpose: To identify dimensions underlying WISC subtest scores for two populations of disabled readers, and a standardization sample
  Subjects: 1) 240 disabled readers, mean age 12, mean IQ 96 (reported by Schiffman & Clements, 1966); 2) 71 disabled readers, age 9 to 10-11 years, IQ 85-127 (data reported by Beck, 1955); 3) Wechsler's (1949) standardization sample of 200 male and 200 female 10½-year-olds

Variables: 11 WISC subtest scores
Method: Principal components condensation with varimax
 rotation, independently for each sample
Results: Three factors emerged in each analysis; factors
 I Verbal and II Spatial-performance appear in all three,
 but the third (unlabeled) factor differed between samples.
Conclusions: These results are discussed in relation to
 other factor-analytic findings, and in the context of
 Bannatyne's categories.
VanHagen, J., & Kaufman, A. S. Factor analysis of the WISC-R
 for a group of mentally retarded children and adolescents.
 Journal of Consulting and Clinical Psychology, 1975, 43,
 661-667.
Purpose: To identify the factorial structure of the WISC-R
 for a group of mentally retarded individuals (MRs)
Subjects: 80 MRs aged 6-16 years
Variables: 12 scaled WISC-R subtest scores
Method: 1) Principal components condensation with varimax
 rotation, 2) principal factors condensation with varimax
 rotation and 3) with oblimax and biquartimin rotation
Results: Two factors emerged in the first analysis: I
 Verbal comprehension; II Perceptual organization. A
 two-factor principal factors solution produced the same
 two factors; however, a distractibility factor (III)
 emerged in the three-factor solution. The oblimax solu-
 tions were virtually identical, but only the two-factor
 biquartimin solution was interpretable.
Wallbrown, F. H., Blaha, J., Wherry, R. J. Sr., & Counts, D. H.
 An empirical test of Myklebust's cognitive structure
 hypothesis for 70 reading-disabled children. Journal of
 Consulting and Clinical Psychology, 1974, 42, 211-218.
Purpose: To identify WISC factor structure for normal and
 reading-disabled children
Subjects: 70 reading-disabled children aged 7-13 years;
 Wechsler's (1949) 7½-, 10½-, and 13½-year-old normals
Variables: 12 WISC subtest scores, plus CA
Method: Hierarchical condensation, independently for dis-
 abled readers and normals
Results: One general (I General) and two group factors (II
 Verbal-educational ability; II Spacial-perceptual) emerged
 for normals. For disabled readers, one general (I General),
 two group (II Verbal-educational ability; III Spacial-
 perceptual), and four additional primary factors (IV
 Freedom from distractibility; V Preponderance of spacial-
 perceptual component; VI Verbal concept formation; VII
 Quasi-specific) emerged.
5. Emotional Disorder
 Proger, B. B., Mann, L., Bayuk, R. J. Jr., Burger, R. M.,
 Cross, L. H., & Green, P. A. Factorial structure of
 the Illinois Test of Psycholinguistic Abilities.
 Psychological Reports, 1973, 32, 931-935.
 Purpose: To identify relationships between the WISC and
 Illinois Test of Psycholinguistic Abilities (ITPA) for
 emotionally disturbed boys

Subjects: 94 emotionally disturbed boys, average age 121.3
months
Variables: 11 WISC and 10 ITPA subtest scores
Method: Principal components condensation with varimax
rotation
Results: Five factors emerged: I Content of verbal com-
munication; II Automatic level of communicative organi-
zation; III Representational level WISC subtests; IV
Expressive category of representational level; V Associ-
ative phase of psycholinguistic process.
D. Counseling and Guidance
Centra, J. A., & Rock, D. College environments and student
academic achievement. American Educational Research
Journal, 1971, 8, 623-634.
Purpose: To identify dimensions of the Questionnaire on
Student and College Characteristics (QSCC, reported by
Centra in 1970 ETS bulletin)
Subjects: 214 colleges
Variables: 77 QSCC item scores
Method: Unspecified condensation with unspecified rotation
Results: Eight factors emerged: I Faculty-student inter-
action; II Activism; III Curriculum flexibility; IV
Unchallenging; V Cultural facilities; VI Restrictiveness;
VII Non-academic emphasis; VIII Laboratory facilities.
Conclusions: Factor scores for five of these factors (I-V)
were used in the present study.
Elterich, K. W. Jr., & Gable, R. K. The factorial dimensions
of the Comparative Guidance and Placement Tests for
three freshman curricula groups. Educational and
Psychological Measurement, 1972, 32, 1061-1067.
Purpose: To identify dimensions underlying the Comparative
Guidance and Placement Battery (CGP)
Subjects: 116 transfer (T), 141 business-technical (BT),
and 93 general studies (GS) freshman community college
students
Variables: 11 interests and 8 cognitive scales scores from
CGP
Method: Image analysis condensation with obliquimax rota-
tion, independently for each group
Results: Six meaningful BT factors emerged: I General
aptitude; II Health-creativity; III Business interest;
IV Interest in history and current events; V Scientific-
technology; VI Mathematics interest and ability. Six
meaningful factors emerged for T: I Verbal aptitude;
II Biology interest; III Business interest; IV Fine arts;
V Inductive-reasoning and perceptual speed; VI Scientific-
technology. Seven interpretable factors emerged for GS:
I Verbal aptitude; II Biology interest; III Social
science-business interest; IV Business interest; V
Scientific-technology; VI Fine arts; VII Non-verbal
aptitude.

Grimaldi, J., Loveless, E., Hennessy, J., & Prior, J. Factor
analysis of 1970-71 version of the Comparative Guidance
and Placement Battery. Educational and Psychological
Measurement, 1971, 31, 959-963.
Purpose: To identify factors underlying the Comparative
Guidance and Placement Battery (CGP), 1970-71 version
Subjects: 1637 community college freshmen
Variables: Scores on 8 cognitive and 11 interests scales
from CGP
Method: Principal components condensation with varimax
rotation
Results: Six factors emerged: I Scholastic aptitude; II
Biology and health interest; III Practical interest;
IV Interest in technological science; V Perceptual-
reasoning; VI (Interest in the arts?).
E. Personnel
Atwood, L. E., & Crain, S. Changes in faculty attitudes
toward university role and governance. Journal of
Experimental Education, 1973, 41, 1-9.
Purpose: To identify 1) dimensions of faculty attitudes
toward university role and governance, and 2) opinion
patterns
Subjects: 255 faculty, stratified across 5 academic ranks
(and including administrative and service personnel);
2 subgroups of 109 were used for Q analysis.
Variables: 18 items applicable to respondent's institution,
11 items applicable to higher education as a whole, all
from a survey extremely similar to that of Atwood and
Starck (1970)
Method: Principal axes condensation with orthogonal varimax rotation, independently for 1) issues (R-type) and
2) opinion types (Q-type)
Results: Three issues factors emerged: I Administrative-
faculty control; II Concern for social issues; III Control orientation-restrictions on access to higher
education institutions. Three opinion-type factors
emerged: I Traditional academic; II Strong conservative
with power-control orientation; III Social concern
orientation.
Conclusions: The 1972 results are compared with 1969
results, and changes discussed.
Arney, W. R. Emergent implications of people's perceptions
of changes in higher education. Journal of Experimental
Education, 1973, 42, 1-11.
Purpose: To identify dimensions (in addition to a priori
item sets) of 118 changes-in-higher-education statements
Subjects: 381 persons connected with higher education
Variables: A subset of scores on a questionnaire containing
118 change statements
Method: Principal components condensation with varimax
rotation
Results: One factor emerged: I Academic flexibility.
Conclusions: Ten canonical analyses are reported in the
major portion of this paper.

Butler, D. C., Gocka, E. F., Hartley, J. A., & Pinneau, S. R.
Analysis of factor variance: Two cases. <u>Psychological
Reports</u>, 1972, <u>31</u>, 267-279.
Purpose: To identify dimensions underlying attitudes of
Headstart paraprofessionals in training
Subjects: 192 Headstart paraprofessionals-in-training (to
be teacher aides or teacher assistants) from 4 training
centers treated as separate Ss on pre- and posttest data
for "total" analysis; subdivided into 16 groups for
"within" analyses on the basis of position, center, and
pre- vs. posttest
Variables: 92 attitude inventory item scores (administered
before and after training)
Method: Principal components condensation with varimax
rotation, independently for total group and for 16 subgroups
Results: Twenty-seven unlabeled factors emerged; 30 factor
matches were reported across analyses.
Dziuban, C. D., & Moser, R. P. Some functional dimensions of
large city public school research. <u>Journal of Educational
Research</u>, 1973, <u>66</u>, 471-478.
Purpose: To identify dimensions of public school research
bureau performance functions
Subjects: 37 of the largest public school research bureaus
in the U. S.
Variables: Binary scores on 27 bureau functions
Method: Principal components condensation with normal
orthogonal varimax rotation followed by oblique rotation
Results: Five functional dimensions emerged: I Supportive
services; II Applied research functioning; III Psycho-
metric activity; IV Administrative research; V Demographic
research.
Isherwood, G. B., & Hoy, W. K. Bureaucratic structure recon-
sidered. <u>Journal of Experimental Education</u>, 1972, <u>41</u>,
47-50.
Purpose: To identify dimensions of the concept bureaucracy
Subjects: 13 secondary schools (rated by approximately 75%
of their teachers)
Variables: 48 School Organizational Inventory item scores
Method: Principal-factor condensation with varimax rotation
Results: Four factors emerged: I Organizational control-
role; II Organizational control methods; III Impersonality;
IV Uninterpreted.
Conclusions: The results tended to support the Parsons-
Douldner notion of organizational control.
1. Teachers and Teacher Training
Aleamoni, L. M., & Spencer, R. E. The Illinois Course Evalua-
tion Questionnaire: A description of its development
and a report of some of its results. <u>Educational and
Psychological Measurement</u>, 1973, <u>33</u>, 669-684.
Purpose: To identify dimensions underlying Illinois Course
Evaluation Questionnaire (CED) items

Subjects: 1) 1200; 2) 1,319; 3 and others) unspecified
numbers of college students
Variables: 1) over 1,000; 2) 450; 3) 50 CEQ item scores
Method: Principal components condensation with varimax
rotation, for all subject-variable sets
Results: Six factors emerged in the initial and all sub-
sequent analyses: I General course attitude; II Method
of instruction; III Course intent; IV Interest and at-
tention; V Instructor; VI Specific items.
Bledsoe, J. C., Brown, I. D., & Strickland, A. D. Factors
related to pupil observation reports of teachers and
attitudes toward their teacher. Journal of Educational
Research, 1971, 65, 119-126.
Purpose: To identify dimensions underlying the Pupil
Observation Report (POSR)
Subjects: 554 student teachers
Variables: 38 POSR item scores
Method: Unspecified factor analysis
Results: Five factors emerged: I Friendly, cheerful, ad-
mired; II Knowledgeable, poised; III Interesting, pre-
ferred; IV Strict control; V Democratic procedure.
Conclusions: POSR factor scores were used in the main in-
vestigation, as were scores from A Scale for Measuring
Attitude Toward Any Teacher.
Brophy, J. E., Coulter, C. L., Crawford, W. J., Evertson,
C. M., & King, C. E. Classroom Observation Scales:
Stability across time and context and relationships
with student learning gains. Journal of Educational
Psychology, 1975, 67, 873-881.
Purpose: To identify relationships among Classroom Obser-
vation Scales (COS) subscales
Subjects: 165 second- and third-grade teachers
Variables: 12 COS scale scores
Method: Principal axes condensation without rotation
Results: Four factors emerged: I Qualitative aspects of
classroom verbal interaction; II Positive affect and
enthusiasm; III Teacher-initiated problem solving,
related to level of cognitive behavior; IV Uninterpreted.
Coats, W. D., Swierenga, L., & Wickert, J. Student perceptions
of teachers - a factor analytic study. Journal of Educa-
tional Research, 1972, 65, 357-360.
Purpose: To identify dimensions underlying student percep-
tions of teacher effectiveness
Subjects: 42,810 seventh- through twelfth-graders, rating
1,427 teachers
Variables: 12 Teacher Image Questionnaire scores
Method: Principal axes condensation, probably with rotation
Results: Three factors emerged: I Teacher charisma or
popularity; II Structure centered; III Student centered.
Cohen, M. W., Mirels, H. L., & Schwebel, A. I. Dimensions of
elementary school student teacher concerns. Journal of
Experimental Education, 1972, 41, 6-10.

Purpose: To identify dimensions of elementary school
student teacher concerns
Subjects: 139 student teachers
Variables: 122 item scores from a questionnaire designed
to tap a wide range of student teacher concerns
Method: Hierarchical factor condensation with varimax
rotation
Results: One major factor (I Concern with being an effective teacher) accounted for 26% of the total variance,
but seven others also emerged: II Class control; III
Pupil concerns; IV Need for support; V Evaluation by
supervisors; VI Self-adequacy in the classroom; VII
Managing interpersonal and administrative problems;
VIII Social desirability.
Coletta, A. J., & Gable, R. K. The content and construct
validity of the Barth Scale: Assumptions of open
education. Educational and Psychological Measurement,
1975, 35, 415-425.
Purpose: To identify dimensions of teachers' perceptions
of their role and the process of children's learning
as assessed by the Barth Scale (BS)
Subjects: 78 open and 113 traditional elementary school
teachers
Variables: 28 Barth Scale item scores
Method: Principal components condensation with obliquimax
rotation
Results: Seven factors emerged: I Curricular flexibility;
II Intellectual development; III Evaluating the child;
IV Learning through involvement; V Learning facilitators;
VI Evaluating the child's work; VII Learning through
exploration.
Cook, M. A., & Richard, H. C. Dimensions of principal and
supervisor ratings of teacher behavior. Journal of
Experimental Education, 1972, 41, 11-14.
Purpose: To identify dimensions of principal and supervisor
ratings of teacher behavior
Subjects: 236 teachers, each rated by their supervisor and
their supervisor's supervisor
Variables: 23 teacher-competence item scores each from the
two raters (46)
Method: Principal components condensation with orthogonal
varimax rotation
Results: Three factors emerged: I Supervising teacher;
II Principal; III Uninterpreted.
Conclusions: The rating scales generated data more reflec-
tive of raters' viewpoint than of teachers' classroom
behavior.
Costin, F. Empirical test of the "teacher-centered" versus
"student-centered" dichotomy. Journal of Educational
Psychology, 1971, 62, 410-412.
Purpose: To identify dimensions underlying the Survey of
Classroom Behavior (SCB)

Study One
Subjects: 201 college students, rating behavior in 1 of
9 discussion sections
Variables: 30 SCB item scores
Method: Principal axes condensation with varimax rotation
Results: Four factors emerged: I Student involvement; II
Teacher support; III Negative affect; IV Teacher control.
Conclusions: Twenty SCB items were selected for further use.
Study Two
Subjects: 425 college students, rating behavior of 1 of 14
discussion sections
Variables: 20 SCB item scores
Method: same as above.
Results: same as above.
Covert, R. W., & Mason, E. J. Factorial validity of a Student
Evaluation of Teaching instrument. Educational and Psy-
chological Measurement, 1974, 34, 903-905.
Purpose: To identify dimensions underlying the Student
Evaluation of Teaching instrument (SET)
Subjects: 3,573 students, evaluating 254 graduate and un-
dergraduate classes
Variables: 17 5-point SET item scores
Method: Principal axes condensation with varimax rotation
Results: Three factors emerged: I Method and style of
teaching; II Student's perception of self as it relates
to the course; III Materials and resources.
Doyle, K. O. Jr., & Whitely, S. E. Student ratings as cri-
teria for effective teaching. American Educational
Research Journal, 1974, 11, 259-274.
Purpose: To identify dimensions underlying 49 specific
instructor characteristics as rated on the Student
Opinion Survey (SOS)
Subjects: 174 introductory French students, rating 1 of
12 instructors
Variables: 49 SOS item scores
Method: Common-factor condensation with varimax rotation
Results: Five factors emerged: I Attitudes toward students;
II Expositional skills; III Motivation of interest; IV
Stimulation of thinking; V Generalization of content.
Conclusions: Dwyer's factor extension procedure was used
to examine relationships with satisfaction ratings,
learning-oriented items, and background information.
Emmer, E. T., & Peck, R. F. Dimensions of classroom behavior.
Journal of Educational Psychology, 1973, 64, 223-240.
Purpose: To identify dimensions underlying, and relation-
ships between, several classroom observation instruments.
Subjects: 28 fifth- and eighth-grade classrooms, on 138
occasions
Variables: Scores on 1) 20 codes from the Observation
Schedule and Record (OScAR5); 2) 23 codes from Fuller
Affective Interaction Records (FAIR); 3) 22 Cognitive
Components System (CCS); 4) 8 Coping Analysis Schedule

for Educational Settings (CASES); 5) 28 Dyadic System
(DS); and 6) 29 of the factors extracted in 1-5
Method: Principal axes condensation with varimax rotation
independently for each instrument and the combined instruments
Results: Eight OScAR5 factors emerged: I Student-idea
oriented versus teacher-idea oriented; II Convergent
evaluative vs. divergent teacher behavior; III Teacher-
initiated problem solving; IV Positive teacher affect;
V Procedural interaction; VI Desisting; VII Teacher
informing vs. student statements; VIII Controlling.
Nine FAIR factors emerged: I Students present vs. routine; II Criticizing-resisting; III Teacher presents vs.
teacher responds; IV Expansive vs. restrictive teacher
behavior; V Student-clarifying activity; VI Teacher
candor; VII Teacher support vs. student information
seeking; VIII Student-initiated discussion; IX Tangen-
tial teacher behavior. Eight CCS factors emerged: I
Conceptual; II Description vs. inferential behavior;
III Explanation; IV Teacher presents; V Association,
drill; VI Pupil-to-pupil description; VII Higher cognitive level student behavior; VIII Description inter-
change: student solicits-teacher responds. Three CASES
factors emerged: I Attention vs. routine activity; II
Passivity; III Inappropriate vs. appropriate social
behavior. Ten DS factors emerged: I Convergent response
opportunities; II Private pupil-teacher contacts; III
No pupil response; IV Incorrect pupil response; V Criticism of wrong response; VI Teacher seeks the correct
response; VII Criticism of pupil work; VIII Opinion
question; IX Open question-correct answer; X Self-
reference question-teacher negates. Eleven factors
emerged when factors from all five instruments were
considered: I Teacher-initiated problem solving; II
Restrictive vs. expansive teaching; III Pupil presentation of ideas; IV Negative affect; V Teacher presentation vs. pupil recitation; VI Divergent vs. convergent
evaluative teacher behavior; VII Teacher-controlling
behavior; VIII Teacher support for correct response;
IX Self-referent pupil questions; X Teacher openness;
XI Pupil unresponsiveness.
Field, T. W., Simpkins, W. S., Browne, R. K., & Rich, P.
Identifying patterns of teacher behavior from student
evaluations. Journal of Applied Psychology, 1971, 55,
466-469.
Purpose: To identify dimensions underlying students' perceptions of teacher behaviors
Subjects: 57 Australian college students enrolled in a
team-taught course
Variables: Scores on 18 teacher-evaluation items, pooled
for ratings of ideal teacher and each member of the
5-teacher team

Method: Principal axes condensation with varimax rotation
Results: Six factors, similar to Isaacson et al.'s (1964) for a longer instrument, emerged: I Skill; II Overload; III Structure; IV Feedback; V Group interaction; VI Rapport.
Conclusions: Multiple discriminant analyses were also reported.
Finkbeiner, C. T., Lathrop, J. S., & Schuerger, J. M. Course and instructor evaluation: Some dimensions of a questionnaire. Journal of Educational Psychology, 1973, 64, 159-163.
Purpose: To identify dimensions underlying students' attitudes toward courses and instructors
Subjects: 1) 1,616 students at university "academic centers," 2) 6,352 students at the main campus, (3) 1,971 students at uncertain location
Variables: 48 item scores from the Course and Instructor Evaluation Questionnaire
Methods: Principal components condensation with varimax rotation, by sample
Results: Five factors emerged for the first two samples (and reportedly were replicated with the third): I General course attitude; II Attitude toward exams; III Attitude toward method; IV Good guy; V Work load.
French-Lazovik, G. Predictability of students' evaluations of college teachers from component ratings. Journal of Educational Psychology, 1974, 66, 373-385.
Purpose: To identify predictor dimensions of student evaluations of college teachers
Subjects: 1) students of 133 faculty members at one university (A), and 2) 6,120 students of 144 faculty members at a second (B)
Variables: Scores on 1) 41 and 2) 16 items of a student evaluation instrument
Method: Principal components condensation, independently for A and B
Results: Eight A and five B factors emerged but were not labeled.
Frey, P. W. Comparative judgment scaling of student course ratings. American Educational Research Journal, 1973, 10, 149-154.
Purpose: To identify dimensions of a course evaluation instrument
Subjects: 599 college students, each enrolled in two or more of four courses evaluated
Variables: 15 item-scores from a course-evaluation instrument
Method: Principal components condensation with varimax rotation
Results: Five factors emerged: I Teacher's presentation; II Work load; III Grading procedure; IV Accessibility; V Uninterpreted.
Conclusions: Evaluations for the four courses were compared.

Frey, P. W., Leonard, D. W., & Beatty, W. W. Student ratings
of instruction: Validation research. American Educa-
tional Research Journal, 1975, 12, 435-447.
  Purpose: To identify dimensions of the Endeavor Instruc-
    tional Rating Form (EIRF)
  Subjects: 1) 421 calculus (college) students; 2) 218
    educational psychology students
  Variables: 21 item scores from the EIRF
  Method: Principal-factor condensation with varimax rota-
    tion, independently for the two groups
  Results: Seven factors emerged in both analyses: I Clarity
    of presentations; II Work load; III Personal attention;
    IV Class discussion; V Organization-planning; VI Grading;
    VII Student accomplishment.
Fulcher, D. G., & Anderson, W. T. Jr. Interpersonal dis-
    similarity and teaching effectiveness: A relational
    analysis. Journal of Educational Research, 1974, 68,
    19-25.
  Study One
  Purpose: To identify dimensions of 115 semantic differen-
    tial (SD) scales used to rate teachers, and thereby
    reduce the number of scales for Study Two
  Subjects: 288 business administration students, rating 1
    of 2 teachers
  Variables: 115 SD scale scores
  Method: Unspecified factor analysis
  Results: An unspecified number of factors emerged; 68 SD
    scales loading .45 or higher were selected for Study Two
  Study Two
  Purpose: To identify dimensions of student-teacher per-
    ceived interpersonal dissimilarity
  Subjects: 195 business administration students, rating
    themselves and 1 of 3 teachers
  Variables: 68 SD scale scores
  Method: Unspecified condensation with varimax rotation
  Results: Twelve factors emerged: I Stage presence; II
    Morality; III Formality; IV Structure; V Stage fright;
    VI Authoritarianism; VII Empathy; VIII Liberalism; IX
    Practicality; X Subjectivity; XI "Sugar daddy"; XII
    Maturity.
  Conclusions: Interpersonal dissimilarity factor scores
    were examined relative to measures of perceived teacher
    effectiveness.
Granzin, K. L., & Painter, J. J. A new explanation for
    students' course evaluation tendencies. American
    Educational Research Journal, 1973, 10, 115-124.
  Purpose: To identify dimensions of potential predictors
    of students' course evaluations
  Subjects: 627 undergraduate, 10 graduate students
  Variables: Scores from questionnaire items related to
    demographics (7 items), commitment (2), static attitude
    (7) variables
  Method: Principal components condensation with varimax
    rotation

Results: Six factors emerged: I Student's maturity; II
Student's sex; III Student's performance; IV Course's
centrality; V Course's specific values; VI Course's
work requirement.
Conclusion: A regression analysis was also performed.
Greenwood, G. E., Bridges, C. M. Jr., Ware, W. B., & McLean,
J. E. Student evaluation of college teaching behaviors.
Journal of Educational Measurement, 1974, 11, 141-143.
Purpose: To identify dimensions of college teaching be-
haviors
Subjects: 328 college students and 554 faculty
Variables: 60 Student Evaluation of College Teaching
Behaviors (SECTB) item scores
Method: Principal axes condensation without rotation,
independently for students, faculty, and combined
groups
Results: Eight factors were identified for the combined
sample: I Facilitation of learning; II Obsolescence
of presentation; III Commitment to teaching; IV Evaluation; V Voice communication; VI Openness; VII Currency
of knowledge; VIII Rapport.
Gross, D. E., & Kaplan, R. M. Teacher attitudes toward im-
plementing career education in the classroom. Journal
of Educational Research, 1975, 69, 106-108.
Purpose: To identify dimensions underlying questionnaire
items related to career education attitudes
Subjects: 373 elementary and intermediate school teachers
Variables: 10 career education attitude items from a
questionnaire
Method: Principal components condensation with orthogonal
varimax rotation
Results: Four factors emerged: I Global attitude toward
career education; II-IV Unlabeled.
Hansen, L. H., Borgatta, E. F., & Lambert, P. Teacher perceptions of intra-occupational status relationships
among elementary, junior high, and senior high school
teaching positions. Journal of Experimental Education,
1971, 40, 51-56.
Purpose: To identify dimensions of teacher perceptions
of intra-occupational status relationships among
elementary (ES), junior high (JHS), and senior high
school (HS) teaching positions
Subjects: 20 teachers, stratified across two levels of
ES and JHS and HS
Variables: 63 item scores from the Questionnaire on
Teacher Attitudes Toward the Profession
Method: Principal factor condensation with quartimax
rotation
Results: Seven factors emerged: I Status; II Training (a);
III Time; IV Training (b); V Predictor; VI Graduate
education; VII Salary.
Conclusions: Factor scores for this pilot sample and a
second sample were examined relative to several other
variables.

Hartley, E. L., & Hogan, T. P. Some additional factors in
    student evaluation of courses. American Educational
    Research Journal, 1972, 9, 241-250.
  Purpose: To identify dimensions of two student-evaluation-
    of-courses measures
  Subjects: Approximately 450 college students
  Variables: Scores on a 30-item "traditional" course eval-
    uation measure and a 26-item scale measuring student
    self-development through the course
  Method: Principal components condensation with varimax
    rotation
  Results: Seven factors were considered interpretable: I
    Overall evaluation; II Student-teacher interaction; III
    Load or difficulty; IV Structure or organization; V
    General cognitive development; VI Field-specific develop-
    ment; VII Relevance.
  Conclusions: There appears to be little in common between
    the traditional student evaluation instrument (Factors
    I-IV) and student ratings of course-affected self-
    development (Factors V-VII).
Holmes, D. S. The Teaching Assessment Blank: A form for the
    student assessment of college instructors. Journal of
    Experimental Education, 1971, 39, 34-38.
  Purpose: To identify dimensions underlying the Teaching
    Assessment Blank (TAB)
  Subjects: 1,648 college students, rating 1 of 7 faculty
    members
  Variables: Scores on the 22 class-related evaluation items
    of the TAB
  Method: Principal axes condensation with varimax rotation
  Results: Four factors emerged: I Quality of instructor's
    presentations; II Evaluation process and student-
    instructor interactions; III Degree to which students
    were stimulated and motivated by instructors; IV Clarity
    of the tests.
  Conclusions: The four statistical dimensions, which have
    been replicated with a second sample, essentially agree
    with the four logical ones designed into the TAB.
Jaeger, R. M., & Freijo, T. D. Some psychometric questions
    in the evaluation of professors. Journal of Educational
    Psychology, 1974, 66, 416-423.
  Purpose: To identify dimensions of two evaluation instru-
    ments for college teaching
  Subjects: 358 undergraduate and graduate students of 21
    faculty members
  Variables: 1) Paired item scores from a 17-item standard
    and a 17-item revised student evaluation of college
    teaching instrument; 2) 17 item scores from the revised
    form
  Method: Principal components condensation, independently
    for the two data sets
  Results: Factors emerging from the paired-item analysis
    were not labeled; five factors were extracted, the first

accounting for 68% of the variance for the standard
form and 60% for the revised form. Five rotated factors emerged in the second analysis: I Judged managerial effectiveness of the professor; II Acquisition
(and professor's role therein) of new educational values;
III Perceptions of progress; IV Behavioral; V Acquisition
(and professor's role therein) of new concepts.

Keaveny, T. J., & McGann, A. F. A comparison of behavioral
expectation scales and graphic rating scales. Journal
of Applied Psychology, 1975, 60, 695-703.
Purpose: To identify dimensions underlying behavioral-
expectation and graphic rating scales for college
teaching
Subjects: 183 sophomore, junior, senior, and graduate
students, rating one of four teachers
Variables: 13 performance dimension scores each for 1)
graphic, and 2) behaviorally anchored rating scales
Method: Unspecified condensation with varimax rotation,
independently for graphic and behaviorally anchored
scales
Results: Seven and nine "non-trivial" but unlabeled factors emerged for the graphic and behaviorally anchored
scales, respectively.

Keith, L. T., Tornatzky, L. G., & Pettigrew, L. E. An
analysis of verbal and nonverbal classroom teaching
behaviors. Journal of Experimental Education, 1974, 42,
30-38.
Purpose: To identify dimensions of teacher-interns' verbal
and nonverbal classroom behaviors
Subjects: 43 teacher-interns in a federally-funded program
Variables: Ratings (from videotapes) on 57 categories of
classroom behaviors
Method: Cluster analysis (oblique)
Results: Three clusters emerged: I Positive task relevant
teacher-pupil interaction; II Observation and group
interaction; III Teacher disapproval and pupil misbehavior.
Conclusions: Disapproval tended to be communicated verbally
and explicitly, whereas approval was communicated pas-
sively and nonverbally.

Kennedy, W. R. Grades expected and grades received - their
relationship to students' evaluations of faculty performance. Journal of Educational Psychology, 1975, 67,
109-115.
Purpose: To identify dimensions underlying a teacher evaluation instrument
Subjects: 549 college freshmen
Variables: Scores on 78 teacher evaluation items
Method: Principal components condensation with varimax
rotation
Results: Three interpretable factors emerged: I Instructional methods; II Interpersonal relationships with
students; III Content competency.

Linn, R. L., Centra, J. A., & Tucker, L. Between, within, and total group factor analyses of student ratings of instruction. Multivariate Behavioral Research, 1975, 10, 277-288.
Purpose: To identify group factor structures of student ratings of instruction (Centra, ETS report, 1972)
Subjects: 9700 students from 437 classrooms
Variables: 31 The Student Instructional Report item scores
Method: Principal axes condensation with varimax and promax rotation
Results: Six factors emerged: I Teacher-student relationship; II Course objectives and organization; III Lectures; IV Reading assignments; V Course difficulty and workload; VI Examinations.
Conclusions: The present report details approximation of the between group covariance matrix.
McDowell, E. E. The semantic differential as a method of teacher evaluation. Journal of Educational Research, 1975, 68, 330-332.
Purpose: To identify dimensions of a teacher evaluation instrument
Subjects: 51 college freshmen
Variables: Scores on semantic differential scales for the concepts: teacher credibility (13 scales), content (12), delivery (13), feedback (13)
Method: Unspecified condensation with varimax rotation, independently for each concept
Results: Five unlabeled factors were extracted in each analysis.
McKeachie, W. J., Lin, Y., & Mann, W. Student ratings of teacher effectiveness: Validity studies. American Educational Research Journal, 1971, 8, 435-445.
Purpose: To identify dimensions of students' ratings of teacher effectiveness
Subjects: 286 introductory economics students
Variables: Unspecified number (more than 14) student-ratings-of-instruction items
Method: Unspecified factor analysis
Results: Six factors emerged: I Skill; II Structure; III Feedback; IV Rapport (warmth); V Change in beliefs; VI Value of the course.
Conclusions: Correlations between these factors and attitude sophistication changes were reported. This was one of five studies reported in the paper.
Miskel, C., & Heller, L. The Educational Work Components Study: An adapted set of measures for work motivation. Journal of Experimental Education, 1973, 42, 45-50.
Purpose: To identify the factorial structure of the Educational Work Components Study (EWCS), a form of the Work Components Study modified for use in schools
Subjects: 550 returns from 153 college seniors, 42 graduate students, 118 public school administrators, and 432 public school teachers

Variables: 1) 66, and 2) 56 EWCS item scores
Method: R-factor condensation with orthogonal varimax and
     oblique maxplane rotation
Results: On the basis of six- and seven-factor solutions
     for the 66-item set, 10 items were eliminatèd. Six
     factors which can be used to measure work motivation
     in educational organizations emerged for the remaining
     56 items: I Potential for personal challenge and
     development; II Competitiveness desirability and reward
     of success; III Tolerance for work pressure; IV Conser-
     vative security; V Willingness to seek reward in spite
     of uncertainty vs. avoidance of uncertainty; VI Sur-
     round concern.
Conclusions: Borgatta's seventh factor, Responsiveness
     to new demands, collapsed into other factors.
Pfeiffer, M. G., Lehmann, W., & Scheidt, U. Methodological
     investigation of perceived structure of college teach-
     ing across two cultures. Perceptual and Motor Skills,
     1972, 35, 619-626.
Purpose: To identify dimensions of students' perceptions
     of college teaching in two cultures
Subjects: 14 upper division students and 10 Ph.D. pro-
     fessors from an American college, and 18 students and
     10 professors from a German university
Variables: Similarity judgments for 30 teaching activities
Method: Unspecified condensation with varimax rotation,
     independently for four groups and by nationality
Results: Eight factors emerged in all six analyses: I
     Teacher (lecturing) dynamism; II Classroom administra-
     tion; III Information dissemination; IV Knowledge dis-
     semination; V Advisory guidance; VI Environmental
     regulation; VII Teacher-student feedback; VIII Control
     of student behavior.
Poole, C. The influence of experiences in the schools on
     students' evaluations of teaching practice. Journal
     of Educational Research, 1972, 66, 161-164.
Purpose: To identify dimensions of practice teaching
     experience
Subjects: 523 teachers college students
Variables: 35 item scores from a questionnaire concerning
     practice teaching
Method: Principal components condensation with varimax
     rotation
Results: Six factors emerged: I Experience of a well-
     organized, supportive situation; II Experience of
     criticism; III Good working relationship with other
     staff; IV Lack of support; V Good working relationship
     with fellow student; VI Good informal working relation-
     ship with T and the children.
Price, J. An instrument for measuring student teacher morale.
     Journal of Educational Measurement, 1971, 8, 47-48.
Purpose: To identify dimensions of student teacher morale

Subjects: 299 student teachers who had just completed
student teaching
Variables: 100 Purdue Teacher Opinionaire (PTO) and 50
experimental item scores
Method: Principal components condensation with varimax
rotation
Results: Fourteen factors emerged: I Rapport with supervising teacher; II Rapport with principal; III Rapport
with university supervisor; IV Community support of
education; V Student teaching load; VI Rapport with
students; VII Rapport with other teachers; VIII Satisfaction with housing; IX Professional preparation; X
School facilities and services; XI Curriculum issues;
XII Satisfaction with student teaching; XIII Student
teacher status; XIV Teacher salary.
Conclusions: The instrument was revised to include 57
PTO and 43 experimental items, representing 12 dimensions. Factor XIV was deleted, and Factors XII and
XIII were combined to "Teaching as a profession."
Rugg, E. A., & Norris, R. C. Student ratings of individualized faculty supervision: Description and evalua-
tion. American Educational Research Journal, 1975,
12, 41-53.
Purpose: To identify dimensions of faculty-supervision
evaluations
Subjects: 125 psychology graduate students
Variables: Scores on 1) 42 descriptive ratings of faculty
supervision, and 2) 9 evaluative (satisfaction) ratings
Method: Principal axes condensation with varimax and
promax rotation, independently for descriptive and
evaluative ratings, and higher-order factor analysis
for descriptive ratings
Results: Ten first-order descriptive factors emerged:
I Respect for students; II Structure and guidance;
III Research productivity; IV Research methods expertise; V Interpersonal rapport; VI Stimulating
teaching; VII Supervisor accessibility; VIII Subject
matter expertise; IX Faculty maturity; X Communications training. Two interpretable second-order factors
emerged: I Teaching style; II Academic qualifications.
Two evaluative dimensions emerged: I Supervisor evaluation; II Experience evaluation.
Sherman, B. R., & Blackburn, R. T. Personal characteristics
and teaching effectiveness of college faculty. Journal
of Educational Psychology, 1975, 67, 124-131.
Purpose: To identify dimensions of teaching effectiveness
as assessed by a semantic differential (SD)
Subjects: 1,500 liberal arts college students, rating
108 faculty
Variables: 30 SD ratings
Method: Principal axes condensation with orthogonal rota-
tion
Results: Four factors emerged: I Personal potency; II
Pragmatism; III Amicability; IV Intellectual competency.

310

Short, B. G., & Szabo, M. Secondary school teachers' knowledge of and attitudes toward educational research. Journal of Experimental Education, 1974, 43, 75-78.
 Purpose: To identify dimensions of an attitudes toward educational research instrument (Short, unpublished dissertation, 1971)
 Subjects: Unspecified
 Variables: Scores on 35 semantic differential scales representing 10 concepts
 Method: Unspecified factor analysis
 Results: Eighteen scales were selected to measure eight concepts: I Research; II Education; III Science; IV Medicine; V Educational research; VI Scientific research; VII Medical research; VIII Statistics.
Sockloff, A. L., & Papacostas, A. C. Uniformity of faculty attitude toward effective teaching in lecture/discussion courses. Journal of Educational Research, 1975, 12, 281-293.
 Purpose: To identify dimensions of perceived effective lecture/discussion teaching for faculty from different disciplines
 Subjects: 387 faculty from 6 disciplines
 Variables: 64 item scores related to lecture/discussion teaching
 Method: Unspecified condensation with biquartimin rotation for each discipline
 Results: Four common factors were interpreted: I Intellectual stimulation; II Organization; III Interpersonal relationship; IV Difficulty level.
 Conclusions: Factor structures were compared by discipline and by faculty rank.
Sontag, M., & Pedhazur, E. Dimensions of educational attitudes: Factorial congruence of two scales. Journal of Educational Measurement, 1972, 9, 189-198.
 Purpose: To identify factors common to Kerlinger's Educational Attitudes Scale (ES-VII) and Oliver and Butcher's Survey of Opinions about Education (SOE)
 Subjects: 356 education graduate students
 Variables: 30 ES-VII and 33 SOE item scores
 Method: Principal axes condensation with oblique promax rotation, followed by second-order factoring
 Results: Seven first-order factors emerged: I Career training; II Moral vs. religious training; III Corporal punishment; IV Individualized instruction; V Subject matter and knowledge; VI Expansion of education; VII School as a social force. Two second-order factors emerged: I ES-VII Traditionalism vs. SOE Tendermindedness (bipolar); II ES-VII Progressivism and SOE Radicalism.
St. John, N. Thirty-six teachers: Their characteristics and outcomes for black and white pupils. American Educational Research Journal, 1971, 8, 635-648.
 Purpose: To identify dimensions of teacher characteristics (to be used further as an independent variable)

Subjects: 36 sixth-grade teachers
Variables: 13 Ryans Scale (semantic differential) scores
Method: Principal components condensation with orthogonal
varimax rotation
Results: Three factors emerged: I Child-oriented; II
Task-oriented; III Fair.
Tobias, S., & Hanlon, R. Attitudes toward instructors, social
desirability, and behavioral intentions. Journal of
Educational Psychology, 1975, 67, 405-408.
Purpose: To identify dimensions underlying students' ratings
of instructors
Subjects: 158 college students from 7 classes
Variables: 32 item scores from a teacher evaluation instru-
ment
Method: Principal components condensation with varimax
rotation
Results: Six interpretable factors emerged: I Instructor
skill; II Evaluation and feedback; III Friendly class-
room atmosphere; IV Classroom organization; V Standards;
VI Assignments.
Wilson, R. C., Dienst, E. R., & Watson, N. L. Characteris-
tics of effective college teachers as perceived by
their colleagues. Journal of Educational Measurement,
1973, 10, 31-37.
Purpose: To identify the dimensions faculty members
associate with good teaching.
Subjects: 119 university faculty
Variables: Scores on 67 items related to a good teacher's
out-of-classroom behavior
Method: Principal components condensation with varimax
rotation
Results: Five factors emerged: I Research activity and
recognition; II Participation in the academic community;
III Intellectual breadth; IV Relations with students;
V Concern for teaching.
Conclusions: These results were related to peer nominations
and other faculty characteristics.
Woog, P. C. A Q study of elementary school teachers' assign-
ments of educational priorities and their practice.
Journal of Experimental Education, 1973, 42, 88-96.
Purpose: To identify dimensions of elementary teachers'
priorities of behaviorally-stated objectives of instruc-
tion
Subjects: 20 secondary and 50 elementary teachers, 15
community members, and 16 school administrators.
Variables: Scores on 100 items (from a Q-sort) of a
behavioral-objectives questionnaire
Method: Unspecified factor analysis
Results: Three factors emerged: I Affective; II High-
cognitive; III Low-cognitive combined with tool-skill.
Conclusions: The 100-item Q-sort was investigated further
in relation to classroom observations.

312

Yee, A. H., & Fruchter, B. Factor content of the Minnesota
    Teacher Attitude Inventory. American Educational
    Research Journal, 1971, 8, 119-133.
    Purpose: To identify dimensions of the Minnesota Teacher
    Attitude Inventory (MTAI)
    Subjects: 368 experienced intermediate grade teachers
    Variables: 150 MTAI item scores
    Method: Principal components condensation with normal
    varimax rotation
    Results: Five factors emerged: I Children's irresponsible
    tendencies and lack of self-discipline; II Conflict
    between teachers' and pupils' interests; III Rigidity
    and severity in handling pupils; IV Pupils' independence
    in learning; V Pupils' acquiescence to the teacher.
F. School Learning and Achievement
    Goldman, R. D., & Sexton, D. W. Archival experiments with
    college admission policies. American Educational
    Research Journal, 1974, 11, 195-201.
    Purpose: To identify dimensions underlying high school
    grades in 12 content categories
    Subjects: 475 students from five colleges
    Variables: College GPA, plus high school GPAs in 12 con-
    tent categories
    Method: Principal-factor condensation, apparently without
    rotation
    Results: A single factor, accounting for 59% of the
    variance, emerged: I General.
    Conclusions: Total high school average was compared with
    academic high school average as a predictor of college
    GPA.
    Schoenfeldt, L. F., & Brush, D. H. Patterns of college
    grades across curricular areas: Some implications for
    GPA as a criterion. American Educational Research
    Journal, 1975, 12, 313-321.
    Purpose: To identify patterns of college grades
    Subjects: 1,904 college students from one entering class
    Variables: Scores on GPA in 12 college areas, SAT (2
    ‚ scores), high school GPA
    Method: Unspecified condensation with varimax and oblimin
    rotation
    Results: Three factors emerged: I General academic
    achievement (GPA factor); II Grades independent of
    achievement and aptitude; III Tested aptitude.
    Simpson, C. K., & Boyle, D. Esteem construct generality and
    academic performance. Educational and Psychological
    Measurement, 1975, 35, 897-904.
    Purpose: To identify dimensions of self-esteem related to
    academic (exam) performance
    Subjects: 78 male and 81 female college students
    Variables: 9 scores: Tennessee Self-Concept Scale,
    Rosenberg Self-Esteem Test, QSE (single-item global),
    intellectual esteem, educational esteem, 2 measures
    of task (midterm) - specific self-esteem, GPA, grade
    Method: Principal components condensation with varimax
    rotation   313

Results: Two factors emerged: I Academic performance;
II Global self-esteem.
Solomon, D., Hirsch, J. G., Scheinfeld, D. R., & Jackson,
J. C. Family characteristics and elementary school
achievement in an urban ghetto. Journal of Consulting
and Clinical Psychology, 1972, 39, 462-466.
Purpose: To identify dimensions underlying school achieve-
ment for urban ghetto fifth-graders
Subjects: 149 urban black ghetto fifth-graders
Variables: Average grades (fourth and fifth grade) for
reading, conduct, and all academic subjects; California
Achievement Test scores for reading, language, and
mathematics; L-T verbal and nonverbal IQ at grade 5;
Kuhlman-Anderson IQ at grade 3
Method: Principal axes condensation without rotation
Results: One unlabeled factor emerged.
Yee, L. Y., & LaForge, R. Relationship between mental
abilities, school class, and exposure to English in
Chinese fourth graders. Journal of Educational Psy-
chology, 1974, 66, 826-834.
Purpose: To identify dimensions underlying several social
class and exposure to English variables
Subjects: 53 American-born Chinese fourth-graders
Variables: Scores on 3 social class and 19 exposure to
English variables
Method: Principal components condensation with varimax
rotation
Results: Eight factors emerged: I Average percentage
of English spoken at home by S's parents, siblings,
and older relatives; II Free time; III Distance from
Chinatown; IV Formal instruction outside elementary
school; V, VIII (Representative of few Ss); VI Unin-
terpretable; VII (Overlap with previous factors).
1. Prediction
Adkins, P. L., Holmes, G. R., & Schnackenberg, R. C. Fac-
tor analyses of the de Hirsch predictive index.
Perceptual and Motor Skills, 1971, 33, 1319-1325.
Purpose: To identify dimensions underlying the de Hirsch
Predictive Reading Index (PRI)
Subjects: 50 5½-year-olds
Variables: 1) 10 PRI scores; 2) plus sex, race, and IQ
Method: Principal components condensation, apparently
without rotation, for each data set
Results: Only one factor emerged in each analysis: I
Visual discrimination.
White, W. F., & Simmons, M. First-grade readiness predicted
by teachers' perception of students' maturity and stu-
dents' perception of self. Perceptual and Motor Skills,
1974, 39, 395-399.
Purpose: To identify dimensions underlying the I Feel -
Me Feel test for first-graders
Subjects: 165 first-graders
Variables: 40 I Feel - Me Feel item scores

Method: Principal components condensation with varimax rotation

Results: Ten unlabeled factors emerged; the first accounted for 33% of the total variance, while the other nine for less than 4% each.

2. Overachievement and Underachievement

Burns, G. W., & Watson, B. L. Factor analysis of the revised ITPA with underachieving children. Journal of Learning Disabilities, 1973, 6, 371-376.

Purpose: To identify the factorial structure of the ITPA for a sample of underachieving children

Subjects: 90 underachieving children, age 5-1 to 9-11 years

Variables: 12 ITPA subtest scores

Method: Unspecified condensation with orthogonal rotation

Results: Five factors emerged: I General auditory language; II Visual language; III Expressive language; IV General language/closure; V Memory/expressive language.

Conclusions: The implications for use of ITPA profiles in planning intervention programs were discussed.

Riedel, R. G., Grossman, J. H., & Burger, G. Special Incomplete Sentences Test for underachievers: Further research. Psychological Reports, 1971, 29, 251-257.

Purpose: To identify dimensions underlying the Special Incomplete Sentences Test (SIST)

Subjects: 106 high school students (33 high achievers, 65 underachievers, 8?)

Variables: 69 SIST item scores

Method: Principal axes condensation with oblimax rotation

Results: Eight factors emerged: I General; II Achievement attitudes; III Reaction to future tasks; IV Interpersonal relations; V Reaction to evaluation; VI Reaction to perceived success and failure; VII Confidence in ability; VIII Academic skills.

G. Curriculum and Programs

Permut, S. E. Toward a factor analytic definition of academic relevance. Educational and Psychological Measurement, 1974, 34, 837-848.

Purpose: To identify dimensions underlying the concept of academic relevance

Subjects: 67 introductory advertising students

Variables: 10 semantic differential ratings for each of 10 "concepts" (topics) in advertising

Method: Principal components condensation with varimax and binormamin rotation, independently by concept

Results: Four factors emerged across concepts: I Easy-difficult; II Useful, good, relevant; III Interesting-uninteresting and practical-impractical; IV Interesting, active, necessary, successful.

Redburn, F. S. Q factor analysis: Applications to educational testing and program evaluation. Educational and Psychological Measurement, 1975, 35, 767-778.

Purpose: To identify dimensions underlying urban interns' social attitudes before and after the internship experience

Subjects: 20 urban interns (in public agencies)
Variables: 51-item Q-sort, administered before and after
internship (treated as separate individuals)
Method: Unspecified Q condensation with varimax rotation
Results: Six factors emerged: I Excitement about constructive change through careers in urban government;
II Political liberalism; III Alienated perspective on
local affairs; IV Uninterpreted; V Insider perspective
on local affairs; VI Viewpoint of the outwardly mobile,
those seeking to leave behind the city and its problems.
Spinivasan, V., & Weinstein, A. G. Effects of curtailment
on an admissions model for a graduate management program. Journal of Applied Psychology, 1973, 58, 339-346.
Purpose: To identify dimensions of predictor variables
used in a corrected-curtailment admissions model for
a graduate management program
Subjects: 40 graduate management program students
Variables: 37 predictor (of academic success) variable
scores
Method: Unspecified factor analysis
Results: Nine factors emerged: I Academic excellence;
II Undergraduate school excellence; III Aptitude for
graduate management study; IV Marital status; V Excellence in science; VI Extracurricular social activity;
VII Excellence in engineering; VIII Maturity; IX Extracurricular sports activity.
1. Teaching Methods and Teaching Aids
Ayers, J. B. Elementary school teachers' attitudes toward
instructional television. Journal of Experimental
Education, 1972, 41, 1-4.
Purpose: To identify dimensions underlying items, from
an instrument measuring attitudes toward instructional
television, which correlated with years of teaching
experience
Subjects: 142 elementary school teachers, variously
divided into ten subgroups
Variables: Scores on 25 attitude-toward-instructional-
television items which correlated with years of
teaching experience
Method: General factor condensation with varimax rotation,
independently for ten subgroups
Results: A four-factor solution was selected as most interpretable for all analyses: I Teaching Process; II
Threat of instructional television; III Problems in the
use of instructional television; IV Learning process.
Janzen, H. L., & Hallworth, H. J. Demographic and biographic
predictors of writing ability. Journal of Experimental
Education, 1973, 41, 43-53.
Purpose: To identify dimensions of (computer graded) linguistic ability
Subjects: 387 college students in educational psychology
classes
Variables: 22 scores obtained by computer-grading a 750+
word essay

Method: Principal components condensation with varimax
    rotation
Results: Five factors emerged: I Mechanical accuracy;
    II Fluency; III Sentence complexity; IV Opinionation
    in writing; V Emotionality in writing.
Conclusions: Factor scores were subsequently used as
    dependent variables in further analyses.
Slotnick, H. B. Toward a theory of computer essay grading.
    Journal of Educational Measurement, 1972, 9, 253-263.
Purpose: To identify dimensions underlying computer-
    graded essay measures
Subjects: 326 high school student papers
Variables: 34 attribute scores graded by the computer
Method: Principal components condensation with varimax
    and oblimax rotation
Results: Six factors emerged in both orthogonal and
    oblique solutions: I Fluency; II Misspelling; III
    Diction; IV Sentence structure; V Punctuation; VI
    Paragraph development.
Conclusions: Factors scores were used in further inves-
    tigation.

XI. PERSONNEL AND INDUSTRIAL PSYCHOLOGY

A. Vocational Choice and Guidance
    Aldag, R. J., & Brief, A. P. Some correlates of work values.
        Journal of Applied Psychology, 1975, 60, 757-760.
        Purpose: To identify dimensions of work values
        Subjects: 131 hourly employees of a manufacturing firm
        Variables: 8 item scores from Blood's (1969) scale
        Method: Principal components condensation with varimax
        rotation
        Results: Two factors emerged: I proProtestant Ethic;
        II nonProtestant Ethic.
    Cohen, D. Differentiating motivations underlying vocational
        choice. Journal of Educational Research, 1971, 64, 229-
        234.
        Purpose: To identify dimensions underlying expressed
        motivation for curricular choice of male college
        students in two disciplines
        Subjects: Unspecified numbers of N.Y.U. male seniors
        majoring in education or business
        Variables: 1) 122, 2) 96, 3) unspecified number of, ratings
        from a 289-item questionnaire motivation list
        Method: Three unspecified factor analyses; the third in-
        cluded all items found to discriminate between
        groups
        Results: The third analysis extracted four factors:
        I "Profile" of prospective teachers; II "Profile" of
        prospective businessmen; III-IV Uninterpreted.
        Conclusions: On the basis of these results and those of
        discriminant analysis, 89 items were retained.
    Gable, R. K., & Pruzek, R. M. Super's Work Values Inventory:
        Two multivariate studies of inter item relationships.
        Journal of Experimental Education, 1971, 40, 41-50.
        Purpose: To identify dimensions underlying Super's Work
        Values Inventory Form A (WVI)
        Subjects: 200 tenth-graders
        Variables: 45 item scores from the WVI
        Method: Image analysis, with normal orthogonal varimax
        rotation
        Results: Ten factors, of the thirteen extracted, were
        interpreted: I Security-economic; II Supervisory
        relations; III Altruism; IV Achievement; V Esthetic-
        creative; VI Intellectual stimulation; VII Management
        prestige; VIII Variety; IX Independence; X Way of life.
        Conclusions: These results were compared with those from
        a Guttman analysis.
    Harrington, T. F., Lynch, M. D., & O'Shea, A. J. Factor
        analysis of twenty-seven similarly named scales of the
        Strong Vocational Interest Blank and the Kuder Occupa-
        tional Interest Survey, Form DD. Journal of Counseling
        Psychology, 1971, 18, 229-233.
        Purpose: To identify the factor structure of 27 similarly-
        named Strong Vocational Interest Blank (SVIB) and Kuder
        Occupational Interest Survey (OIS) scales

318

Subjects: 175 male college students
Variables: 1) 27 SVIB scale scores, 2) 27 OIS, 3) 54
    (combined)
Method: Principal components condensation with varimax
    rotation, independently by instrument and combined
Results: Four OIS factors emerged: I Technical and
    skilled; II Medical-scientific and social service; III
    Business and persuasive; IV Scientific and verbal, or
    femininity-masculinity, or good impression. Five SVIB
    factors emerged: I Physical science vs. social science;
    II Biological science or medical vs. business detail;
    III Persuasive and verbal, or use of language to in-
    fluence and persuade; IV Femininity vs. masculinity, or
    aesthetic vs. objective; V Technical interest vs. social
    service. Seven factors emerged in the combined analysis;
    the five SVIB factors (I-V) remained intact and two OIS
    factors also emerged: VI General; VII Personal service.
Kennedy, P. W., & Dreger, R. M. Development of criterion
    measures of overseas missionary performance. Journal
    of Applied Psychology, 1974, 59, 69-73.
Purpose: To identify dimensions underlying behavioral
    attributes relevant to the missionary enterprise as
    measured by the Missionary in Action checklist (MINA)
Subjects: 137 missionaries, rated by a total of 430 peers
    and self
Variables: 135 5-point MINA item scores
Method: Principal axes condensation with varimax and
    promax rotation
Results: Eleven factors emerged: I Person who is very
    understanding and accepting of people and ideas; II
    Person who is insensitive to events around him, has
    few, if any, close friends, and has few personal skills
    for coping with social and professional situations; III
    Individual's ability to organize his time and energy in
    carrying out his professional responsibilities; IV Open-
    ness to and acceptance of changes in people and social
    situations; V Individual's philosophy of life and way
    in which this philosophy affects his personal and
    professional activities; VI Leadership abilities; VII
    Commitment to Christ and efforts and abilities to share
    this faith with others; VIII Humility and dedication;
    IX Capacities to adjust to cultural demands and to re-
    late to people in cultures different from his own; X
    Concern about people with special needs, such as the
    poor, the blind, and the physically handicapped; XI
    Family and home relationships.
Lorr, M., & Suziedelis, A. A dimensional approach to the
    interests measured by the SVIB. Journal of Counseling
    Psychology, 1973, 20, 113-119.
Purpose: To identify dimensions underlying the SVIB
Subjects: 1 & 2) 488 men-in-general; 3) 488 men-in-general
Variables: 1) Scores on 100, 100, 100, 99-item subsets of
    the SVIB; 2) Scores on 63 "subtests" identified in (1);
    3) 198 items for second-order factoring

Method: Principal components condensation with varimax
    rotation, independently for each data set
Results: A total of 63 (unnamed) "subtests" were identified
    via the first four analyses. Fourteen first-order fac-
    tors emerged in the fifth analysis: I Scales and
    business management; II High-status professional; III
    Leading and directing; IV Extrovertive-competitive games;
    V Religious activities; VI Social welfare activities;
    VII Liberal nonconformist; VIII Art activities; IX Quan-
    titative science; X Mechanical activities; XI Outdoor
    work; XII Outdoor sports; XIII Military activities; XIV
    Risk and change. Five second-order factors emerged
    (second subsample, 198 items) in the final analysis:
    I People-related activities; II Mechanical and symbol
    manipulation; III Personal expression and the arts; IV
    Outdoor activities; V Risk-change (or sensation-seeking).
McLaughlin, G. W., & Butler, R. P. Perceived importance of
    various job characteristics by West Point graduates.
    Personnel Psychology, 1973, 26, 351-358.
  Purpose: To identify dimensions underlying West Point
      graduates' perceptions of the importance of job char-
      acteristics
  Subjects: 605 West Point graduates
  Variables: Scores on 31 need-fulfillment and extrinsic-
      intrinsic job characteristics items from a questionnaire
  Method: Principal axes condensation with varimax and pro-
      max rotation (hierarchical solution)
  Results: Six first-order factors emerged: I Self-esteem;
      II Safety/security; III Family and community involvement;
      IV Interpersonal relationships; V Army opportunities;
      VI Reputation. Two orthogonal second-order factors
      emerged: I Narrow or specific job aspects (everyday
      phenomena); II Concern for general aspects of work
      (growth, contribution, freedom-responsibility).
McLaughlin, G. W., & Butler, R. P. Use of weighting and data
    on alternative careers to predict career retention and
    commitment. Journal of Applied Psychology, 1974, 59,
    87-89.
  Purpose: To identify dimensions of job characteristics
      satisfaction for military careers (for military,
      civilian, and difference satisfaction data)
  Subjects: 470 U.S. Military Academy graduates
  Variables: Ratings on 1) 31 job characteristics in a mili-
      tary career; 2) 31 for military and 31 for civilian
      satisfaction; 3) military-civilian difference
  Method: Unspecified factor and cluster analysis, indepen-
      dently for three types of scores
  Results: Seven unlabeled clusters emerged for the classi-
      cal (first set of) scores and for difference scores;
      fifteen clusters emerged in the second analysis.
Manhardt, P. J. Job orientation of male and female college
    graduates in business. Personnel Psychology, 1972, 25,
    361-368.

Purpose: To identify dimensions of job orientation in
college graduates in business
Subjects: 569 college graduates (since 1966) in business
Variables: 5-point importance ratings for 25 job characteristics
Method: Principal components condensation with varimax
rotation
Results: Three factors emerged: I Long-range career objectives (higher for males); II Comfortable working
environment and pleasant interpersonal relations (higher
for females); III Autonomy and self-actualization
(negligible sex difference).

Nafziger, D. H., & Helms, S. T. Cluster analyses of interest
inventory scales as tests of Holland's occupational
classification. Journal of Applied Psychology, 1974,
59, 344-353.
Purpose: To identify occupation dimensions underlying the
Strong Vocational Interest Blank (SVIB), Minnesota
Vocational Interest Inventory (MVII), and Kuder Occu-
pational Interest Survey (OIS)
Subjects: Samples reported in the manuals: SVIB (Campbell,
1971), MVII (Clark & Campbell, 1965), OIS (Kuder, 1971)
Variables: 72 SVIB, 21 MVII, 23 (males) or 21 (females)
OIS scores
Method: Cluster analysis, independently for the three
instruments
Results: Two main clusters each emerged for SVIB (I Pro-
fessional occupations; II Residual occupations), MVII
(I Oriented toward people; II Oriented toward objects),
and OIS males (I Object, data, and thing oriented; II
Science, social, and business oriented) and females
(I Science, social, and business oriented; II "Feminine"
occupations). Subclusters were interpreted in Holland's
terms.

Parnicky, J. J., Kahn, H., & Burdett, A. B. Standardization
of the VISA (Vocational Interest and Sophistication
Assessment) technique. American Journal of Mental
Deficiency, 1971, 75, 442-448.
Purpose: To identify dimensions underlying the VISA, a
reading-free test of vocational interest and sophistication (and compare with pilot data)
Subjects: 1,686 test and 1,021 retest males, and 1,421
test and 973 retest females, all retardates trained (for
work placement) at institutions, schools, or workshops
Variables: 75 interest item scores for males and 53 for
females - test and retest data
Method: Unspecified condensation with varimax rotation,
independently for test and retest data by sex and location of training
Results: Seven factors emerged for all analyses for males,
and were similar to pilot results: I Farm/grounds; II
Food service; III Garage; IV Industry; V Laundry; VI
Maintenance; VII Materials handling; Four factors
emerged for all analyses for females: I Business/clerical;

II Food service; III Housekeeping; IV Laundry.
Robey, D. Task design, work values and worker response: An
     experimental test. Organizational Behavior and Human
     Performance, 1974, 12, 264-273.
     Purpose: To identify dimensions underlying preferences for
     job aspects
     Subjects: 126 college students
     Variables: 14 item scores from Friedlander's (1965) job
     aspects measure
     Method: Unspecified condensation with orthogonal varimax
     rotation
     Results: Three factors emerged: I Concern for extrinsic
     job aspects; II Concern for intrinsic job aspects; III
     Concern for advancement.
     Conclusions: The present factor structure was similar to
     Friedlander's. Scores on I and II were used in a further
     experiment with a subsample of these subjects.
Stewart, L. H. Relationships between interests and personality
     scores of occupation-oriented students. Journal of
     Counseling Psychology, 1971, 18, 31-38.
     Purpose: To identify clusters of occupation-oriented students
     on the bases of their 1) Interest Assessment Scale (IAS)
     and 2) Omnibus Personality Inventory (OPI) profiles
     Subjects: 2,458 junior college students enrolled in 43
     trade and technical curricula
     Variables: 1) 8 IAS scale scores, 2) 7 OPI scale scores
     Method: Key cluster analysis, independently by instrument
     Results: Five clusters emerged from IAS data, six from OPI
     data; all were unlabeled.
Summers, G. F., Burke, M., Saltiel, S., & Clark, J. P. Sta-
     bility of the structure of work orientations among high
     school students. Multivariate Behavioral Research, 1971,
     6, 35-50.
     Purpose: To identify dimensions underlying the Work Com-
     ponents Study (WCS) with high school students
     Subjects: 359 male and 364 female students in two 4-year
     rural high schools
     Variables: 7 WCS scale scores
     Method: Cluster analysis, by sex
     Results: Seven clusters emerged: I Potential for personal
     challenge and development; II Responsiveness to new
     demands; III Competitive desirability; IV Tolerance for
     work pressure; V Conservative security; VI Willingness
     to seek reward in spite of uncertainty vs. avoidance of
     uncertainty; VII Surround concern. Subsequent analyses
     by grade revealed essentially the same structure, al-
     though Factors I and III had some tendency to collapse.
Wollack, S., Goodale, J. G., Wijting, J. P., & Smith, P. C.
     Development of the Survey of Work Values. Journal of
     Applied Psychology, 1971, 55, 331-338.
     Purpose: To identify the factorial structure of the Survey
     of Work Values (SWV); data reported by Goodale, 1969
     (unpublished manuscript)

Subjects: 495 employees from 7 occupational groups
Variables: 67 SWV attitude item scores
Method: Principal components condensation with quartimin rotation
Results: Six factors emerged: I Intrinsic values; II Organization-man ethic; III Upward striving; IV Social status of job; V Conventional ethic; VI Attitude toward earnings.

Wood, D. A. Effect of worker orientation differences on job attitude correlates. Journal of Applied Psychology, 1974, 59, 54-60.
Purpose: To identify dimensions underlying job involvement (JI)
Subjects: Unspecified number of incumbent semiskilled and skilled male papermakers
Variables: Scores on 20 JI items
Method: Principal components condensation apparently without rotation
Results: Five factors emerged: I Work attraction; II Failure sensitization; III Work commitment; IV Job preeminence; V Work identification.

B. Selection and Placement

Alper, S. W. Racial differences in job and work environment priorities among newly hired college graduates. Journal of Applied Psychology, 1975, 60, 132-134.
Purpose: To identify dimensions underlying job/work environment priorities among newly hired college graduates
Subjects: 70 black and 179 white newly hired college graduates
Variables: Importance ratings (relative to accepting employment) for 22 job and company items
Method: Principal factors condensation with varimax rotation
Results: Two factors emerged: I Future or growth orientation of the individual; II Hygienic or extrinsic factors.
Conclusions: Factor scores were examined as a function of race.

Arvey, R. D., & Mussio, S. J. Determining the existence of unfair test discrimination for female clerical workers. Personnel Psychology, 1973, 26, 559-568.
Purpose: To identify dimensions underlying biographical and background variables for female clerical workers
Subjects: 266 female clerical (civil service) workers
Variables: Age, income, father's education and occupation, number of jobs father held, size of city,where early life was spent, education, high school standing, and Environmental Participation Index score
Method: Principal components condensation with varimax rotation
Results: Three factors emerged: I Cultural, or socioeconomic; II Age; III Uninterpretable.
Conclusions: Scores on Factor I were used to define culturally advantaged and disadvantaged subsamples.

Cecil, E. A., Paul, R. J., & Olins, R. A. Perceived importance of selected variables used to evaluate male and female job applicants. Personnel Psychology, 1973, 26, 397-404.
   Purpose: To identify perceived importance dimensions in evaluations of male and female job applicants
   Subjects: 118 graduate and undergraduate management students, 58 rating a "female" and 60 a "male" applicant
   Variables: Perceived importance ratings on 50 attributes which could be used to evaluate a job applicant
   Method: Unspecified condensation with unspecified rotation
   Results: Four factors emerged: I Motivation - ability; II Personality - appearance; III Interpersonal relations; IV Skills - education.
   Conclusions: Factors I and III were found to relate to male applicants, Factors II and IV to females.
Mayer, S. E., & Bell, A. I. Sexism in ratings of personality traits. Personnel Psychology, 1975, 28, 239-249.
   Purpose: To identify dimensions underlying males' and females' ratings of (fictitious) male and female job applicants
   Subjects: 75 male and 75 female college students, each rating 3 male and 3 female "job applicants"
   Variables: Scores on 26 Checklist for Describing Job Applicants items
   Method: Principal axes condensation with orthogonal rotation, independently for 4 (sex of rater x sex of ratee) matrices
   Results: Four factors emerged for females rating both females and males, whereas six emerged for both male rater analyses. The factors were discussed but not named (except an independent Masculinity-femininity factor for each analysis).
Rosenbaum, B. L. Attitude toward invasion of privacy in the personnel selection process and job applicant demographic and personality correlates. Journal of Applied Psychology, 1973, 58, 333-338.
   Purpose: To identify dimensions of job applicants' attitude toward invasion of privacy in the personnel selection process
   Subjects: 1,392 job applicants (mostly for sales or middle management positions)
   Variables: Scores on 66 items from an invasion of privacy questionnaire
   Method: Principal components condensation with varimax rotation
   Results: Five factors emerged: I Family and background influences; II Personal history data; III Interests and values; IV Financial management data; IV Social adjustment.
C. Training
Deppe, A. H. Performance criteria in diver training. Perceptual and Motor Skills, 1971, 32, 718.

Purpose: To identify dimensions underlying behavioral
    elements assessed in diver training
Subjects: 71 male Scuba diver trainees
Variables: Scores on 29 behavioral (diving) elements
Method: Principal factors condensation with varimax
    rotation
Results: Four factors emerged: I Diving skill; II Intel-
    lectual ability; III Task-orientation; IV Emotional
    maturity.

Gordon, M. E., & Cohen, S. L. Training behavior as a pre-
    dictor of trainability. Personnel Psychology, 1973,
    26, 261-272.
Purpose: To identify dimensions underlying 11 learning
    (welding) tasks
Subjects: 58 trainees in a welding program
Variables: Time to completion scores on 11 learning tasks
    in the welding program
Method: Principal axes condensation without rotation
Results: Three factors emerged: I Complicated joining
    operations in overhead position vs. bead weld opera-
    tions; II-III Not interpreted.

Harris, J. G. Jr. Prediction of success on a distant Pacific
    island: Peace Corps style. Journal of Consulting and
    Clinical Psychology, 1972, 38, 181-190.
Purpose: To identify dimensions underlying final board and
    staging ratings for Peace Corps Volunteers
Subject: 1) 44, 2) 98, 3) 145 Peace Corps Volunteers
Variables: Scores on 1) 10 final board ratings, 2) 7
    staging ratings, 3) 6 final board ratings
Method: Principal axes condensation with varimax rotation,
    independently for each S-variable set
Results: Three factors emerged in the first analysis: I
    Final board personality; II Staging interview; III
    Language competence. Two factors, identical to I and
    II above, emerged in the second analysis; the third
    yielded two factors like I and III above.

Kraut, A. I. Prediction of managerial success by peer and
    training-staff ratings. Journal of Applied Psychology,
    1975, 60, 14-19.
Purpose: To identify dimensions underlying peer-rated
    characteristics for middle-level managers-in-training
    and executives-in-training
Subjects: 185 middle-level managers and 99 higher-level
    executives, at conclusion of one-month training course
Variables: Peer-ratings on 13 characteristics (related to
    later success)
Method: Principal components condensation with varimax
    rotation, independently for managers and executives
Results: Two factors emerged in both analyses: I Impact;
    II Tactfulness.

Mencke, R. A., & Cochran, D. J. Impact of a counseling out-
    reach workshop on vocational development. Journal of
    Counseling Psychology, 1974, 21, 185-190.

325

Purpose: To identify dimensions underlying attitudinal
    measures related to vocational development
Subjects: 141 male college students
Variables: Unspecified number of attitude-to-vocational-
    development item scores
Method: Unspecified condensation with orthogonal rotation
Results: An unspecified number of uninterpreted factors
    emerged; they were of little relevance to the experi-
    mental hypotheses.
Stone, L. A., Bassett, G. R., Brosseau, J. D., deMers, J. L.,
    & Stiening, J. A. Training staff's multidimensional
    perceptions of a class of Medex (physician's extension)
    trainees: A method of grading. Perceptual and Motor
    Skills, 1973, 36, 395-402.
Purpose: To identify dimensions of judged similarity
    among MEDEX trainees
Subjects: 6 MEDEX training staff members
Variables: Similarity judgments for 19 MEDEX trainees,
    during months 1 and 15 of training
Method: 1) Inverse principal components condensation
    without rotation, by time; 2) principal components
    condensation with varimax rotation, by time
Results: For Month 1, two judge factors emerged, although
    most judges had high loadings on the first factor; for
    Month 15, one factor emerged, defined by all judges. For
    the second set of analyses, six Month 1 factors emerged:
    I Personality characteristics (bipolar); II "Early" per-
    ception of those entering the program with a "high degree
    of medical expertise vs. coming in with minimal medical
    background;" III Solid citizens; IV Mode of dress; V
    American Indian trainee; VI Youthfulness. Four factors
    emerged for Month 15: I Quality (bipolar); II Verbosity
    contrast; III Conventional vs. unconventional personal
    style; IV Age.
Stone, L. A., & Brosseau, J. D. Cross-validation of a system
    for predicting training success of Medex trainees.
    Psychological Reports, 1973, 33, 917-918.
Purpose: To identify dimensions underlying judged similar-
    ity of MEDEX trainees
Subjects: 4 MEDEX training program staff members
Variables: Similarity ratings of unspecified number of
    MEDEX trainees
Method: Principal components condensation with orthogonal
    rotation
Results: An unspecified number of factors emerged; Factor
    I (Quality of trainee within the program) accounted for
    25% of the variance after rotation.
Stone, L. A., & Brosseau, J. D. Another utilization of MESA
    in the evaluation-grading of a class of Medex trainees.
    Perceptual and Motor Skills, 1974, 38, 275-278.
Purpose: To identify dimensions underlying judged simi-
    larity of MEDEX trainees

326

Subjects: 4 MEDEX training staff members, half-way through training

Variables: Similarity judgments of 26 MEDEX trainees

Method: 1) Inverse, 2) principal components condensation with varimax rotation

Results: A single judge dimension emerged. Six trainee factors emerged: I Quality of trainee in program; II Male Indianness; III Femininity exhibited by male stimuli; IV Person obnoxiousness; V Low profileness; VI Having unusual training sites.

D. Task and Work Analysis

Andrulis, R. S. Construct validation of a standardized achievement test. Educational and Psychological Measurement, 1973, 33, 499-503.

Purpose: To identify dimensions underlying an achievement test for life insurance personnel

Subjects: 5,834 students in the chartered Life Underwriters program

Variables: Scores on 100 items of Test 1 for Course Area I

Method: Principal components condensation with varimax rotation

Results: Five factors, of twenty-four extracted, were identifiable: I Mathematical definitions; II Calculation; III Problem-solving; IV Comprehension of insurance definitions; V Comprehension and application of insurance principles within a mathematics domain.

Einhorn, H. J. Expert judgment: Some necessary conditions and an example. Journal of Applied Psychology, 1974, 59, 562-571.

Purpose: To identify patterns (of histological cues reported) of judges' (pathologists') expert judgment

Subjects: 193 biopsy slides

Variables: Ratings on 9 histological characteristics plus global severity, made by 3 medical pathologists

Method: Unspecified condensation with varimax rotation, by rater

Results: Four unlabeled factors emerged in each analysis.

London, M., & Kumoski, R. J. A study of perceived job complexity. Personnel Psychology, 1975, 28, 45-56.

Purpose: To identify dimensions of RNs' perceived job complexity (PJC)

Subjects: 153 female RNs

Variables: 33 PJC item scores

Method: Hierarchical factor analysis (principal factors condensation, minres correction, varimax and hierarchical rotation)

Results: Two secondary (I Task demands and situational constraints; II Control and authority) and one general factor (III Overall job complexity) emerged.

McCormick, E. J., Jeanneret, P. R., & Mecham, R. C. A study of job characteristics and job dimensions as based on the Position Analysis Questionnaire (PAQ). Journal of Applied Psychology, 1972, 56, 347-368.

Purpose: To identify dimensions of job elements and job
    attributes as assessed by the Position Analysis Ques-
    tionnaire (PAQ)
Subjects: 1) 100, 268, 268, 536 jobs, 2) 67 job attributes,
    all rated by 62 rater pairs
Variables: Scores on 178 PAQ job elements
Method: Principal components condensation with unspecified
    rotation, by job subsamples and for job attributes
Results: Twenty-seven job data factors could be grouped
    under five dimensions: Information input (I Visual in-
    put from devices/materials; II Perceptual interpretation;
    III Information from people; IV Visual input from distal
    sources; V Evaluation of information from physical
    sources; VI Environmental awareness; VII Awareness of
    body movement/posture), Mediation processes (VIII Deci-
    sion making; IX Information processing), Work output
    (X Machine/process control; XI Manual control/coordination
    activities; XII Control/equipment operation; XIII General
    body activity; XIV Handling/manipulating activities; XV
    Use of finger-controlled devices vs. physical work; XVI
    Skilled/technical activities), Interpersonal activities
    (XVII Communication of decisions/judgments; XVIII Job-
    related information exchange; XIX Staff-related activities;
    XX Supervisor-subordinate relationships; XXI Public-
    related contact), Work situation and job context (XXII
    Unpleasant/hazardous physical environment; XXIII Personally
    demanding situations), Miscellaneous aspects (XXIV Busi-
    nesslike work situations; XXV Attentive/discriminating
    work demands; XXVI Unstructured vs. structured work;
    XXVII Variable vs. regular work schedule). Twenty-one
    job attribute factors could be grouped under six cate-
    gories: Information input (I Visual input from devices/
    materials; II Perceptual input from processes/events;
    III Evaluation of visual input; IV Nonvisual input; V
    Physical/environmental awareness; VI Verbal/auditory
    input/interpretation), Mediation process (VII Use of
    job-related knowledge; VIII Information processing),
    Work output (IX Manual control/coordination activities;
    X Control/equipment operation; XI General body/handling
    activities), Interpersonal activities (XII Interpersonal
    communications; XIII Serving/entertaining/ XIV Signal/
    code communications), Work situation and job context
    (XV Unpleasant/hazardous physical environment; XVI Per-
    sonally demanding situations), Miscellaneous aspects
    (XVII Attentive/discriminating work demands; XVIII Un-
    structured/responsible work activities; XIX Paced/regular
    work activities; XX Businesslike work situations; XXI
    Merit income).
Mobley, W. H., & Ramsay, R. S. Hierarchical clustering on the
    basis of inter-job similarity as a tool in validity
    generalization. Personnel Psychology, 1973, 26, 213-225.
    Purpose: To identify job similarity clusters

Subjects: 126 hourly-rated entry, production, and maintenance jobs in two chemical plants, rated by experienced
supervisory personnel
Variables: Ratings of 5 job characteristics and 15 human
attributes potentially important to successful performance
Method: Principal components condensation with varimax
rotation
Results: Four factors emerged: I Mechanical-dexterity;
II Reasoning-problem solving; III Semantic-communicative;
IV Responsibility-complexity.
Pfeiffer, M. G., Kuennapas, T., & Fastiggi, C. F. Common
elements approach to multidimensional similarity analysis
among job tasks. Perceptual and Motor Skills, 1973, 36,
3-12.
Purpose: To identify intellective complexity dimensions
underlying classroom tasks of college professors
Subjects: 11 psychology-student raters
Variables: Intellective complexity scale scores for 30
classroom tasks
Method: Principal components condensation with varimax
rotation
Results: Eight unlabeled factors emerged; these results
were compared with those from previous studies.
Schneider, B., & Hall, D. T. Toward specifying the concept
of work climate: A study of Roman Catholic diocesan
priests. Journal of Applied Psychology, 1972, 56, 447-
455.
Purpose: To identify dimensions of Roman Catholic diocesan
priests' work activities and climate
Subjects: 97 (subsample), and 373 Roman Catholic diocesan
priests
Variables: Scores on 1) 35 work activities items, 2) 35
work climate items
Method: Principal components condensation with varimax
rotation, independently for activities and climate,
first with N=97 followed by N=373
Results: Four activities factors emerged: I Parochial;
II Administration; III Community involvement; IV Personal
development. Four climate factors emerged: I Superior
effectiveness; II Work challenge and meaning; III Personal acceptance; IV Supportive autonomy.
Sheposh, J. P., Abrams, A. J., & Licht, M. H. Assessment by
technical personnel of the role of the change advocate.
Journal of Applied Psychology, 1975, 60, 483-490.
Purpose: To identify dimensions of attitudes toward a
change advocate ("system specialist") among Naval personnel
Subjects: 83 technicians and 9 officers from various USN
ship types
Variables: 15-point importance ratings for 25 traits
relevant to a change advocate
Method: Unspecified condensation with varimax rotation

Results: Five factors emerged: I Credible communicator;
II Constructive team member; III Realistic perspective;
IV Interpersonally adept; V Competent technician.
Stone, L. A., & Bassett, G. R. The Medex medical occupation
concept in subjective medical occupation-profession
multidimensional space. Psychological Reports, 1972,
31, 167-174.
Purpose: To identify dimensions underlying subjective
medical occupation-professions similarities
Subjects: 13 MEDEX training staff members
Variables: Similarity ratings of 16 medical occupation-
professions
Method: Principal components condensation with orthogonal
rotation
Results: Five factors emerged: I Amount of training -
degree of responsibility - prestige; II Therapist; III
Nurse; IV Technician; V Hospital nursing staff authority.
E. Performance and Job Satisfaction
Aiken, W. J., Smits, S. J., & Lollar, D. J. Leadership behav-
ior and job satisfaction in State rehabilitation
agencies. Personnel Psychology, 1972, 25, 65-73.
Purpose: To identify dimensions of leadership behavior
and job satisfaction perceived as operative by rehabilitation counselors
Subjects: 230 rehabilitation counselors from 31 state
agencies
Variables: Scores on 4 Leadership Behavior Description
Questionnaire subscales, 4 Relationship Inventory subscales, 8 Job Satisfaction Inventory subareas
Method: Principal components condensation with normalized
varimax rotation
Results: Eight factors emerged: I Interpersonal aspects
of the work environment; II Involvement aspects of the
job; III Leader directed aspects of the job; IV Financial
reward aspects of the job; V Exertive aspects of the job;
VI Autonomous functioning aspects of the job; VII-VIII
Not interpreted.
Barth, R. T., & Vertinsky, I. The effect of goal orientation
and information environment on research performance: A
field study. Organizational Behavior and Human Performance, 1975, 13, 110-132.
Purpose: To identify dimensions of work-related goal orien-
tation
Subjects: 89 scientists (limnologists)
Variables: 18 item scores from the goal-orientation section
of a questionnaire
Method: Unspecified condensation with oblimin rotation
Results: Four factors emerged: I Leader-administration;
II Entrepreneur; III Scientist-basic; IV Scientist-
applied.
Conclusions: Relationships of scores on these factors with
several other variables were also reported.

Campbell, J. P., Dunnette, M. D., Arvey, R. D., & Hellervik, L. V. The development and evaluation of behaviorally based rating scales. Journal of Applied Psychology, 1973, 57, 15-22.
Purpose: To identify dimensions underlying behaviorally based rating of job performance
Subjects: 537 department managers, each rated by store manager and assistant manager
Variables: Summated ratings, and scaled expectations; of performance on 9 behavior dimensions (scale scores)
Method: Principal factors condensation with varimax rotation, independently for each rater x method matrix
Results: Nine factors, each defined by one variable, emerged in each analysis, although VI-IX were generally only weakly defined.
Dickinson, T. L., & Tice, T. E. A multitrait-multimethod analysis of scales developed by retranslation. Organizational Behavior and Human Performance, 1973, 9, 421-438.
Purpose: To identify dimensions of 1) ratings of fire-fighters, and 2) ratings, tenure, and absenteeism
Subjects: 149 firefighters
Variables: 1) Scores on skill at getting along with others, dedication, and ability to apply learning, derived from each of three methods (peer nominations, peer checklist rating, supervisor checklist rating, 2) plus tenure and absenteeism
Method: Restricted maximum-likelihood condensation
Results: Seven ratings factors emerged: I General ability or reputation; II-III Extent of ratee ordering based on retranslation dimensions; IV-VII Extent ratees were ordered consistently by the methods. Seven factors also emerged when tenure and absenteeism were included.
Friedlander, F. Performance and orientation structures of research scientists. Organizational Behavior and Human Performance, 1971, 6, 169-183.
Purpose: To identify dimensions of scientists' research success
Subjects: 178 full-time research scientists from the six largest Naval R&D labs
Variables: Scores on 13 measures of productive output, quality/characteristics of research, and length/quality of relationship with organization, plus age
Method: Principal components condensation with varimax rotation
Results: Three factors emerged: I Research orientation; II Professional orientation; III Local organization.
Conclusions: Differences on Factors I and II were found between six disciplinary subgroups.
Gavin, J. F., & Ewen, R. B. Racial differences in job attitudes and performance: Some theoretical considerations and empirical findings. Personnel Psychology, 1974, 27, 455-464.

Purpose: To identify job satisfaction dimensions for
blacks and whites
Subjects: 81 black and 390 white, male blue-collar workers
from 14 major cities; data collected and factor analysis
reported by Katzell, et al. (1970, unpublished manuscript.)
Variables: 53 job satisfaction item scores
Method: Principal axes condensation with varimax rotation
Results: Five interpretable factors emerged: I Advance-
ment; II Job and company; III Supervision; IV Coopera-
tion between coworkers and supervisors; V Pay and
working conditions.
Gluskinos, U., & Brennan, T. F. Selection and evaluation pro-
cedure for operating room personnel. Journal of Applied
Psychology, 1971, 55, 165-169.
Purpose: To identify dimensions of evaluation (performance)
ratings for operating room (OR) personnel
Subjects: 163 OR personnel, rated by their supervisors
Variables: Scores on 42 behavioral (OR) performance items
Method: Principal components condensation, apparently
without rotation
Results: Four factors emerged: I General technical knowl-
edge; II Attention to detail; III Patient awareness;
IV Social interaction.
Goodale, J. G., & Aagaard, A. K. Factors relating to varying
reactions to the 4-day workweek. Journal of Applied
Psychology, 1975, 60, 33-38.
Purpose: To identify dimensions underlying a questionnaire
designed to tap job-related changes, leisure-related
changes, and general satisfaction with the 4-day workweek
Subjects: 434 clerical and 40 supervisory accounting
division personnel (approximately 90% of whom had
worked for the company prior to the change to 4-day
workweek 1 year before)
Variables: 63 item scores from a questionnaire dealing
with job- and leisure-related changes and general
satisfaction with the 4-day workweek
Method: Principal components condensation with varimax
rotation
Results: Four factors emerged: I General satisfaction/
good life; II Activity changes; III Work changes; IV
Financial complications.
Guion, R. M., & Elbert, A. J. Factor analyses of work-
relevant need statements in two populations. Multivariate
Behavioral Research, 1973, 8, 41-62.
Purpose: To identify dimensions underlying a set of work-
relevant need statements for two populations
Subjects: 1) 100 college students anticipating future em-
ployment, 2) 105 employed college graduates, all males
Variables: Scores on 105 work-relevant need statements
Method: Principal axes condensation with varimax rotation,
by sample

Results: Eleven student factors emerged: I Physical working conditions vs. responsibility; II Need for advancement; III Need for maintenance of moral values; IV Need
to be in authority; V Need for security; VI Creativity
needs; VII Job content; VIII Job context; IX Need for
consideration; X Need for precision; XI Not interpreted.
Twelve factors emerged for graduates, including II-VI
above; others were as follows: I Need for recognition;
VII Need for activity; VIII Need for good working conditions; IX Need for autonomy; X Need for reward; XI
Need for social status; XII Need to fulfill responsibility.

Guion, R. M., & Landy, F. J. The meaning of work and the
motivation to work. Organizational Behavior and Human
Performance, 1972, 7, 308-339.

Study One:
Purpose: To identify clusters of semantic differential
(SD) ratings related to the meaning of work
Subjects: 1,426 employed persons total, from unspecified
number of occupational and organizational categories
Variables: 31 SD ratings related to the meaning of work,
plus unspecified number of SD ratings of management
and spare time
Method: Unspecified number of cluster analyses; by occu-
pational and organizational categories
Results: Three clusters emerged persistently across
analyses: I Stimulus value; II Autonomy; III Structure.

Study Two:
Purpose: To identify dimensions of need strengths and need
satisfactions for graduating seniors and employed college
graduates, respectively (data collected by Elbert, unpublished M.A. thesis, 1967)
Subjects: Unspecified numbers of 1) graduating college
seniors, and 2) employed college graduates
Variables: Scores on 105 job items responded to in terms
of 1) need strengths, or 2) need satisfactions
Method: Unspecified factor analysis, independently for
senior-strengths and employed-satisfactions
Results: Seven common factors emerged: I Imaginative,
unique ways of doing things; II Something to do most of
time; III Tell people what to do; IV Religious beliefs
not interfered with; V Chance to advance; VI Proper
recognition; VII Adequate working conditions.
Conclusions: The results from both these studies were considered in designing the main study reported in this
paper.

Herman, J. B., & Hulin, C. L. Managerial satisfactions and
organizational roles: An investigation of Porter's
need deficiency scales. Journal of Applied Psychology,
1973, 57, 118-124.
Purpose: To identify dimensions underlying the Porter
Need Satisfaction Questionnaire (PNSQ)

Subjects: 1) 174 supervisory personnel (4 levels) from
a manufacturing plant; 2) unspecified number of hospital personnel (Sonsoni & Johnson, unpublished paper);
3) unspecified number of government agency personnel;
4) unspecified number of personnel from a large retail
corporation
Variables: Raw scores, and need-, have-, and importance-
weighted scores, on 13 PNSQ need deficiency items
Method: Principal axes condensation with varimax and
oblimax rotation, independently for raw and weighted
scores within each sample
Results: In all cases, a one-factor solution was most
appropriate.
Hundal, P. S. A study of entrepreneurial motivation: Comparison of fast- and slow-progressing small-scale
industrial entrepreneurs in Punjab, India. Journal of
Applied Psychology, 1971, 55, 317-323.
Purpose: To identify dimensions of job-related attitudes
and motivation for slow- and fast-progressing small-
scale Indian entrepreneurs
Subjects: 92 fast-progress and 92-slow-progress small-
scale industrial entrepreneurs in Ludhiana, India
Variables: Scores on achievement motivation, job satisfaction, job prestige, value patterns, aspirations,
interests, attitudes toward money and labor unions
(12 scores)
Method: Principal axes condensation with varimax rotation
by group
Results: Four factors emerged in both analyses: I
Achievement motivation; II Self-image; III Living well;
IV Quasi-specific (different for fast and slow groups).
Inn, A., Hulin, C. L., & Tucker, L. Three sources of criterion
variance: Static dimensionality, dynamic dimensionality,
and individual dimensionality. Organizational Behavior
and Human Performance, 1972, 8, 58-83.
Purpose: To identify static, dynamic, and individual
dimensionality of criterion variance
Subjects: 184 reservations agents from a domestic airline
Variables: Scores on 11 performance measures (e.g., amount
of revenue of telephone calls per shift), collected for
each of 5 consecutive months
Method: Three-mode factor condensation with varimax rotation (static: 920x11; dynamic: 2024x5; individual:
184x55 matrix)
Results: Three static factors emerged: I Speed and accuracy; II Customer rapport; III Sales ability. Three
time (dynamic) factors emerged: I Steady performance
level over time; II Performance increase over time;
III Little variance months 1-4 but gain during last
month. Four individual factors emerged: I Level of
performance; II Increased customer rapport and decreased
sales; III Temporary decrease in sales with increased
rapport; IV Sales ability.

Johannesson, R. E. Some problems in the measurement of organizational climate. Organizational Behavior and Human Performance, 1973, 10, 118-144.
Purpose: To identify dimensions common to work attitude and organizational climate measures
Subjects: 499 employees of one company (two locations)
Variables: 1) 78 item scores from the SRA Employee Inventory, 90 perceptual organizational climate (OC) item scores, and 71 JDI item scores; 2) 8 SRA, 6 OC, and 8 JDI cluster scores
Method: Cumulative communality cluster analysis, independently for each set of item scores and the combined cluster scores
Results: Eight SRA clusters emerged: I Management interest; II Working conditions; III Employee benefits; IV Supervision; V People I work with; VI Work itself; VII Pay; VIII Supervision-interpersonal. Six OC clusters emerged: I Degree of organization; II Pressure; III Rewards; IV Friendly team spirit; V Rewards - management criticism; VI Pay. Eight JDI clusters emerged: I Promotions; II Supervision - interpersonal; III Work - intrinsic; IV Pay; V Co-workers - technical competence; VI Supervision - technical competence; VII Work - extrinsic; VIII Co-workers - interpersonal. Five clusters emerged from the analysis of cluster scores: I Pay; II Work content; III Supervision; IV Work context; V People.
Conclusions: There appears to be some overlap between measures of work attitudes and those of organizational climate.
Karson, S., & O'Dell, J. W. Performance ratings and personality factors in radar controllers. Journal of Clinical Psychology, 1971, 27, 339-342.
Purpose: To identify relationships between the 16PF and radar controllers' performance ratings
Subjects: 264 radar controllers
Variables: Scores on the usual 16 16PF scales plus a "Motivational Distortion" scale, and Career Potential and Performance scales from the FAA's Employee Appraisal Record for Nonsupervisory Employees form (EAR)
Results: Eight factors emerged: I Anxiety vs. dynamic integration; II Subduedness vs. independence; III EAR; IV Pathemia vs. cortertia, or cortical alertness; V Remainder of invia vs. axvia; VI Intelligence; VII Superego, or obsessiveness-complusiveness vs. sociopathic deviancy; VIII Rebelliousness.
Conclusions: The two criterion measures loaded on a virtually private factor.
Kesselman, G. A., Hagen, E. L., & Wherry, R. J. Sr. A factor analytic test of the Porter-Lawler expectancy model of work motivation. Personnel Psychology, 1974, 27, 569-579.
Purpose: To identify dimensions of work motivation
Subjects: 76 female telephone company employees

Variables: Scores on 4 expectancies and role measures,
4 effort and performance measures, 6 satisfaction
measures
Method: Principal factors and minimum residual condensation with hierarchical varimax rotation
Results: A general factor (I General), three group factors
(II Expectancy-behavior; III Behavior-satisfaction; IV
Satisfaction-expectancy), and two subfactors (composing
first group factor: V Hourly worker modification; VI
Piece rate modification) emerged.
Klimoski, R. J., & London, M. Role of the rater in perfor-
mance appraisal. Journal of Applied Psychology, 1974,
59, 445-451.
Purpose: To identify dimensions of supervisory, peer,
and self rating of performance for RNs
Subjects: 153 RNs
Variables: Scores on 18 individual characteristics, total
rating, and overall effectiveness rating, each for self-,
peer-, and supervisor rating (i.e., 60x60 matrix)
Method: Hierarchical factor analysis (multiple group condensation with varimax hierarchical rotation)
Results: One general (I General) and five subgeneral factors (II Self-bias; III Supervisor bias; IV Peer bias;
V Job competence; VI Effort) emerged.
Landy, F. J. Motivational type and the satisfaction-perfor-
mance relationship. Journal of Applied Psychology, 1971,
55, 406-413.
Purpose: To identify 1) dimensions of a satisfaction in-
ventory (SI), 2) dimensions of peer-rated motivation
for engineers, and 3) types or clusters of engineers
Subjects: 175 engineers
Variables: Scores on 1) 21 7-point SI items, 2) peer
ratings on 7 motivation-to-work dimensions, 3) motivation dimensions (high, moderate, low standard score)
Method: 1) key cluster analysis, 2) principal factors
condensation with quartimax and varimax rotation, 3)
O-type cluster analysis
Results: Five SI clusters emerged: I Advancement; II
Ethical principles; III Creativity; IV Pay; V Working
conditions. Three motivation factors emerged: I Professional identification; II Team attitude; III Task
concentration. Thirteen engineer clusters, accounting
for 171 Ss, emerged but were not labeled.
Landy, F. J. A procedure for occupational clustering.
Organizational Behavior and Human Performance, 1972,
8, 109-117.
Purpose: To identify dimensions of "My job" perceptions
for a heterogeneous group of employees, and to identify
clusters of employee groups
Subjects: 1,346 employees from 46 organizations, who could
be subgrouped into nine occupational groupings and five
organizational types
Variables: 1) 31 semantic differential ratings for the
concept "My Job," and 2) similarity matrix (by occupation)

Method: 1) Cluster analysis (V-analysis), independently
by group/type, and 2) linkage analysis and graphic
cluster analysis
Results: Three clusters were common to all group/types:
I Stimulus value; II Structure imposed by work; III
Degree of autonomy. Three group clusters and two
isolate groups emerged in the similarity matrix analysis:
I Supervisors and administrators, teachers and professionals; II Tellers and nurses; III Technicians and
clerical; IV Executives; V Semiskilled workers.
Mayfield, E. C. Value of peer nominations in predicting life
insurance sales performance. Journal of Applied Psy-
chology, 1972, 56, 319-323.
Purpose: To identify dimensions of peer-nominated predic-
tions of life insurance sales performance
Subjects: 117 inexperienced life insurance agents, at the
end of a 3-week training course
Variables: Peer nomination scores on 12 work- and socially-
oriented items
Method: Principal components condensation with varimax
rotation
Results: Three factors emerged: I Technical skill; II
General sociability; III Ability to form close personal
relationships.
McCarrey, M. W., & Edwards, S. A. Organizational climate
conditions for effective research scientist role performance. Organizational Behavior and Human Performance,
1973, 9, 439-459.
Purpose: To identify dimensions 1) underlying research
scientist role performance, 2) underlying role perfor-
mance-experience, and 3) of organizational climate related
to role performance
Subjects: 72 biological scientists employed in 4 Canadian
federal labs
Variables: 1) 9 performance measure scores; 2) 9 performance
plus 4 experience criterion measure scores; 3) scores on
subsets of 131 organizational climate items related to
one of three performance components identified in (1)
Method: Principal components condensation with orthogonal
rotation, independently for each set of scores
Results: Three factors emerged in the role performance
only analysis: I Creative productivity; II Quality
and originality of published work; III Impact of published work on that of others. Four factors emerged
in the role performance-plus-experience analysis: I
Creative productivity; II Experience; III Quality and
originality of published work; IV Impact of published
work on that of others. Four factors, of ten extracted
in the analysis of climate items related to Creative
Productivity, discriminated high from low performers:
I Ease of written communication and orientation toward
increased management contact; II Cool autonomous work-
group atmosphere; III Lack of customer contact; IV

Global supervision, planning, and role diversity. Three
factors, of the ten extracted from climate items related
to the Quality and originality factor, so discriminated:
I Achievement orientation (cosmopolitan-local combination); II Supervisory flexibility; III Technician under
utilization, consultative, and goal setting activities.
Four factors, of ten extracted from climate items related to Impact factor, so discriminated: I High management priorities orientation and local hierarchical
orientation; II Low customer contact, effective project
forecast; III Criticality regarding advancement; IV
High external controls regarding punctuality.

Mowday, R. T., Porter, L. W., & Dubin, R.   Unit performance,
situational factors, and employee attitudes in spatially
separated work units.   Organizational Behavior and
Human Performance, 1974, 12, 231-248.
Study One:
Purpose: To identify dimensions of perceived sources of
organizational attachment
Subjects: 411 female clerical bank workers from 37 branches
Variables: Scores on 12 items regarding sources of organ-
izational attachment
Method: Principal axes condensation with varimax rotation
Results: Four dimensions of perceived influence emerged:
I Aspects of the branch; II Organizational policies;
III Aspects of the organization; IV Branch location.
Study Two:
Purpose: To identify dimensions of branch-bank performance
Subjects: 37 branches, rated by regional executive per-
sonnel
Variables: Ratings on employee relations, marketing,
operations, loan performance, leadership effectiveness
Method: Principal axes condensation with varimax rotation
Results: Two factors emerged: I Branch performance; II
Branch manager's loan performance.

Murdy, L. B., Sells, S. B., Gavin, J. F., & Toole, D. L.
Validity of personality and interest inventories for
stewardesses.   Personnel Psychology, 1973, 26, 273-278.
Purpose: To identify dimensions of stewardesses' perfor-
mance
Subjects: 113 (female) stewardesses
Variables: Scores on 34 performance variables derived
from personal files
Method: Principal components condensation with varimax
rotation
Results: Six factors were extracted: I Infractions; II
Commendations; III Says unable to work because of acci-
dents; IV Sick days; V-VI Not interpreted.
Conclusions: Four performance variables (corresponding to
Factors I-IV) were selected for use in investigation of
interests and personality.

Nandy, A.   Motives, modernity, and entrepreneurial competence.
Journal of Social Psychology, 1973, 91, 127-136.

338

Purpose: To identify dimensions of entrepreneurial competence
Subjects: 67 entrepreneurs from 2 Indian subcultures
(initiator of industrial unit who survived in business
for at least 5 years)
Variables: Scores on ratings (after 1 year's observation)
of dynamism of firm, entrepreneur's innovativeness,
growth in profits, growth in firm size, proportion of
"jobbed out" orders, prognosis of future entrepreneurial
status, predisposition to expansion, adaptive capacity,
relationship to workers
Method: Principal components condensation without rotation
Results: Since the first factor accounted for 32% of the
total variance, it was considered appropriate to talk
about a unitary construct of entrepreneurial competence;
five variables were selected to compose an entrepreneurial
competence scale.
Roach, D. E., & Davis, R. R. Stability of the structure of
employee attitudes: An empirical test of factor invariance. Journal of Applied Psychology, 1973, 58, 181-
185.
Purpose: To identify dimensions underlying employee attitudes for samples 10 years apart in time
Subjects: 1) 4,052 employees (heterogeous) in 1956, 2)
4,882 in 1966
Variables: Scores on 98 attitudinal and 5 classification
items
Method: Centroid (first-order) and multiply group (second-
order) condensation with hierarchical rotation, by
sample
Results: Seventeen 1956 first-order factors emerged: I
Physical working conditions; II Justice and interest of
management; III Pride in company; IV Intrinsic job
satisfaction; V Co-workers; VI Immediate supervision;
VII Job security; VIII Freedom from work rules; IX
Setting up and enforcing job standards; X Confidence in
ability of management; XI Downward communications; XII
Employee benefits; XIII Infrequent benefits; XIV Opportunities for relief from boredom; XV Job demands; XVI
Development and advancement; XVII Salary. Four second-
order (I Need fulfillment from immediate work environment; II Physical working conditions; III Impersonal
future rewards; IV Immediate personal rewards) and two
third-order factors (I Personal need fulfillment; II
Corporate reward system) were grouped under a general
factor. For 1966 data the results were essentially the
same, except a new second-order factor (V Corporate-
provided need fulfillment) emerged, 1956 Factor XIII
failed to emerge, and 1956 second-order Factor II became
first-order for 1966.
Roberts, K. H., Walter, G. A., & Miles, R. E. A factor
analytic study of job satisfaction items designed to
measure Maslow need categories. Personnel Psychology,
1971, 24, 205-220.

Purpose: To identify dimensions underlying job satisfaction items designed to measure Maslow need categories
Subjects: 380 managers from 5 hierarchical levels of a
high-technology manufacturing firm (subgrouped into 4
levels for "now" analysis)
Variables: Scores on 12 items designed to meet Maslow
need categories, 4 general satisfaction items, and pay,
each computed as 1) deficiency scores, 2) "now" scores,
3) "should be" scores, and 4) importance scores
Method: Unspecified factor analysis, independently for
deficiency, "should be," importance, and 4 management
levels within "now" scores
Results: Four difference factors emerged: I-II Not interpretable; III Prestige and fulfillment; IV Recognition and growth. Five "should be" factors emerged: I
Uninterpretable in terms of Maslow categories; II
General satisfaction items; III Prestige and esteem;
IV Self-actualization; V In the know, or inside dopester.
Four importance factors emerged: I Uninterpretable in
terms of Maslow categories; II General satisfaction
items; III Esteem (or at least related); IV Self-actualization. Four "now" factors emerged, apparently for all
4 management-level subgroups: I Growth and recognition
1; II General satisfaction items; III In the know; IV
Growth and recognition 2.
Smith, P. C., Smith, O. W., & Rollo, J. Factor structure
for blacks and whites of the Job Descriptive Index and
its discrimination of job satisfaction. Journal of
Applied Psychology, 1974, 59, 99-100.
Purpose: To identify dimensions underlying the Job Descrip-
tive Index (JDI) for blacks and whites
Subjects: 217 white and 107 black civil service, 110 pre-
dominantly white bank employees
Variables: 72 JDI item scores
Method: Principal component condensation with varimax
rotation, independently for 3 S groups
Results: Seven factors emerged for all groups: I Pay;
II Promotion; III Co-workers; IV Work 1; V Work 2
(white civil service only - unspecified for others);
VI Supervision 1; VII Supervision 2.
South, J. C. Early career performance of engineers - its
composition and measurement. Personnel Psychology,
1974, 27, 225-243.
Purpose: To identify dimensions of engineers' early career
performance
Subjects: 124 supervisors, rating an "outstanding" (83),
"average" (85), or "poor" (81) young engineer
Variables: 297 5-point checklist item scores
Method: Unspecified condensation with graphic and orthogonal
rotation
Results: Six factors emerged: I Communication; II Relating
to others; III Administrative ability; IV Motivation; V
Technical knowledge and ability; VI Self-sufficiency.

Spector, P. E. Relationships of organizational frustration
with reported behavioral reactions of employees. _Journal
of Applied Psychology_, 1975, 60, 635-637.
Purpose: To identify dimensions of employees' responses
to organizational frustration
Subjects: 82 employees from a wide range of jobs (mostly
mental health)
Variables: Scores on 35 response-to-frustration questionnaire items
Method: Principal components condensation with varimax
rotation
Results: Six factors, of ten extracted, were interpretable:
I Aggression against others; II Sabotage; III Wasting
of time and materials; IV Interpersonal hostility and
complaining; V Interpersonal aggression; VI Apathy
about the job.
Waters, L. K., & Roach, D. A factor analysis of need-ful-
fillment items designed to measure Maslow need categories.
Personnel Psychology, 1973, 26, 185-190.
Purpose: To identify dimensions underlying a set of need-
fulfillment items frequently used to measure Maslow need
categories
Subjects: 101 male managerial level personnel
Variables: Scores on 16 need-fulfillment items, overall
job satisfaction rating, and job grade
Method: Principal axes condensation with varimax and
hierarchical rotation
Results: Two subgeneral (I Higher order need fulfillment;
II Lower order need fulfillment) and two group factors
(III Esteem or prestige; IV Participation in the
management process) emerged.
Weitzel, W., Pinto, P. R., Dawis, R. V., & Jury, P. A. The
impact of the organization on the structure of job
satisfaction: Some factor analytic findings. _Personnel
Psychology_, 1973, 26, 545-557.
Purpose: To identify dimensions of self-reported job
satisfaction
Subjects: Salaried employees from companies of a corporation: 106 corporate staff, 224 retailing staff, 80
real estate-construction, 489 retailing, 200 discount
sales company staff
Variables: 28 4-item satisfaction self-report scales from
the Triple Audit Opinion Survey
Method: Hierarchical (principal factor and minimum residual) condensation with varimax rotation, independently
for each company
Results: One general (I General satisfaction), two subgeneral factors (II Satisfaction with the job; III
Satisfaction with the organization), and four first-
order factors (I Satisfaction with personal progress
and development; II Satisfaction with compensation; III
Satisfaction with the organizational context; IV Satisfaction with superior-subordinate relationships) emerged
for all five companies.

Conclusions: Differences between companies were examined.
Wild, R., & Kempner, T. Influence of community and plant
   characteristics on job attitudes of manual workers.
   Journal of Applied Psychology, 1972, 56, 106-113.
   Purpose: To identify dimensions of job attitudes for a
      large sample of female manual electronics workers
   Subjects: 2,543 female manual electronics workers from 10
      geographically dispersed plants
   Variables: 46 item scores from a questionnaire section
      concerning work, supervision, induction and training,
      wages, social relations, management, and physical working
      conditions
   Method: Principal components condensation with varimax
      rotation
   Results: Twelve factors emerged: I Pay; II Self-actualiza-
      tion; III Induction; IV Recognition; V Training; VI
      Physical effort and conditions; VII Supervision; VIII
      Working conditions; IX Mental vs. physical work; X
      Control of work vs. output required; XI Social peer
      relations; XII Responsibility for quality.
F. Management and Organization
   Alderfer, C. P. Effect of individual, group, and intergroup
      relations on attitudes toward a management development
      program. Journal of Applied Psychology, 1971, 55, 302-
      311.
      Purpose: To identify dimensions of attitudes toward a
         management development program, as assessed by the
         Management Development Questionnaire (MDQ)
      Subjects: 22 trainees, 80 officers, 62 male and 99 female
         employees of a bank
      Variables: 24 MDQ item scores
      Method: Principal components condensation with varimax
         rotation
      Results: Six program (I Overall evaluation of the program;
         II Degree to which trainees received undeserved prefer-
         ential treatment and misused their privileges; III
         Degree to which trainees were perceived as a closed
         cliquish group; IV Degree to which the program was per-
         ceived to be a barrier to one's own career development;
         V Degree to which trainees were seen as exclusively
         graduates of Ivy League colleges; VI Degree to which
         trainees were not moving fast enough in the organiza-
         tion and were barriers to each other in the organization)
         and two additional factors (VII Career promise; VIII
         Pay satisfaction) emerged.
   Bass, B. M., Valenzi, E. R., Farrow, D. L., & Solomon, R. J.
      Management styles associated with organizational, task
      personal and interpersonal contingencies. Journal of
      Applied Psychology, 1975, 30, 720-729.
      Purpose: To identify dimensions of management styles
      Subjects: 178 subordinates and managers from manufacturing
         organizations, 80 from a USAR unit, 67 from social ser-
         vice and volunteer organizations

Variables: A total of 288 item scores, separated into 6 (pilot) questionnaire sections
Method: Principal components condensation with varimax rotation, independently for each section
Results: Two leader styles factors emerged: I Initiating structure; II Consideration. Results for the remaining analyses are inferred (on the basis of scales used in the final instrument) as follows: Five organizational inputs factors emerged: I Constraints; II Clarity; III Warmth; IV Order; V External influences. Five task inputs factors emerged: I Clear objectives; II Routine; III Discretionary opportunities; IV Complexity; V Managerial activity. Four personal inputs factors emerged: I Fair; II Assertive; III Egalitarian; IV Introspective. Nine within-systems relations factors emerged: I Boss power; II Subordinate power; III Boss information; IV Subordinate information; V Long-term objectives; VI Structure; VII Intragroup harmony; VIII Interdependence; IX Commitment to group. Three system output factors emerged: I Work-unit effectiveness; II Satisfaction with supervision; III Satisfaction with job.

Blanz, F., & Ghiselli, E. E. The mixed standard scale: A new rating system. Personnel Psychology, 1972, 25, 185-199.
Purpose: To identify dimensions underlying the mixed-standard merit rating scale
Subjects: 100 middle-management personnel, rated by 23 superordinate (higher-level management) personnel
Variables: 18 trait scores from the mixed-standard merit rating scale
Method: Unspecified condensation with unspecified rotation
Results: Four factors emerged: I Work efficiency; II adjustment; III Carefulness and responsibility; IV Intelligence and social behavior.

Butterfield, D. A., & Farris, G. F. The Likert Organizational Profile: Methodological analysis and test of System 4 theory in Brazil. Journal of Applied Psychology, 1974, 59, 15-23.
Purpose: To identify dimensions of management systems used by Brazilian development banks, as measured by the Likert Organizational Profile (LOP)
Subjects: 256 employees of (13) Brazilian development banks
Variables: 20 LOP item scores, ideal form
Method: Principal components condensation with varimax rotation
Results: Six factors, not consistent with the hypothesized dimensions, emerged: I Leadership; II Resistance; III Guidance; IV Informed decision making; V Dispersion of goal setting and control; VI Motivation and communication.

Csoka, L. S. Relationship between organizational climate
    and the situational favorableness dimension of Fiedler's
    contingency model. Journal of Applied Psychology, 1975,
    60, 273-277.
  Purpose: To identify dimensions of organizational climate
    in Army units
  Subjects: 487 U.S. Army personnel
  Variables: Unspecified number of (item or) scale scores
    from the Organizational Climate Scale
  Method: Principal components condensation with varimax
    rotation
  Results: Four factors emerged: I Conflict and inconsistency;
    II Formalization and structure; III Communication flow;
    IV Tolerance of error.
DeVries, D. L., & Snyder, J. P. Faculty participation in
    departmental decision-making. Organizational Behavior
    and Human Performance, 1974, 11, 235-249.
  Purpose: To identify items measuring perceived departmental
    formalization
  Subjects: 387 university junior and senior faculty members
    (46 departments)
  Variables: 11 item scores from a questionnaire segment
    dealing with formalization of departmental procedures
  Method: Unifactor principal component condensation without
    rotation
  Results: Nine items loaded above .40 on the factor; two
    items were consequently dropped from further analyses.
Eden, D., & Leviatan, U. Implicit leadership theory as a
    determinant of the factor structure underlying super-
    visory behavior scales. Journal of Applied Psychology,
    1975, 60, 736-741.
  Purpose: To identify dimensions underlying students'
    implicit leadership theories
  Subjects: 235 undergraduate and graduate students (in
    Israel): 1) 99 with little or no work experience, 2)
    119 with 4 or more years of work experience, 3) 46
    claiming "random" answers.
  Variables: Scores on 12 leadership items from Survey of
    Organizations
  Method: Principal components condensation with varimax
    rotation
  Results: Four factors emerged for the first two analyses:
    I Support; II Work facilitation; III Interaction facili-
    tation; IV Goal emphasis. Five "random" factors emerged:
    I Support; II Work facilitation; III Interaction facili-
    tation; IV Goal emphasis; V Encourage effort (singlet).
Fox, W. A., Hill, W. A., & Guertin, W. H. Dimensional analysis
    of the Least Preferred Co-Worker Scales. Journal of
    Applied Psychology, 1973, 57, 192-194.
  Purpose: To identify dimensions underlying Fiedler's Least
    Preferred Co-worker (LPC) scales for three samples
  Subjects: 1) 114 Internal Revenue Service (IRS) tax ex-
    aminers, 2) 147 Marines, 3) 186 English managers

Variables: Scores on 1 & 3) original 16 LPC items, 2) 24
    LPC items
Method: Principal axes condensation with varimax (and
    oblique) rotation, independently for the three S-
    variable sets
Results: Four IRS factors emerged: I Hostile-ineffective;
    II Remote-rejecting; III Intense; IV Boring-ineffective.
    An additional two factors were extracted for Marines and
    English managers: V Hesitant; VI Not discussed.
Gannon, M. J., & Hendrickson, D. H. Career orientation and
    job satisfaction among working wives. Journal of Applied
    Psychology, 1973, 57, 339-340.
Purpose: To identify dimensions of career orientation among
    working wives
Subjects: 69 retail clerks and office workers, all working
    wives
Variables: Scores on 7 career orientation items
Method: Principal components condensation with orthogonal
    rotation
Results: Two factors emerged: I Job involvement; II Work
    vs. family orientation.
Gannon, M. J., & Paine, F. T. Unity of command and job
    attitudes of managers in a bureaucratic organization.
    Journal of Applied Psychology, 1974, 59, 392-394.
Purpose: To identify dimensions underlying characteristics
    of managerial and organizational effectiveness
Subjects: 181 GSA managers who reported to only one superior,
    123 who reported to two or more
Variables: 50 attitude-to-managerial-effectiveness item
    scores
Method: Principal components condensation with orthogonal
    rotation
Results: Eleven factors emerged: I Selection based on
    ability; II Initiative-flexibility; III Role conflict
    and pressure; IV Role ambiguity or clarity; V Employee
    utilization; VI Coordination; VII People are not
    satisfied with performance appraisals (singlet); VIII
    Adequacy of delegation; IX Employee effort; X Planning
    activity; XI Employee development.
Gavin, J. F. Organizational climate as a function of per-
    sonal and organizational variables. Journal of Applied
    Psychology, 1975, 60, 135-139.
Purpose: To identify dimensions underlying 1) organizational
    climate perceptions, 2) biographical data, 3) organiza-
    tional data for managerial-level employees; 4) to iden-
    tify clusters of managers
Subjects: 140 managerial level bank employees
Variables: Scores on 1) 106 organizational climate per-
    ception items, 2) 38 biographical data items, 3) 20
    organizational variables, 4) biographical factors, and
    organizational factors identified in (2) and (3)
Method: 1) and 3) Principal axes, 2) principal components
    condensation, all with varimax rotation; 4) Wolfe's (1970)
    cluster analysis, independently for biographical and
    organizational data
345

Results: Six climate factors emerged: I Clarity and efficiency of organizational structure; II Hindrance;
III Rewards; IV Esprit; V Managerial trust and consideration; VI Challenge and risk. Six biographical
factors emerged: I Financial responsibility; II Stability; III School adjustment; IV Social and physical
well-being; V Vocational choice; VI Childhood environment. Five organizational factors emerged: I Clerical
and repetitive work operations; II Professional and
technical operations; III Tenure and status in department; IV Restricted work area; V Business office routine.
Three clusters of managers (I Atonic; II Work-oriented;
III Socially-oriented) emerged using biographical
scores, and three (I Line functions-lower levels; II
Line functions-upper levels; III Staff functions) using
organizational data.

Kirchhoff, B. A. A diagnostic tool for management by objectives. Personnel Psychology, 1975, 28, 351-364.

Purpose: To identify dimensions underlying the Managerial
Style Questionnaire (MSQ)

Subjects: 389 managers (not first-line) or professional
employees from a variety of organizations

Variables: 47 MSQ item scores

Method: Principal factors condensation with quartimax
rotation

Results: Four factors emerged: I, III, IV Not interpreted;
II Goal use.

Kirton, M. J., & Mulligan, G. Correlates of managers' attitudes toward change. Journal of Applied Psychology,
1973, 58, 101-107.

Purpose: To identify dimensions underlying a questionnaire
measure of managers' attitudes toward appraisal schemes,
his responsibilities, his position relative to his
talents, and competing management styles

Subjects: 1) 80 managers, 2) 100 managers

Variables: 1) 62, 2) 26 questionnaire item scores

Method: Principal components condensation with varimax
rotation, by subject-variable set

Results: Although eight factors, of 19 extracted, were
interpretable, only three from the first analysis were
discussed: I Conservative-radical; II Content-discontent; III Confidence level. In the second analysis,
three factors, of nine extracted, were interpretable:
I Content-discontent; II Confidence level; III Assessment scheme attitude.

Lawler, E. E. III, & Suttle, J. L. Expectancy theory and
job behavior. Organizational Behavior and Human Performance, 1973, 9, 482-503.

Purpose: To identify dimensions of expectancy-attitude
items related to job behavior.

Subjects: 69 retail store department managers

Variables: Scores on 38 expectancy-attitude items related
to job behavior

Method: Principal components condensation with varimax
rotation
Results: Three factors emerged: I Internally-mediated
rewards; II Externally-mediated rewards; III Negatively-
valued outcomes.
Lawler, E. E. III, Hall, D. T., & Oldham, G. R. Organizational
climate: Relationship to organizational structure, pro-
cess and performance. Organizational Behavior and Human
Performance, 1974, 11, 139-155.
Purpose: To identify dimensions of organizational climate
perceived by R&D lab scientists
Subjects: 291 R&D lab scientists (from 21 labs)
Variables: Scores on 15+ semantic differential scales
related to organizational climate (number unspecified)
Method: Unspecified condensation with varimax rotation
Results: Five factors emerged: I Competent/potent; II
Responsible; III Practical; IV Risk-oriented; V Impul-
sive.
Conclusions: These scores were examined relative to data
gathered from directors of the labs; need satisfaction
data was also gathered.
Mahoney, T. A., Frost, P., Crandall, N. F., & Weitzel, W.
The conditioning influence of organization size upon
managerial practice. Organizational Behavior and Human
Performance, 1972, 8, 230-241.
Purpose: To identify dimensions of organizational charac-
teristics
Subjects: (Managers of unit supervisors from) 386 units
(section, department, division) within 19 firms; data
collected by Mahoney, 1967 and/or Mahoney & Weitzel,
1969
Variables: Scores on 114 measures of specific organiza-
tional characteristics
Method: Unspecified factor analysis
Results: Eighteen factors emerged: I Flexibility; II
Development; III Cohesion; IV Democratic supervision;
V Reliability; VI Delegation; VII Bargaining; VIII
Results emphasis; IX Staffing; X Decentralization; XI
Planning; XII Cooperation; XIII Performance-support-
utilization; XIV Communication; XV Initiation; XVI
Supervisory control; XVII Conflict; XVIII Supervisory
backing.
Morrison, R. F., & Sebald, M. Personal characteristics dif-
ferentiating female executive from female nonexecutive
personnel. Journal of Applied Psychology, 1974, 59,
656-659.
Purpose: To identify dimensions underlying a biographical
questionnaire (BQ)
Subjects: 39 pairs of working women (executive, nonexecutive), matched on age, education, years of employment,
type of organization
Variables: Scores on 104 items from a biographical questionnaire

347

Method: Principal components condensation with varimax rotation

Results: Six factors emerged: I Leadership; II Self-esteem; III Supervisory role perceptions; IV Early career socialization process; V Social skills; VI Independence.

Paine, F. T., & Gannon, M. J. Job attitudes of supervisors and managers. Personnel Psychology, 1973, 26, 521-529.

Purpose: To identify dimensions underlying job attitudes of supervisors and managers

Subjects: 317 managers and 404 supervisors from a nation-wide government organization

Variables: 50 item scores from a job attitude questionnaire

Method: Principal components condensation with orthogonal rotation

Results: Eight factors emerged: I Equity of rewards; II Adequacy of work force; III Goal clarity; IV Level of commitment or initiative in the organization; V Adequacy of performance appraisal; VI Autonomy; VII Planning and coordination of work; VIII Skill utilization in the organization.

Palmer, W. J. Management effectiveness as a function of per-sonality traits of the manager. Personnel Psychology, 1974, 27, 283-295.

Purpose: To identify dimensions of first-line managers' performance ratings of their managers

Subjects: 306 first-line managers, each rating their manager's performance

Variables: Overall 5-point performance rating, plus an unspecified number (8 or more) of other item scores from the same opinion survey

Method: Unspecified factor analysis

Results: An unspecified number of factors emerged; the performance rating loaded on a factor associated with Manager-employee interpersonal relationships.

Payne, R. L., & Pheysey, D. C. G. G. Stern's Organizational Climate Index: A reconceptualization and application to business organizations. Organizational Behavior and Human Performance, 1971, 6, 77-98.

Purpose: To identify dimensions underlying the scales of the Business Organization Climate Index (BOCI), an instrument adapted from the Organizational Climate Index for use in the British business climate

Subjects: 120 British junior managers (from 100+ companies)

Variables: 24 BOCI scale scores

Method: Principal components condensation without rotation

Results: Two factors, of the five extracted, were con-sidered interpretable: I Organizational progressiveness; II Normative control.

Conclusions: These data were discussed in relation to Stern's work.

Pinder, C. C., & Pinto, P. R. Demographic correlates of managerial style. Personnel Psychology, 1974, 27, 257-270.

Purpose: To identify clusters of managers with similar
inbasket performance
Subjects: 200 managers who were graduates from one business
school; data gathered by England (1967, 1968; England
& Keaveny, 1970; Keaveny, 1940)
Variables: 15 inbasket test scores
Method: Hierarchical grouping analysis
Results: Three clusters of managers emerged: I One man
show, or autocratic; II Decisive and efficient but
capable of dealing effectively with others in the
organization; III Consultative, thoughtful, courteous.

Pinto, P. R., & Pinder, C. C. A cluster analytic approach
to the study of organizations. Organizational Behavior
and Human Performance, 1972, 8, 408-422.

Purpose: To identify dimensions of (behavioral) effectiveness in organizational units
Subjects: 227 organizational units from a variety of
industries
Variables: Scores on 18 dimensions of organizational
behavior, 5 demographic characteristics, and an overall
supervisory effectiveness rating
Method: Hierarchical grouping cluster analysis
Results: Eight clusters emerged: I Planning, cooperation,
reliability, productivity-support-utilization; II Low
scores on all behavioral dimensions; III Supervisory
control and backing; IV High staffing and low on development, cohesion, democratic supervision, reliability,
productivity-support-utilization; V High on most behavioral dimensions and on effectiveness rating; VI Jagged
or "relief" cluster; VII Low behavioral dimension and
effectiveness scores; VIII High on flexibility, cohension,
supervisory control and democracy, delegation, bargaining, planning, productivity, communication, and conflict
but low on development, results emphasis, staffing,
decentralization, and supervisory backing.

Prien, E. P., & Ronan, W. W. An analysis of organization
characteristics. Organizational Behavior and Human
Performance, 1971, 6, 215-234.

Purpose: To identify organization dimensions for a sample
of small metal working firms
Subjects: (Top executive from) 107 small metal working
firms
Variables: 38 measures of organization characteristics
from an interview
Method: Principal components condensation with varimax
rotation
Results: Nine factors emerged: I Standardization of
individual roles; II Change, products, and technology;
III Succession; IV Specialization; V Marketing strategy;
VI Standardization of individual recognition; VII Organization site; VIII Marketing-technology interface with
criterion of organization effectiveness; IX Quality
production.

Conclusion: These results were compared with those from
    other studies.
Roberts, K. H., & O'Reilly, C. A. III. Measuring organiza-
    tional communication. _Journal_ _of_ _Applied_ _Psychology_,
    1974, _59_, 321-326.
  Purpose: To identify dimensions of communication between
    and across organizations
  Subjects: 86 mental health workers
  Variables: Scores on 1) 51 communication and 9 noncommuni-
    cation items, 2) 9 noncommunication items
  Method: 1) V-type cluster analysis; 2) principal components
    condensation with varimax rotation
  Results: Ten clusters emerged in the first analysis: I
    Trust; II Influence; III Mobility; IV Desire for inter-
    action; V Directionality-upward; VI Directionality-
    downward; VII Directionality-lateral; VIII Accuracy;
    IX Summarization; X Gatekeeping. Three factors emerged
    in the second analysis, which was apparently replicated
    with 95 military subjects: I Trust in superior; II
    Perceived influence of superior; III Subordinate's
    mobility aspirations.
Ronan, W. W., Latham, R. P., & Kinne, S. B. III. Effects of
    goal setting and supervision on worker behavior in an
    industrial situation. _Journal_ _of_ _Applied_ _Psychology_,
    1973, _58_, 302-307.
  Purpose: To identify dimensions of producers' supervisory
    practices
  Subjects: 292 pulpwood producers
  Variables: Scores on 16 supervisory practices items
  Method: Principal components condensation with varimax
    rotation
  Results: Three factors emerged: I Employee-production-
    centered style of supervision which included goal
    setting; II Production-centered style of supervision;
    III Employee-centered style of supervision which did
    not include goal setting.
Ronan, W. W., & Prien, E. P. An analysis of organizational
    behavior and organizational performance. _Organizational_
    Behavior _and_ _Human_ _Performance_, 1973, _9_, 78-99.
  Purpose: To identify dimensions of characteristic and
    performance variables for units of an industrial plant
  Subjects: 64 units, or departments, of an industrial
    plant
  Variables: Scores on 8 characteristic and 10 performance
    variables
  Method: 1) Principal components condensation with varimax
    and 2) graphic rotation; 3) hierarchical factor con-
    densation
  Results: The three solutions yielded approximately equiva-
    lent seven-factor solutions: I Extent of manufacturing
    a product particularly; II Effectiveness or management
    strategy; III Degree of unit build-up with either no
    pressure to avoid or inadequate recording of tardiness,

with low insurance claims; IV Job discontent, job withdrawal, or "conflict" phenomena; V Salaried discontent;
VI Absenteeism contingent on hourly/salaried employee
mix; VII Priority for intervention (employee relations).
Sank, L. I. Effective and ineffective managerial traits
   obtained as naturalistic descriptors from executive
   members of a super-corporation. Personnel Psychology,
   1974, 27, 423-434.
Purpose: To identify dimensions underlying effective and
   ineffective managerial traits
Subjects: 145 male middle managers: 65 describing an effective manager, 80 an ineffective manager
Variables: Protocol response categorization-derived
   descriptive ratings: 45 for effective, 29 for ineffective managers
Method: Chain cluster analysis
Results: Two large clusters, each defined by an unspecified
   number of subclusters emerged: I Effective managerial
   traits (Subclusters: achievement-oriented traits, firm
   and fair, honest with empathy); II Ineffective managerial
   traits (subcluster: deficiency in character).
Schneider, B. Organizational climate: Individual preferences
   and organizational realities revisited. Journal of
   Applied Psychology, 1975, 60, 459-465.
Purpose: To identify clusters of life insurance agencies
   on the basis of old-agent perceptions (Schneider, 1974
   unpublished manuscript)
Subjects: Unspecified number of old (experienced) life
   insurance agents
Variables: Scores on 6 Agency Climate Questionnaire
   factors
Method: Cluster analysis
Results: Four agency clusters emerged: I Conflict; II
   Typical agency; III Theory Y/System 4; IV Disaster.
Schwyhart, W. R., & Smith P. C. Factors in the job involve-
   ment of middle managers. Journal of Applied Psychology,
   1972, 56, 227-233.
Purpose: To identify dimensions of middle managers' job
   involvement (JI)
Subjects: 1) 149 male middle managers from one company,
   58 from four others; 2) 100 Ss randomly chosen from
   pooled groups for replication sample
Variables: Scores on 20 JI items
Method: Principal components condensation with varimax
   rotation, by group and for replication sample
Results: In the initial analyses, six factors emerged for
   the first group, and seven for the second. Three factors were replicated in the second analysis: I Job
   ambition; II Job centrality; III Job conscientiousness.
   (Three additional factors for which only one item replicated were also named: IV Striving for perfection;
   V Disinterest in nonjob activities; VI Job interest for
   reasons other than pay.)

Shiflett, S. C. Stereotyping and esteem for one's least
    preferred co-worker. Journal of Social Psychology,
    1974, 93, 55-65.
    Purpose: To identify dimensions underlying Least Preferred
    Coworker (LPC) and Most Preferred Coworker (MPC) scales
    Subjects: 107 male U.S. Army trainees
    Variables: Scores on 16 LPC and 16 MPC scales
    Method: Principal axes condensation, apparently without
    rotation and independently for LPC and MPC scores
    Results: Two interpretable factors were reported: I
    Interpersonal relations; II Task orientation.
Sims, H. P. Jr., & LaFollette, W. An assessment of the
    Litwin and Stringer Organization Climate Questionnaire.
    Personnel Psychology, 1975, 28, 19-38.
    Purpose: To identify dimensions underlying the Organization
    Climate Questionnaire (OCQ)
    Subjects: 997 employees (janitors to RNs) of a major medical
    complex
    Variables: 50 item scores from the OCQ (some questions
    modified to setting)
    Method: Principal axes condensation with varimax rotation
    Results: Six factors emerged: I General affect tone
    toward other people in the organization; II General
    affect tone toward management/organization; III Policy
    and promotion clarity; IV Job pressure and standards;
    V Openness of upward communication; VI Risk in decision
    making.
Sims, H. P. Jr., & Szilagyi, A. D. Leader reward behavior
    and subordinate satisfaction and performance. Organi-
    zational Behavior and Human Performance, 1975, 14,
    426-438.
    Purpose: To identify dimensions of the Leader Reward
    Behavior instrument (previously extracted and reported
    by Johnson in unpublished doctoral dissertation, 1973)
    Subjects: Unspecified
    Variables: 17 Leader Reward Behavior item scores
    Method: Unspecified factor analysis
    Results: Two factors emerged: I Positive reward behavior;
    II Punitive reward behavior.
Spautz, M. E. A new scale for theories X and Y. Australian
    Journal of Psychology, 1975, 27, 127-141.
    Purpose: To identify dimensions underlying the Realistic
    Employee Motivation Questionnaire (REMQ)
    Subjects: 209 male business administration graduate
    students
    Study One:
    Variables: 20 REMQ item scores
    Method: Unspecified factor analysis (apparently without
    rotation
    Results: One general (I Theory X vs. Y attitude) and five
    specific factors (II-VI Uninterpreted) emerged.
    Study Two:
    Variables: 20 REMQ item scores, F-Scale, D-Scale, LOQ
    structure, LOQ Consideration scale scores

Method: Unspecified condensation with orthogonal rotation
Results: Eight factors emerged: I Theory X (employee
    motivation); II Dogmatism-authoritarianism-structure;
    III Role of management; IV Internal vs. external locus
    of control of employees' behavior; V Inconsideration;
    VI Not interpreted; VII Traditional management func-
    tions; VIII Superior-subordinate interpersonal relations.
Study Three:
Purpose: To identify dimensions of a peer rating form, the
    Professional Development Evaluation (PDE, Spautz, un-
    published)
Subjects: 25 PDE item scores
Method: Unspecified condensation with orthogonal rotation
Results: Five factors emerged: I Competence; II Liberality;
    III Employee orientation; IV Industry; V Decisiveness.
Stephenson, R. W., & Ward, J. H. Jr. Computer-assisted dis-
    cussions to help a policy group assign weights to criterion
    weightings. Personnel Psychology, 1971, 24, 447-461.
Purpose: To identify clusters of items relevant to pro-
    motion policy
Subjects: 10 policy board members, 12 advisors, and "one or
    two" customers in a research-and-development organization
Variables: 7-point ratings of actual and ideal importance
    for 112 promotion policy related items
Method: Key cluster analysis with direct oblique rotation
Results: Ten clusters, used in later procedures, emerged:
    I Effective performance in a management advisory capacity;
    II Effective performance as a person who initiates change;
    III Likeability; IV Technical knowledge; V Inspires con-
    fidence; VI Effective performance in areas of functional
    expertise; VII Professional knowledge; VIII Knowledge
    of supervisory practices; IX Sensitivity to job-related
    problems and involvement therein; X Timeliness in pro-
    cessing personnel actions.
Szilagyi, A. D., & Sims, H. P. Jr. Cross-sample stability
    of the Supervisory Behavior Description questionnaire.
    Journal of Applied Psychology, 1974, 59, 767-770.
Purpose: To identify dimensions underlying the Supervisory
    Behavior Description (SBD) with a predominantly female
    population (for comparison with previous results)
Subjects: 1,161 employees of a major medical center (80%
    female)
Variables: 48 (?) SBD item scores
Method: Principal axes condensation with varimax rotation
Results: Two factors emerged: I Consideration; II
    Initiating structure.
Conclusion: These results are compared with those of
    Fleishman (1953) and of Tscheulin (1973).
Templer, A. J. Self-perceived and others-perceived leadership
    style using the Leader Behavior Description Questionnaire.
    Personnel Psychology, 1973, 26, 359-367.
Purpose: To identify dimensions of self-perceived and
    others-perceived leadership style using the Leader Behavior
    Description Questionnaire (LBDQ)

Subjects: 1) 49 male middle-management personnel from one
South African firm, 2) 60 of the same subjects, 3) 49
subjects rated by supervisors
Variables: 1) 12 LBDQ scores each for self-rating and for
supervisor rating, 2) 12 self-rated LBDQ scale scores,
3) 12 supervisor-rated LBDQ scale scores
Method: 1) Maximum likelihood condensation with varimax
rotation; 2 & 3) principal-factor condensation with
varimax and direct quartimin rotation
Results: Two of the seven factors extracted were interpreted
for the combined analysis of self- and supervisor-rated
scores: I General self-rating; II General supervisor
rating. Three factors emerged for self-ratings alone;
they did not appear to reflect any expected groupings
of scales and were not labeled. Three factors, likewise
uninterpreted, also emerged for supervisor ratings alone.
Tscheulin, D. Leader behavior measurement in German industry.
Journal of Applied Psychology, 1973, 57, 28-31.
Purpose: To identify dimensions of German supervisory
behavior
Subjects: 183 employees, describing their (44) immediate
supervisors, in West Germany
Variables: Scores on 48 Supervisory Behavior Description
Questionnaire (German translation)
Method: Principal axes condensation with varimax rotation
Results: Two factors emerged: I Consideration; II
Initiating structure.
Conclusions: These results were compared with those for one
of Fleishman's American samples.
Waters, L. K., Roach, D., & Batlis, N. Organizational climate
dimensions and job-related attitudes. Personnel Psy-
chology, 1974, 27, 465-476.
Purpose: To identify dimensions underlying perceptually-
based organization climate ratings
Subjects: 105 radio and TV station employees
Variables: 22 organization climate scale scores (from 3
instruments used by previous investigators)
Method: Principal factors condensation with varimax
rotation
Results: Five factors emerged: I Effective organizational
structure; II Work autonomy vs. encumbered by nonproduc-
tive activities; III Close impersonal supervision; IV
Open challenging environment; V Management and peer support, or employee centered orientation.
G. Special Environments
Worchel, S., & Mitchell, T. R. An evaluation of the effectiveness of the culture assimilator in Thailand and
Greece. Journal of Applied Psychology, 1972, 56, 472-
479.
Purpose: To identify dimensions underlying self-evaluated
performance and adjustment in a foreign culture
Subjects: 51 AFAG airmen and USOM civilian advisors sta-
tioned in Thailand (and receiving either a Thai culture
assimilator or Thai essay)

Variables: Scores on 17 self-evaluation performance and
adjustment items
Method: Unspecified condensation with unspecified rotation
Results: Five factors emerged: I Affective; II Training;
III Thai cooperation; IV Productivity; V Job pressure.
H. Advertising and Consumer Psychology
Farley, J. U., Howard, J. A., & Weinstein, D. The relation-
ship of liking and choice to attributes of an alternative
and their saliency. Multivariate Behavioral Research,
1974, 9, 27-35.
Purpose: To identify attribute dimensions of evaluations
of a subcompact automobile
Subjects: 265 Denver residents who expressed interest in
buying a subcompact
Variables: Scores on 20 attributes of the brand
Method: Principal components condensation with varimax
rotation
Results: Three factors emerged: I Physical aspects of
driving; II Economic considerations of owning and
operating a car; III Personal feelings about driving
the car.
Green, P. E., & Carmone, F. J. The effect of task on intra-
individual differences in similarities judgments.
Multivariate Behavioral Research, 1971, 6, 433-450.
Purpose: To identify dimensions underlying similarity
judgments of bakery-type food items
Subjects: 4 finance and commerce graduate students
Variables: 22 bipolar scale ratings for 20 bakery-type
food items
Method: Principal components condensation with varimax
rotation, for group and by judge
Results: Seven factors emerged for the group: I Caloric-
ness; II Naturalness of flavor; III Softness of texture;
IV Expensiveness; V Complexity of flavor; VI Formality
of serving; VII Perishability. Between 4 and 6 factors
emerged in each of the judge matrices.
Conclusions: INDXCAL results are also reported.
Hunter, T. E. Method s of reordering the correlation matrix
to facilitate visual inspection and preliminary cluster
analysis. Journal of Educational Research, 1973, 10,
51-61.
Purpose: To identify clusters of a marketing orientation
instrument (Lessin, unpublished dissertation, 1968)
Subjects: 109 males
Variables: Scores on 70 marketing orientation items
Method: Hierarchical (pairs of items, pair subclusters,
clusters) cluster analysis
Results: Six clusters emerged: I Other directedness; II
Conformity; III Quickness; IV Sociability; V Emptiness;
VI Manipulation.
Roach, D. E., & Wherry, R. J. Sr. The use of hierarchical
factor analysis in the determination of corporate image

355

dimensions. Educational and Psychological Measurement, 1972, 32, 31-44.
Purpose: To identify corporate image dimensions
Subjects: 472 male heads of households in 14 eastern states
Variables: 139 item scores from a questionnaire concerning advisability of activities (for a multi-line insurance company to engage in)
Method: Wherry-Winer (1953) condensation with hierarchical rotation
Results: Eighteen primary factors emerged: I Competent agency staff; II Product quality involvement; VI Agents contacts (social responsibility); VII Special services (social responsibility); VIII Concern for employee welfare (social responsibility); IX Good claims service (marketing program); X Low rates (marketing program); XI TV and radio sponsership (marketing program); XII Mass marketing (marketing program); XIII International involvement (competitive strategy); XIV Political involvement (competitive strategy); XV Cut-throat competition (competitive strategy); XVI Sound investment policy; XVII Agent role in community affairs (community involvement); XVIII Financial involvement (community involvement). Four second-order factors (I Quality products and services; II Social responsibility; III Marketing program; IV Community involvement) and two third-order factors (I Prestige through quality services or products; II Prestige through self-serving manipulation) also emerged.
Schutz, H. G., & Rucker, M. H. A comparison of variable configurations across scale lengths: An empirical study. Educational and Psychological Measurement, 1975, 35, 319-324.
Purpose: To identify dimensions underlying four forms (lengths) of a food-use questionnaire
Subjects: 15 male and 15 female college students each for the 4 forms
Variables: Scores for rating appropriateness of 10 different foods in ten different situations, on 2-, 3-, 6-, or 7-point scales
Method: Principal components condensation with orthogonal rotation, independently for the four subject-form sets
Results: Three unlabeled factors emerged in all four analyses.
Conclusions: Differences between scale lengths were examined.
I. Engineering Psychology
Richards, J. M. Jr., & Claudy, J. G. Does farm practice adoption involve a general trait? Journal of Applied Psychology, 1973, 57, 360-362.
Purpose: To identify dimensions of adopted farm practices
Subjects: 1) 170 farm owner-operators (data originally collected by Wilkening, 1954); 2) 157 cattle farmers (Copp, 1956)

Variables: 1) 11 farm practices scores; 2) 21 farm practices scores
Method: Principal axes condensation with varimax rotation, by sample-variable set
Results: Three factors emerged for the first analysis: I Insemination practices; II Milking practices; III Use of chemicals. Three factors emerged in the second analysis: I Conservative good practices; II Care of calves; III Receptivity to change.

1. Displays and Controls
   Siegal, A. J., & Fischl, M. A. Dimensions of visual information displays. Journal of Applied Psychology, 1971, 55, 470-476.
   Purpose: To identify perceived dimensions of visual information displays
   Subjects: 17 USAFR officers (rank captain and up) experienced with command context, 15 male college students experienced with displays, 21 male college students with neither kind of experience
   Variables: Similarity judgments for 12 air-defense-oriented visual information displays
   Method: Principal components condensation with equamax rotation, by group
   Results: Seven factors emerged for the USAFR group: I Stimulus numerosity; II Structure scanning; III Cognitive processing ability; IV Critical relationships; V Cue integration; VI Contextual discrimination; VII Primary coding. Factors I, III, IV, V, VI emerged for the trained college students; Factors I, II, III, IV, V, VI emerged for the naive college students.

J. Driving and Safety
   Fhaner, G., & Hane, M. Seat belts: Relations between beliefs, attitude, and use. Journal of Applied Psychology, 1974, 59, 472-482.
   Purpose: To identify dimensions of attitudes toward seat belt usage.
   Subjects: 368 car owners
   Variables: Scores on 52 belief items
   Method: Principal factors condensation with oblique rotation
   Results: Five factors emerged: I Feelings of discomfort; II Worry; III Feelings of risk; IV Effect; V Inconvenience in handling.
   Conclusions: Relationship of factor scores to reported use of seat belts was subsequently examined.

Daly, D. L., 249
Darbes, A., 138
Das, J. P., 14, 26, 270
Daum, J. W., 95
Daut, R. L., 215
Davis, K. E., 51
Davis, R. R., 339
Dawlis, R. V., 341
Day, G. J., 181
Deines, M. M., 218
Delaney, J. O., 62
Deleon, G., 228
Delhees, K. H., 155
Delisser, O., 138
Demaree, R. G., 182
de Mers, J. L., 326
DeMille, R., 102
Dempewolff, J. A., 159
Deppe, A. H., 324
Derogatis, L. R., 216, 240
Detre, T., 222
Detterman, D. K., 10, 21
Deutsch, M., 58
Devine, B., 73
Devito, P. J., 43
De Vries, D. L., 344
De Vries, R., 26
Dexter, W. R., 162
Deyoung, A. J., 280
Diamond, J. J., 286
Dickinson, T. L., 331
Dielman, T. E., 52, 53, 54,
    80, 216
Dienst, E. R., 312
Dies, R. R., 205
Dimascio, A., 243
Dion, K. K., 91
Dion, K. L., 91
Distefano, M. K. Jr., 208
Dixon, P. W., 159
Dobyns, Z. P., 80, 124
Doherty, E. G., 209
Domino, G., 263, 282
Donnerstein, E., 102
Donnerstein, M., 102
Donoghue, R., 255
Dorr, D., 194
Doughtie, E. B., 27, 188, 189
Douty, H. I., 136
Dowd, A. F., 79
Dowds, B. N., 218
Doyle, K. O. Jr., 301
Dregar, R. M., 42, 257, 319
Dressler, D. M., 236

Dubin, R., 338
Dudley, H. K. Jr., 230
Dudzinski, M. L., 20
Du Jovne, B. E., 142
Dunham, J. L., 12
Dunin-Markiewicz, A., 23
Dunnette, M. D., 331
Durojaiye, M. O. A., 288
Duske, E. R., 142, 143
Dwarshuis, L., 222
Dytell, R. S., 37
Dziuban, C. D., 298

Eaves, L. J., 43
Eckert, H. M., 27
Eddy, G. L., 251
Eden, D., 344
Edgerly, J., 109
Edney, J. J., 95
Edwards, A. L., 160
Edwards, C. B. H., 20
Edwards, C. N., 103
Edwards, K. J., 161
Edwards, S., 86
Edwards, S. A., 337
Egeland, B., 44
Egelhoff, C. J., 122
Eichorn, D. H., 27
Einhorn, H. J., 327
Eisenman, R., 138
Ekehammar, B., 103, 108,
    128, 192
Ekman, G., 67
Elbert, A. J., 332
Ellsworth, R., 209
Elterich, K. W. Jr., 296
Elton, C. F., 252, 253
Emery, A., 85
Emmer, E. T., 301
Endicott, J., 217
Endler, N. S., 161
Engelbrektson, K., 141
Epstein, A. S., 55
Epting, F., 141
Erdmann, J. B., 191
Erickson, J., 130, 131
Erlich, H. J., 103
Evan-Wong, L., 153
Evans, D. R., 161
Evans, G. T., 143
Evans, W. J., 286
Evertson, C. M., 299
Ewen, R. B., 331

Gleser, G. C., 196
Globerson, T., 42
Gluskinos, U., 332
Gluskinos, U. M., 275
Gocka, E. F., 298
Goebel, R. A., 258
Goldberg, H. L., 241
Goldberg, J., 164
Goldberg, K., 218
Golding, S. L., 104, 164, 181
Goldman, R. D., 82, 313
Goldschmid, M. L., 204
Goldstein, G., 257
Goldstein, K. M., 276
Goodale, J. G., 322, 332
Gordon, M. E., 325
Gordon, T. P., 175
Gorman, B. S., 125
Gorsuch, R. L., 71, 92
Gottheil, E., 213
Gottlieb, J., 83, 294
Gough, H. G., 67, 143, 165
Graham, J. R., 165, 207, 210, 266
Gralnick, A., 203
Granzin, K. L., 304
Graves, W., 282
Gray, J. E., 166, 210
Green, B. L., 196
Green, D. L., 83
Green, P. E., 355
Green, R. T., 75
Greenberg, G., 77
Greenberg, L. M., 215
Greenberg, R. P., 226
Greenhaus, J. H., 276
Greenwood, G. E., 305
Greever, K. B., 83
Gregson, R. A. M., 5
Greif, E. B., 166
Griffin, C., 185
Griffith, G. M., 184
Grimaldi, J., 297
Grinspoon, L., 228
Gronwall, D. M. A., 2
Gross, D. E., 305
Grossman, J. H., 315
Gruzen, J., 74
Guarnaccia, V. J., 260
Guertin, W. H., 66, 93, 344
Guidubaldi, J., 39
Guilford, J. P., 28, 36, 143, 144
Guion, R. M., 332, 333

Gunderson, E. K. E., 130, 131, 139
Gurel, L., 250
Gurland, B. J., 218
Gurman, A. S., 198
Guthrie, G. M., 218

Haase, R. F., 105
Hagen, E. L., 335
Hagen, K., 120
Hakel, M. D., 125
Kakstian, A. R., 205
Haley, J. V., 127
Hall, D. T., 329, 347
Hall, W. S., 47
Hallahan, D. P., 28, 28
Hallenbeck, C. E., 260
Hallworth, H. J., 316
Halperin, S., 44
Halpin, W. G., 288
Halverson, C. F. Jr., 51
Hamlin, R. M., 245
Hammill, D., 36
Hammond, S. B., 64, 187
Hampe, E., 48
Hampson, S. L., 93
Hamsher, J. H., 152
Hane, M., 357
Hanlon, R., 312
Hansen, J. B., 144
Hansen, L. H., 305
Hanson, D. L., 263
Hanson, R. A., 56
Hanson, R. W., 80
Hanssen, C., 57
Hardy, M., 28, 38
Hare, A. P., 96
Hare, B., 36
Hargreaves, D. J., 44
Hargreaves, W. A., 196
Hariton, E. B., 78
Harrington, T. F., 318
Harris, C. W., 29, 144, 273
Harris, E. L., 10
Harris, J. G. Jr., 325
Harris, M. L., 144
Harris, S., 196
Harrison, P. R., 56
Harrison, R. H., 261
Harrow, M., 249
Hartford, D., 136
Hartley, E. L., 306
Hartley, J. A., 298
Harvey, W. M., 120, 121, 252

Hasleton, S., 71
Hautaluoma, J., 196
Hawley, P., 105
Hayman, J.L., Jr., 47
Haynes, J. R., 114
Heckel, R. V., 200
Hedlund, J.L., 212
Heeler, R. M., 131
Hekmat, H., 173
Heller, L., 308
Hellervik, L.V., 331
Helms, S. T., 321
Henderson, R. W., 71
Hendrick, C., 135
Hendrickson, D. H., 345
Henisz, J. E., 268
Hennessy, J., 297
Hensley, D. R., 167
Hensley, W. E., 167
Herman, J. B., 333
Herman, K. D., 243
Herrenkohl, R. C., 126
Heskins, K. J., 115
Hick, T. L., 29
Hill, J. A., 203
Hill, K., 118
Hill, W. A., 344
Hirsch, J. G., 314
Hirschberg, M. A., 102
Hirschfeld, R., 219
Hodapp, V., 6
Hoffman, H., 233, 234, 235
Hogan, R., 60, 166
Hogan, R., 60, 166
Hogan, T. P., 181, 306
Hogarty, G. E., 211
Horberg, A., 83, 90
Holland, T. R., 237
Hollandsworth, J. G. Jr., 164
Hollenbeck, G. P., 29, 33
Hollos, M., 30
Holmes, D. S., 306
Holmes, G. R., 200, 219, 227
Holt, N., 237
Holzman, P. S., 5
Hooke, O., 174
Hopkins, K. D., 30
Horn, J. L., 149, 167
Hornstra, R. K., 267
Houlihan, K. A., 59
Hourany, L., 126
House, E. R., 270
Houtz, J. C., 28, 38
Howard, J. A., 355

Howard, K. F., 205
Howarth, E., 167, 183
Hoy, W. K., 298
Hrycenko, I., 96
Hubbell, M., 263
Huberty, C. J., 56
Hulin, C. L., 333
Hummel, T. J., 204
Humphreys, L. G., 30
Hundal, P. S., 334
Hunt, D., 287
Hunter, S., 220, 226
Hunter, S. H., 230
Hunter, T. E., 355
Hurt, M. Jr., 71
Hyde, J. S., 144

Iker, H. P., 113
Inn, A., 334
Innes, J. M., 137
Insua, A. M., 191
Isbister, A. G., 287
Isherwood, G. B., 298
Itkins, S., 101
Ivinskis, A., 62
Iwawaki, S., 117

Jackson, D. N., 126, 150, 151,
    168, 235
Jackson, J. C., 314
Jackson, R., 287
Jacobs, P. I., 185
Jacobs, P. J., 239
Jacobs, S. S., 149
Jaeger, R. M., 276, 306
Jahn, J. C., 169
Janicki, R. S., 257
Jansen, D. G., 234, 253
Janzen, H. L., 316
Jeanneret, P. R., 327
Jenkins, J. D., 45
Jensema, C. J., 211
Jensen, A. R., 31
Joe, V. C., 168, 169
Johannesson, R. E., 335
Johnson, A. L., 169
Johnson, E. G., 34
Johnson, F. L., 50
Johnson, J. H., 169, 171, 176
Johnson, M., 201
Johnson, R. W., 170
Johnson, W. H., 51
Johnston, R. J., 76
Jones, F. H., 63

Jones, J. M., 84
Jones, R. A., 106, 112, 127
Jordan, R., 109
Josefowitz, N., 73
Judd, L. R., 127
Jury, P. A., 341

Kahn, H., 321
Kanh, M. W., 237
Kaminski, E. P., 154
Kane, R. B., 276
Kaplan, R. B., 82
Kaplan, R. M., 305
Karson, S., 335
Kayzenmeyer, W. G., 45
Kaufman, A. S., 29, 31, 32, 33, 295
Kayser, B. D., 145
Kayton, L., 10
Kazelskis, R., 45
Kear-Colwell, J. J., 220, 221, 234
Keaveny, T. J., 307
Kehle, T., 39
Keith, L. T., 307
Keith-Spiegel, P., 213
Kempner, T., 342
Kendall, P. C., 226
Keniston, K., 82
Kennedy, J. J., 272
Kennedy, P. W., 319
Kennedy, W. R., 307
Kerins, T., 270
Kerlinger, F. N., 84, 91
Kernan, J. B., 84
Kerr, B. A., 294
Kerry, R. J., 222, 242
Kesselman, G. A., 335
Khajavi, F., 173
Khan, S. B., 63, 269
Khatena, J., 148
Kidd, A. H., 170
Kidd, R. M., 170
Kilmann, R. H., 271
Kiloh, L. G., 240
Kimball, C. P., 264
King, C. E., 299
Kinne, S. B. III, 350
Kipperman, A., 244
Kirby, D. M., 70, 106
Kirchholff, B. A., 346
Kirk, S. A., 84
Kirton, M. J., 346
Kleiber, D., 170

Klein, T. W., 43
Klerman, G. L., 222, 243
Klimoski, R. J., 336
Klingler, D. E., 145
Klockars, A. J., 160
Klonoff, H., 33
Knapp, R. R., 172
Knops, L., 11
Knudson, R. M., 104
Kogan, N., 45, 185
Koh, S. D., 10
Kohn, P. M., 72
Kolton, M. S., 222
Koral, J., 222
Korchin, S. J., 183
Korn, S. J., 47
Kothandapani, V., 79
Kourtrelakos, J., 64
Koutstaal, C. W., 4, 11
Kraft, I. A., 189
Kraso, D. J., 107
Kraut, A. I., 325
Krebs, D., 137
Kreitler, H., 46
Kreitler, S., 46
Krug, S. E., 170
Ksionzky, S., 97
Kubiniec, C. M., 125, 131
Kuennapas, T., 329
Kumoski, R. J., 327
Kupfer, D. J., 222

Labouvie, E. W., 46
Labouvie-Vief, G., 64
Lacher, M., 171
Lacorsiere-Paige, F., 22
LaCrosse, M. B., 199
Laffal, J., 112
LaFollette, W., 352
LaForge, R., 314
Lamberd, W. G., 194
Lambert, N. M., 277
Lambert, P., 305
Lamberth, J., 77
Lamiell, J. T., 109
Lamont, J., 223
Lan, H., 15
Landis, D., 47
Landy, F. J., 333, 336
Langbauer, W., 255
Langevin, R., 145
Langhorne, J. E. Jr., 269, 277
Lansdell, H., 258, 259
Lanyon, B. J., 127

Mason, E. J., 301
Massa, E., 36
Mastbaum, N. A., 204
Mathews, A. M., 78
Mathis, W. J., 18
Matter, D. E., 254
Mausner, B., 120
Mayer, S. E., 324
Mayerberg, C. K., 85
Mayfield, E. C., 337
McArthur, D. L., 29
McCallon, E. L., 280
McCandless, B. R., 50
McCarrey, M. W., 86, 337
McConnell, J., 22
McCormick, E. J., 327
McCormick, M., 262
McDowell, E. E., 308
McGann, A. F., 307
McGarty, M., 263
McGettigan, J., 36
McGinnies, E., 87
McGowan, J. R., 36
McGrady, H. J., 294
McKeachie, W. J., 308
McKerracher, D. W., 289, 294
McKinney, J. D., 291
McLain, R., 263
McLaughlin, G. W., 320
McLaughlin, J. A., 34
McLean, J. E., 305
McNeal, K. A., 171
McNeil, T. F., 225
Meadows, C. M., 129
Mecham, R. C., 327
Meddis, R., 129
Meeker, M., 65
Mehra, N., 280
Mehrabian, A., 97, 108
Mehryar, A. H., 173
Meikle, S., 9, 242
Menaker, S. L., 170
Mencke, R. A., 325
Mendels, J., 225, 244
Mendels, G. E., 35
Menne, J. M., 201
Merrill, P. F., 146
Mertens, C., 265
Messick, S., 146, 150
Metcalfe, M., 169
Meyer, R. M., 181
Meyers, C. E., 65
Meyers, L. S., 12

Mezei, L., 73
Michael, J. J., 47, 48, 138, 174
Michael, W. B., 1, 47, 48, 174
Michelli, F. A., 257
Michlin, M. L., 50
Miles, R. E., 339
Miller, A., 253
Miller, J. K., 174
Miller, L., 286
Miller, L. C., 48, 57
Miller, R. G., 219
Millimet, C. R., 226
Minton, H. L., 96
Mintz, J., 201, 206, 256
Miranda, M., 245
Mirels, H. L., 86, 299
Miskel, C., 308
Mitchell, H. E., 77
Mitchell, K. M., 200
Mitchell, K. R., 207
Mitchell, M. C., 21, 242
Mitchell, T. R., 354
Mobley, W. H., 328
Molloy, G. N., 14
Monahan, J., 112
Moorey, B., 136
Moos, R., 228
Moos, R. H., 212, 280
Moran, L. J., 72
Morf, M. E., 151, 168, 187
Morgan, B. B. Jr., 15
Morgan, D. W., 212
Morris, J. D., 93
Morrison, R. F., 347
Morrison, T. L., 269
Morse, S. J., 74
Mos, L., 146
Moser, R. P., 298
Moses, J. J., 133
Mosher, D. L., 77
Mowday, R. T., 338
Muir, W., 246
Mulhern, T., 10
Mullens, B. N., 247
Mulligan, G., 346
Munro, H. P., 167
Munz, D. C., 130
Murdy, L. B., 338
Murphy, D. L., 248
Murphy, J., 265
Mussio, S. J., 323

371

Vincent, J. E., 94
Viney, L. L., 188
Vitale, J. H., 265
Vajtisek, J. E., 230
Von Kulmiz, P., 86
Vreeland, R., 140
Vroegh, K., 76

Wagner, F. R., 133
Wainer, H., 275
Wakefield, J. A. Jr., 27, 188, 189
Walberg, H. J., 281
Waldnop, M. F., 51
Walker, R. E., 180
Walter, G. A., 339
Walters, S. M., 150
Wallbrown, F. H., 24, 39, 295
Wallbrown, J. D., 39
Wanberg, K. W., 167
Wang, A. M., 70
Ward, J. H. Jr., 353
Ware, E. E., 131
Wardell, D., 134, 146
Ware, W. B., 305
Waters, L. K., 89, 341, 354
Waters, C. W., 89
Watson, B. L., 315
Watson, E. P., 278
Watson, J. P., 207
Watson, N. L., 312
Wearing, A. J., 68, 90
Webb, S. C., 140
Webb, W. W., 291
Webberley, M., 13
Wechsler, H., 141
Weckowicz, T. E., 246
WeFring, L. R., 233
Weigl, K., 79
Weinstein, A. G., 316
Weinstein, D., 355
Weinstein, N., 225
Weiss, R. L., 260
Weissman, A., 236
Weissman, H. N., 111, 204
Weissman, M., 243
Weitzel, W., 341, 347
Weldon, D. E., 252
Welkowitz, J., 197
Wentworth, D. R., 285
Wernimont, P. F., 116
Wessman, A. E., 125
Westler, L., 157
Wexley, K., 39

Wherry, R. J., 24, 39, 90, 295, 335, 355
White, K., 13, 15
White, W. F., 22, 48, 49, 314
Whitehead, J. L., 286
Whitelock, P. R., 235
Whiteman, M., 58
Whitley, S. E., 301
Whitney, D. R., 161
Wickert, J., 299
Wiederanders, M. R., 7
Wiegerink, R., 225
Wiesenthal, D. L., 67
Wiggins, N., 230
Wiggins, N. H., 101, 115
Wijting, J. P., 322
Wilcox, A. H., 189
Wild, R., 342
Wilkins, G., 141
Will, D. P. Jr., 182
Williams, A. F., 141
Williams, C., 189
Williams, C. M., 271
Williams, F., 286
Williams, J. D., 230
Williams, J. E., 116
Williams, T. H., 40
Willerman, L., 41
Wilson, G. D., 76
Wilson, H. R., 28, 38
Wilson, R. C., 312
Winkelmann, W., 40
Winter, D. G., 83
Wober, M., 89
Wojtowicz, E. H., 233
Wolff, B. B., 18
Wolff, E., 74
Wollack, S., 322
Wollitzer, A. D., 265
Wood, D. A., 323
Wood, F. B., 116
Woodbury, R., 239
Woods, E. M., 34
Woog, P. C., 312
Worchel, S., 354
Wright, L., 190, 252
Wright, T. L., 190
Wyant, K., 190

Yee, A. H., 313
Yee, L. Y., 314
Yelling, W. F., 90
Yeh, E. G., 281
Yen, W. M., 144